Dodson · Gonzalez   Experiments In Mathematics Using Maple

Springer
Berlin
Heidelberg
New York
Barcelona
Budapest
Hong Kong
London
Milan
Paris
Santa Clara
Singapore
Tokyo

C. T. J. Dodson   E. A. Gonzalez

Experiments In Mathematics Using Maple

With 146 Figures

 Springer

Christopher T. J. Dodson
Department of Mathematics
University of Toronto
200 College Street
Toronto, Ontario M5S 1A4, Canada

Elizabeth A. Gonzalez
University of Toronto Instructional and
Research Computing
200 College Street
Toronto, Ontario M5S 1A4, Canada

CIP data applied for

Die Deutsche Bibliothek - CIP-Einheitsaufnahme

Dodson, Christopher T. J.:
Experiments in mathematics using maple / C. T. J. Dodson ;
E. A. Gonzalez. - Berlin ; Heidelberg ; New York ; Barcelona ;
Budapest ; Hong Kong ; London ; Milan ; Paris ; Santa Clara ;
Singapore ; Tokyo : Springer, 1995
ISBN-13:978-3-540-59284-6   e-ISBN-13:978-3-642-79758-3
DOI: 10.1007/978-3-642-79758-3

NE: Gonzalez, Elisabeth A.:

Mathematics Subject Classification (1991):
26-01, 26A06, 26A09, 51-01, 68Q40

ISBN-13:978-3-540-59284-6

The use of general descriptive names, registered names, trademarks, etc. in this publication does not imply, even in the absence of a specific statement, that such names are exempt from the relevant protective laws and regulations and therefore free for general use.

Cover design: Künkel + Lopka Werbeagentur GmbH, Ilvesheim
Typesetting: Camera-ready output from the authors
SPIN: 10501406   41/3143 - 5 4 3 2 1 0 - Printed on acid-free paper

Preface

We believe that to understand and be comfortable with mathematical concepts and methods, it is necessary to *do* mathematics and, traditionally, doing meant with a pencil and paper. Now, a modest home computer can provide a platform for a computer algebra package like *Maple*, which can perform all of the operations encountered in secondary school mathematics and beyond, and provide graphical representations of functions, including animations. The capability of rendering accurate graphics for mathematical functions greatly enhances the learning experience, and helps intuition work in new situations, before beginning to do the algebra and calculus needed to solve a problem. For example, if you can see that a function has a clear minimum from its graph, then you are more likely to be able to identify its location precisely by analysis.

This book is designed to support the interactive *Maple* worksheets that we have developed for the two senior secondary school years and made available via the Internet at the anonymous ftp site `ftp.utirc.utoronto.ca` (`/pub/ednet/maths/maple`). This takes you from basic algebra, functions and sequences, to calculus and its applications. Thus, the book is a hardcopy version of the worksheets, with additional explanatory text, cross-referencing, appendices containing worked solutions to all exercises, and indexing. On the other hand, the worksheets are live documents and allow addition of your own examples and explorations—just like a traditional exercise book—so you can expand the study of any particular themes that interest you, adding more topics, graphs and animations. Each of our worksheets contains open-ended experiments with some hints for project work. In fact, all mathematics on a computer is experimental mathematics an we hope that you get in the habit of trying to check results by analytic methods! Our worksheets and this book can be used for independent study or in conjunction with any course text that you have.

The software *Maple* is provided in many schools, colleges and universities, and is one of the standard mathematical packages used by working scientists, engineers and teachers; student editions are available at low cost. There is a mailing list for discussions about *Maple*, called The *Maple* Users' Group, which you may join by sending email to `majordomo@daisy.uwaterloo.ca` with the following command in your message: `subscribe maple-list <your email address>`. There is a Usenet newsgroup `sci.math.symbolic` for the discussion of computer algebra systems like *Maple*.

This book has been prepared in LaTeX with *Maple* graphics. You can use your own *Maple* worksheets in a similar way, to edit into a book customized to include your own interests and explorations in mathematics and its applications.

We wish to thank Dr. C.D. Sadleir, who initiated the EDnet Project from which this book arose, and through whom support for our work was obtained. We thank also Dr. M. Peters of Springer-Verlag for his enthusiastic support of the book.

Kit Dodson and Elizabeth Gonzalez
University of Toronto
August 4, 1995

Acknowledgements

The authors wish to thank the following publishers for permission to quote examples from texts as indicated below.

C.H. Edwards, Jr. and D.E. Penney: *Calculus and Analytic Geometry*. 2nd edn. Prentice-Hall, Inc., Englewood Cliffs, New Jersey 1986.

#47	p14
#35	p22
#18, #20	p38
#91	p170
#23	p190
#8	p438
#29	p510

S. Goldenberg and H. Greenwald: *Calculus Applications in Engineering and Science*. D.C. Heath and Company, Toronto 1990.

Application 1.2	p4
Application 1.3	p5
Application 2.1	p13
Application 3.1	p18
Application 3.3	p22
Application 6.4	p63

Contents

II Beginning Calculus

List of Figures

List of Tables

Part I

Pre-Calculus Mathematics

Chapter 1

Introduction to Maple

1.1 Documentation

Welcome to *Maple*. This chapter will help you become familiar with some of the most commonly used commands in the subsequent chapters. For more detailed information, consult the the supporting documents for *Maple* on your computer; note also following sources:

On-Line Tutorial *Maple* has an on-line tutorial which can be run by typing `tutorial();` at the prompt.

Help Command To get information about any command in *Maple,* simply type a question mark followed by the name of the command and press return.

```
> ?evalf
```

This facility describes how the command is used and gives some examples. The Help Browser in the upper right hand corner of the *Maple* window is another source of examples. On your computer you can use the mouse to copy from the Help window into a *Maple* worksheet window; this allows you to copy and modify examples for your own use.

Books There are plenty of books on the use of *Maple.* Several of them are listed in the Bibliography and new information is available over the Internet, as mentioned in the Preface.

1.2 First *Maple* Session

1.2.1 Basic Operations

To calculate the value of $2 + 3$, we simply type the following command.

```
> 2+3;
```

Mathematical Operation	*Maple* Symbol
addition	+
subtraction	−
division	/
multiplication	*
exponentiation	^

Table 1.1. *Maple* symbols for basic mathematical operations

You must put a semi-colon or colon at the end of a command. A semi-colon causes the command to be executed and the result to be displayed. Using a colon prevents the output being displayed on the screen. Most of the basic mathematical commands have easy to remember expressions in the language of *Maple*; comman early errors are to forget the end-of-command semi-colon or to forget a bracket.

Table 1.1 lists the *Maple* symbols needed to do basic mathematical operations. Here are some examples to demonstrate their use.

```
> 4-1;
```
$$3$$

```
> 8*4;
```
$$32$$

```
> 9/3;
```
$$3$$

```
> 3*(8+4);
```
$$36$$

```
> 3^2;
```
$$9$$

```
> (x+3)^2;
```
$$(x + 3)^2$$

To expand this expression, use the **expand** command. Some of the most commonly used commands are demonstrated in Section 1.3.8.

```
> expand((x+3)^2);
```
$$x^2 + 6\,x + 9$$

To assign an expression to a variable name, you must use := as in,

> y:=5-9;

$$-4$$

With *Maple,* the result is always exact, no matter how large the numbers are.

> a:=38975*43076;

$$a := 1678887100$$

> b:=6745/53648;

$$b := \frac{6745}{53648}$$

To get a decimal approximation of the result, use evalf to evaluate,

> evalf(b);

$$.1257269609$$

1.2.2   Solving Equations

We wish to solve the equation $y = 2x - 1$ for the value of x at $y = 0$. One way is to assign the right hand side of the equation to y and use the solve command. The solve command always equates the value of the variable to be solved to zero unless otherwise specified. In this case, it will solve $2x - 1 = 0$.

> y:=2*x-1;

$$y := 2x - 1$$

> solve(y,{x});

$$\left\{ x = \frac{1}{2} \right\}$$

1.3   More Commands

We remind you here that most of the lines of *Maple* commands in this book are already typed for you in the worksheets we provide free over the Internet.

1.3.1   Calling Function Packages

Many of *Maple's* functions are stored in packages and must be called in before they can be used. This is done using the with command. For example, if we want to use the `isolate` command, we must first call in the `student` function package.

> `with(student);`

$$[D, Doubleint, Int, Limit, Lineint, Sum, Tripleint, changevar, combine,$$
$$completesquare, distance, equate, extrema, integrand, intercept,$$
$$intparts, isolate, leftbox, leftsum, makeproc, maximize,$$
$$middlebox, middlesum, midpoint, minimize, powsubs, rightbox,$$
$$rightsum, showtangent, simpson, slope, trapezoid, value]$$

Don't panic if when you call a package you get a message like: `Warning: new definition for isolate`. *Maple* is just letting you know that certain commands have been redefined. You can always use ? to find out more.

1.3.2   Isolating Variables

Suppose we wish to solve for the variable x in terms of y in the equation $y = 2x-1$. One way is to call an isolating function `isolate` from the `student` package. To call the package `student` we use the command `with(student)`, just once for the whole *Maple* session.

First we give the equation $y = 2x - 1$ a name so that we may refer to it easily.

> `eq:=y=2*x-1;`

$$eq := y = 2x - 1$$

> `with(student):`

> `isolate(eq,x);`

$$x = \frac{1}{2}y + \frac{1}{2}$$

For a simple case like this, the equation can be entered into the `isolate` command directly. For more complicated computations, naming the equality is preferable because we wish to do as little typing as possible, so reducing errors of transcription.

> `isolate(y=2*x-1,x);`

$$x = \frac{1}{2}y + \frac{1}{2}$$

1.3.3   Variable Names

It's a good idea to use a descriptive variable name, for example 'velocity' rather than 'v', especially when you are dealing with many variables. A variable name can be up to 499 characters long and contain letters, digits or underscores (_). The first character in a variable name must be a letter.

> variable:=2;
$$variable := 2$$

> variable_name:=2;
$$variable\_name := 2$$

If you wish to use a name which contains other characters or spaces, the name must be in back quotes (').

> 'Have a Nice Day!':=thank_you;
$$Have\ a\ Nice\ Day! := thank\_you$$

1.3.4   Syntax

Maple is case sensitive. For example, y is not the same as Y.

> y:=2:
> y;
$$2$$

> Y;
$$Y$$

Maple does not, however, recognize spaces. For consistency, no spaces are included in the commands in this, and subsequent, chapters.

> 35*81;
$$2835$$

> 35 * 81;
$$2835$$

Round and square brackets are not interchangeable. Round brackets must be used in mathematical expressions.

> 1-(2+3);
$$-4$$

```
> 1-[2+3];
```

$$1 - [5]$$

1.3.5   Assigning Variables

When we solved for x or isolated it on the right hand side of the equation, the expression on the left hand side was not assigned to the name x, so x was still a free symbol variable. Here's how to assign the variable, using the example from Section 1.3.2.

```
> y:=2*x-1;
```

$$y := 2\,x - 1$$

```
> solution:=solve(y,{x});
```

$$solution := \left\{ x = \frac{1}{2} \right\}$$

The value of x at $y = 0$ has been assigned to the name *solution*. x has not yet been assigned the value of $\frac{1}{2}$.

```
> x;
```

$$x$$

To assign it,

```
> assign(solution);
```

```
> x;
```

$$\frac{1}{2}$$

To unassign variables, we tell the symbol x that it should again become a free variable:

```
> x:='x':
```

```
> y:='y':
```

```
> x,y;
```

$$x, y$$

To unassign all variables, and wipe clean *Maple's* memory of them, use the `restart` command.

```
> restart;
```

1.3.6   Ditto

A quotation mark ("), or ditto, is used to refer to the last command activated. For example,

> `solve(y=2*x-1,x);`

$$x = \frac{1}{2}y + \frac{1}{2}$$

> `assign(");`

> `x;`

$$\frac{1}{2}y + \frac{1}{2}$$

Several dittos may be used to refer to commands activated earlier.

> `2+3;`

$$5$$

> `1+";`

$$6$$

> `1-"";`

$$-4$$

Care must be used with dittos because if the series of commands is not activated in the original order, the ditto may refer to a command other than was intended.

1.3.7   Notation of Mathematical Expressions

Several forms of notation can be used when describing mathematical expressions, such as mapping notation, normal notation, function notation, and set builder notation. In *Maple,* expressions are defined using the assignment operation (:=), or by using arrow notation.

Assignment Operation

This type of notation was used in the previous sections of this chapter. Simply, the left hand side of the expression is assigned to a name.

> `y:=x+3;`

$$y := x + 3$$

To determine the value of y for a value of x, either substitute the value of x into the expression (using the **subs** command), or assign the value to x.

```
> subs(x=-3,y);
```
$$0$$

```
> x:=-3;
```
$$x := -3$$

```
> y;
```
$$0$$

Arrow Notation

Maple's arrow notation is similar in form to mapping notation. It is used to define functions. (We shall be looking more closely at functions in subsequent chapters.) To make the arrow symbol, type a minus sign $(-)$ followed by a greater than sign $(>)$. This is simply stating the rule for what must be done to the input variable. For example,

```
> y:=x->x+3;
```
$$y := x \rightarrow x + 3$$

Notice that the value of the function at x is referred to as $y(x)$, not y, which is the *name* of the function.

```
> y;
```
$$y$$

```
> y(x);
```
$$x + 3$$

For substituting values of x into an expression, this is the most useful notation. To determine values of $y(x)$ for any x, insert the value of x into the brackets, as shown below, which culminates in a plot of the relation.

```
> y(-3);
```
$$0$$

```
> set_of_ordered_pairs:=[[-3,y(-3)],[-2,y(-2)],[-1,y(-1)],[0,y(0)],
> [1,y(1)],[2,y(2)],[3,y(3)]];
```
$$set\_of\_ordered\_pairs :=$$
$$[[-3,0],[-2,1],[-1,2],[0,3],[1,4],[2,5],[3,6]]$$

```
> plot(y(x),x=-3..3);
```

To find out more about plotting, see Section 1.4.

1.3.8 Commonly Used Commands

The most important commands when dealing with polynomials are demonstrated
below.

`collect`

```
> d:=x->a*x+3*x-a;
```
$$d := x \rightarrow a\,x + 3\,x - a$$

```
> collect(d(x),x);
```
$$(a + 3)\,x - a$$

```
> collect(d(x),a);
```
$$(x - 1)\,a + 3\,x$$

`normal`

```
> f:=x->(x^2-2*x-3)/(x+1);
```
$$f := x \rightarrow \frac{x^2 - 2\,x - 3}{x + 1}$$

```
> normal(f(x));
```
$$x - 3$$

`expand`

```
> f:=x->(x+3)*(x-2);
```
$$f := x \rightarrow (x + 3)(x - 2)$$

```
> expand(f(x));
```
$$x^2 + x - 6$$

sort

```
> g:=x->x+2*x^2-1;
```
$$g := x \rightarrow x + 2\,x^2 - 1$$

```
> sort(g(x));
```
$$2\,x^2 + x - 1$$

factor

```
> factor(g(x));
```
$$(\,x+1\,)\,(\,2\,x-1\,)$$

simplify

```
> h:=f(x)+g(x);
```
$$h := (\,x+3\,)\,(\,x-2\,) + 2\,x^2 + x - 1$$

```
> simplify(h);
```
$$3\,x^2 + 2\,x - 7$$

solve

```
> solve(h,{x});
```
$$\left\{ x = -\frac{1}{3} + \frac{1}{3}\,\sqrt{22} \right\}, \left\{ x = -\frac{1}{3} - \frac{1}{3}\,\sqrt{22} \right\}$$

```
> evalf(");
```
$$\{\,x = 1.230138587\,\}, \{\,x = -1.896805253\,\}$$

fsolve

```
> fsolve(h,{x});
```
$$\{\,x = -1.896805253\,\}, \{\,x = 1.230138587\,\}$$

op

To extract operands,

```
> op(");
```
$$\mathrm{op}(\,\{\,x = -1.896805253\,\},\{\,x = 1.230138587\,\}\,)$$

```
> op(1,");
```
$$\{\,x = -1.896805253\,\}$$

```
> op(2,"");
```
$$\{\,x = 1.230138587\,\}$$

1.4 Plots

Maple's plots package contains many useful commands for generating a wide range of plots. But first let's look at the simplest command plot, which is contained in the main *Maple* library. (The plots themselves are omitted in this chapter to save space, but you can generate them on your own computer with the code provided.)

In *Maple,* you can plot ordered pairs ...

```
> set_of_ordered_pairs:=[[-3,1],[-2,-1],[-1,-1],[0,0],[1,1],[2,2],
> [3,2]];
```
$$set\_of\_ordered\_pairs :=$$
$$[[-3,1],[-2,-1],[-1,-1],[0,0],[1,1],[2,2],[3,2]]$$

```
> plot(set_of_ordered_pairs);
```
... or expressions over an interval of the x−axis. Note that *Maple* joins up the points with straight lines unless you tell it not to. This is done using the style option, as in the next example.

```
> plot(set_of_ordered_pairs,style=POINT);
```

```
> plot(3*x-7,x);
```

```
> f:=x->3*x-7;
```
$$f := x \rightarrow 3\,x - 7$$

```
> plot(f(x),x);
```

1.4.1   Options

There are many optional features which can be added to plots to make them more informative. Some of these options are shown in the next series of `plot` commands. Run them on the worksheet and try them yourself.

```
> f:=x->2*x+8;
```
$$f := x \rightarrow 2\,x + 8$$

```
> plot(f(x),x);
> plot(f(x),x=-3..3);
> plot(f(x),x=-3..3,y=0..14);
> plot(f(x),x=-3..3,y=0..14,scaling=CONSTRAINED);
> plot(f(x),x=-3..3,y=0..14,scaling=CONSTRAINED,style=POINT);
> plot(f(x),x=-3..3,y=0..14,scaling=CONSTRAINED,style=POINT,
> numpoints=100);
> plot(f(x),x=-3..3,y=0..14,scaling=CONSTRAINED,style=POINT,
> numpoints=100,axes=BOXED);
> plot(f(x),x=-3..3,y=0..14,scaling=CONSTRAINED,style=POINT,
> numpoints=100,axes=BOXED,color=BLUE);
> plot(f(x),x=-3..3,y=0..14,scaling=CONSTRAINED,style=POINT,
> numpoints=100,axes=BOXED,color=BLUE,title='f(x)');
> plot(f(x),x=-3..3,y=0..14,scaling=CONSTRAINED,style=POINT,
> numpoints=100,axes=BOXED,color=BLUE,title='f(x)',
> axesfont=[COURIER,30],labelfont=[COURIER,30],font=[COURIER,30]);
```

(All of the text in the plots in this book were generated using 30 point courier font.)

Features such as the `style`, `axes` type and `projection` may be changed using the pull-down menus at the top of the `plot` window.

1.4.2   Text

Text, other than the title, may be added anywhere in a plot using the `textplot` command. The first and second arguments in the square brackets represent the $x-$ and $y-$coordinates of the text, and the third argument, which must be surrounded by back quotes, contains the text to be added. The `display` command is used to overlay a plot, or series of plots, and text. One way to use the `display` command is to assign a name to each `plot` or `textplot` command and then specify by name the plots to combine. Notice that a colon, rather than a semi-colon, ends the command when plots are named.

```
> with(plots):
> p:=plot(f(x),x=-3..3,scaling=CONSTRAINED,style=POINT,
> numpoints=200):
> t:=textplot([3,14,'f(x)']):
```

```
> display({p,t});
```

The other way is to include the `plot` and `textplot` commands within the `display` command.

```
> display({plot(f(x),x=-3..3,scaling=CONSTRAINED,style=POINT,
> numpoints=200),textplot([3,14,'f(x)'],align={ABOVE,RIGHT})});
```

1.4.3   Implicit Plots

It is often more convenient to plot expressions implicitly, that is, as an equation expressed in two variables, using `implicitplot`.

```
> eq:=x^2-y^2=1;
```

$$eq := x^2 - y^2 = 1$$

```
> with(plots):
> implicitplot(eq,x=-3..3,y=-3..3);
```

To plot this hyperbola explicitly, we must solve for y and then plot each solution branch separately. In this example it is possible to solve for y explicitly, but in general, this may not be the case, as for example if we have $x^3 + y^3 = 3xy$, which gives a curve called the Folium of Descartes.

1.4.4   3D Plots

Plots in three dimensions are made using the `plot3d` command. Two of the three dimensions (x, y, and z) must be specified. To redraw a plot after the `projection`, `axes`, `line style`, or `color` is changed, press the middle mouse button.

```
> with(plots):
> plot3d(sin(x+y),x=-3..3,y=-3..3);
```

This is an example from the help menu under `plot3d`.

```
> c1:= [cos(x)-2*cos(0.4*y),sin(x)-2*sin(0.4*y),y]:
> c2:= [cos(x)+2*cos(0.4*y),sin(x)+2*sin(0.4*y),y]:
> c3:= [cos(x)+2*sin(0.4*y),sin(x)-2*cos(0.4*y),y]:
> c4:= [cos(x)-2*sin(0.4*y),sin(x)+2*cos(0.4*y),y]:
> plot3d({c1,c2,c3,c4},x=0..2*Pi,y=0..10,grid=[25,15],style=patch);
```

1.4.5   Animations

A most important feature of *Maple* is to represent continuous movements using animations. The graph of a function can be animated using the `animate` command. In the example that follows, x is the independent variable and t is the animation parameter; you can think of t as measuring progress in time.

```
> animate(t*x^2,x=-2..2,t=-2..2);
```

The `animate` command takes the same options as other plots.

To animate a series of plots, we use the `display` command with the option `insequence=true`. For example,

```
> display({plot(x,x),plot(2*x,x),plot(3*x,x)},insequence=true);
```

Complicated animations can be quite demanding for computers with slow cpu or small amounts of RAM so it is always wise to begin with a basic skeleton having few steps, and gradually include refinements.

1.5   Preparing Worksheets for Printing

Often you will wish to save and print a worksheet. To do this you will normally want to edit it, adding a title and your name, removing any unwanted input and also including some of your graphics.

All of the plots you produce in your *Maple* worksheet will be generated in a separate window. To paste plots into the body of a worksheet, use the mouse to select Copy in the Edit pull-down menu of the plot window, or type Ctrl-C while the cursor is in the plot window. Then move the cursor to the worksheet and click at the position where you would like the plot to appear. Choose Paste in the Edit pull-down window or type Ctrl-V to insert the plot. Now, after saving, the plot is a permanent part of the worksheet. You can delete the plot as you would any other output.

The procedure for printing varies from one computer to another. Usually, from the File menu you can select Print, to create a file that may be printed.

Chapter 2

Functions

Commands used in this chapter
- @
- `expand`
- `plot`
- `plot[options]`
- `plots`
- `plots[display]`
- `plots[implicitplot]`
- `plots[textplot]`
- `solve`
- `sort`
- `student`
- `student[isolate]`
- `subs`
- `with`

2.1 Relations and Functions

A *relation* is a set of ordered pairs; the set of first elements in each ordered pair is called the *domain*, and the set of second elements is called the *range*. For example, in a family, the property 'being uncle of' is expressed as a relation among the set of family members; A is uncle of B can have more than one B for each choice of A. A *function* is a relation for which each value in the domain corresponds to a unique value in the range. In other words, a vertical line drawn anywhere from the x-axis will intersect the graph of a function at one and only one point. Mathematicians also use the words map or mapping to mean a function. The concept of a function as a well-defined, that is unambiguous, rule is one of the most important in mathematics.

The equation $y = x+3$ defines a function (Figure 2.1(a)). Each value in the domain maps to a unique value in the range. Any non-vertical line is a function. A horizontal line is a constant function.

> `plot(x+3,x=-5..5);`

The equation $y = x$ defines a function (Figure 2.1(b)), called the *identity function*.

> `plot(x,x=-5..5);`

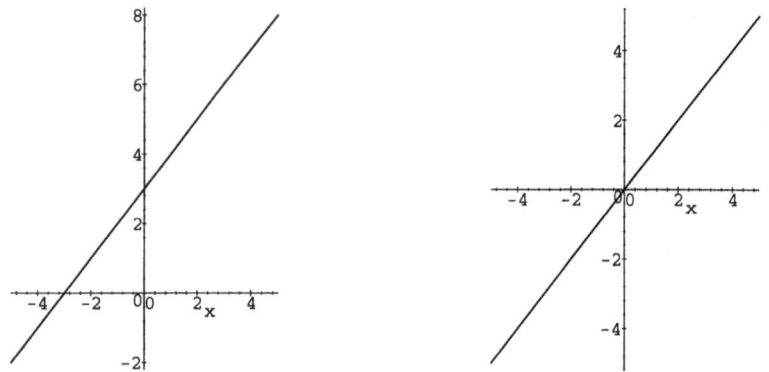

Figure 2.1. (a) The graph of $y = x + 3$, on $[-5, 5]$. (b) The graph of $y = x$, on $[-5, 5]$

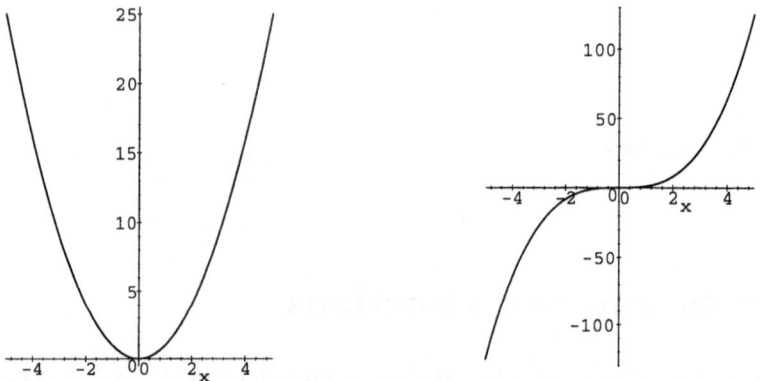

Figure 2.2. (a) The graph of $y = x^2$, on $[-5, 5]$. (b) The graph of $y = x^3$, on $[-1, 1]$

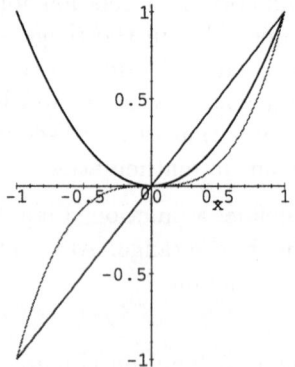

Figure 2.3. The graphs of $y = x$, $y = x^2$, and $y = x^3$, on $[-1, 1]$

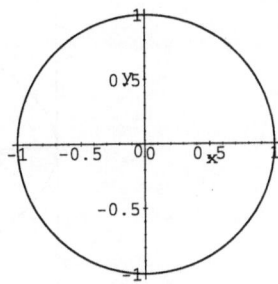

Figure 2.4. The graph of $x^2 + y^2 = 1$, on $[-1, 1]$

The equation $y = x^2$ defines a function (Figure 2.2(a)).

```
> plot(x^2,x=-5..5);
```

The equation $y = x^3$ defines a function (Figure 2.2(b)).

```
> plot(x^3,x=-5..5);
```

Figure 2.3 overlays the above three functions.

```
> plot({x,x^2,x^3},x=-1..1);
```

The circle shown in Figure 2.4 is the graph of a relation, not a function. All values of x, except the x-intercepts, relate to two values of y.

```
> with(plots):
> implicitplot(x^2+y^2=1,x=-1..1,y=-1..1);
```

To prove this, solve the equation for any value in the domain $-1 < x < 1$.

```
> subs(x=0,x^2+y^2=1);
```

$$y^2 = 1$$

```
> solve(",{y});
```

$$\{y = 1\}, \{y = -1\}$$

The equation $y = x^2$ defines a function (Figure 2.5(a)).

```
> plot(x^2,x=-2..2);
```

The equation $y^2 = x$ defines a relation, not a function (Figure 2.5(b)).

```
> implicitplot(y^2=x,x=-9..9,y=-3..3);
```

To prove this, solve the equation for any value of $x > 0$.

```
> subs(x=1,y^2=x);
```

$$y^2 = 1$$

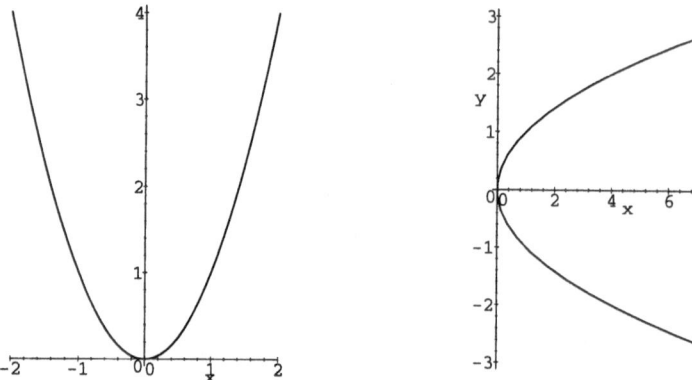

Figure 2.5. (a) The graph of $y = x^2$, on $[-2, 2]$. (b) The graph of $y^2 = x$, on $[0, 9]$

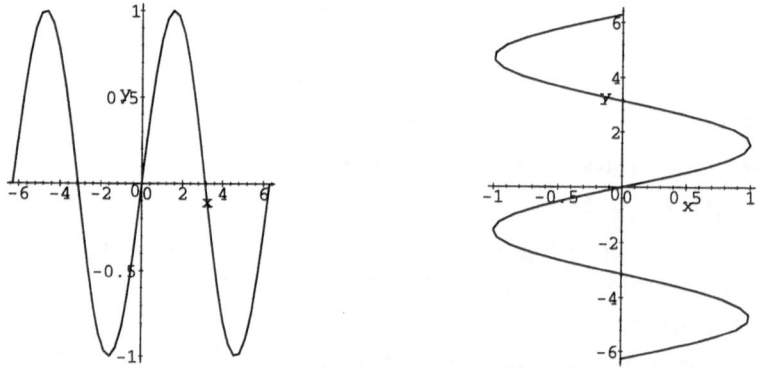

Figure 2.6. (a) The graph of $y = \sin(x)$, on $[-2\pi, 2\pi]$. (b) The graph of $x = \sin(y)$, on $[-1, 1]$

```
> solve(",{y});
```
$$\{\, y = 1 \,\}, \{\, y = -1 \,\}$$

The *inverse* of a function is always a relation but not necessarily a function. We shall investigate inverse functions later in this chapter and find that by restricting domains of functions we can usually find partial inverses.

The sine function is another example of a function whose inverse is not a function. Compare the plots in Figure 2.6.

```
> implicitplot(y=sin(x),x=-2*Pi..2*Pi,y=-2*Pi..2*Pi,
> numpoints=2000);

> implicitplot(x=sin(y),x=-2*Pi..2*Pi,y=-2*Pi..2*Pi,
> numpoints=2000);
```

Note that if we *restrict* $y = \sin(x)$ to domain $[-\frac{\pi}{2}, \frac{\pi}{2}]$ we do have an inverse.

Exercise 2.1   Which of the following relations are functions and why?

   (a) $f = \{(1,1),(3,4),(5,5),(5,6)\}$

   (b) $\frac{x^2}{4} + \frac{y^2}{2} = 1$

   (c) $y = x^3 - 2x^2$

   (d) $\frac{x^2}{4} - \frac{y^2}{2} = 1$

   (e) $y = 1/x$

Experiment 2.1   Use `implicitplot` to plot the graphs of the relations $x^n + y^n = 1$ for $n = 2,4,6,8$. Find subsets of these relations which define functions.

Experiment 2.2   Why do odd powers of x, like $x, x^3, x^5, \ldots$ have inverse functions over the whole real line but even powers of x do not? Look at their graphs.

2.2   Multiplication by a Real Number

Multiplication of a function by a real number may result in a stretch, compression, or reflection of the graph of the function in the y-axis.

   If the constant is greater than one, then the resultant function will be stretched in the y-axis (Figure 2.7(a)).

```
> s:=4*x-3;
```
$$s := 4x - 3$$

```
> r:=3*s;
```
$$r := 12x - 9$$

```
> t:=textplot([[3,27,'3s(x)'],[3,9,'s(x)']],align={ABOVE,RIGHT}):
> p:=plot({s,r},x=-3..3):
> display({p,t});
```

Multiplication of $f(x)$ by -1 gives $-f(x)$ and we see that $f(x)$ and $-f(x)$ give graphs which are mirror images in the x-axis. See Figure 2.7(b).

```
> '-s':=-s;
```
$$-s := -4x + 3$$

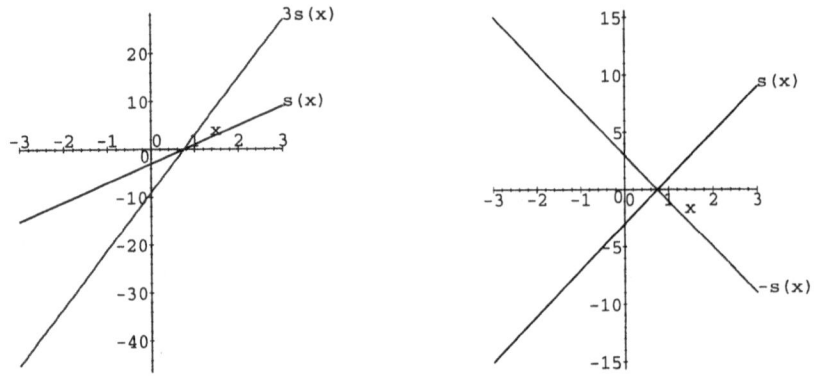

Figure 2.7. (a) Multiplication of a function by 3. (b) Multiplication of a function by -1

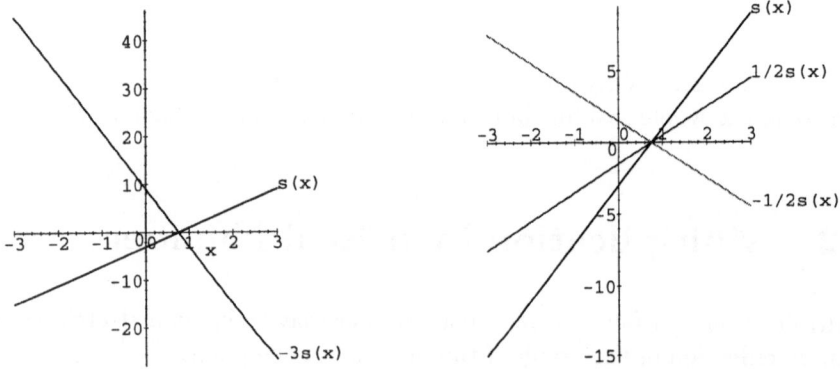

Figure 2.8. (a) Multiplication of a function by -3. (b) Multiplication of a function by $\frac{1}{2}$

```
> t:=textplot([[3,-9,'-s(x)'],[3,9,'s(x)']],align={ABOVE,RIGHT}):
> p:=plot({s,'-s'},x=-3..3):
> display({p,t});
```

If the constant is less than -1, then the graph of the resultant function will be stretched and reflected in the y-axis (Figure 2.8(a)).

```
> v:=-3*s;
```

$$v := -12\,x + 9$$

```
> t:=textplot([[3,-27,'-3s(x)'],[3,9,'s(x)']],align={ABOVE,RIGHT}):
> p:=plot({s,v},x=-3..3):
> display({p,t});
```

If the constant is between $+1$ and -1, the graph of the resultant function will be compressed in the y-axis. If the constant is negative, the graph of the resultant function will also be reflected (Figure 2.8(b)).

```
> u:=1/2*s;
```

$$u := 2\,x - \frac{3}{2}$$

```
> y:=-1/2*s;
```

$$y := -2\,x + \frac{3}{2}$$

```
> t:=textplot([[3,9,'s(x)'],[3,4.5,'1/2s(x)'],[3,-4.5,'-1/2s(x)']],
> align={ABOVE,RIGHT}):
> p:=plot({s,u,y},x=-3..3):
> display({p,t});
```

Exercise 2.2 Given $f(x) = 2x^2 - 4x + 9$, plot $f(x)$ and $3f(x)$, $-10 \le x \le 10$, on the same graph.

Exercise 2.3 Perfect gases expand when heated according to the Ideal Gas Law $PV = nRT$ where P is the pressure, V is the volume, n is proportional to the amount of gas present, R is a constant, and T is the Kelvin temperature. Engineers who design cars use this formula to determine the *explosion point*, that is, the condition when increase in temperature will cause the windshield of a car to pop out. (p5, Goldenberg and Greenwald [7])

(a) Assume the interior of a locked car is airtight, the contents are at one atmosphere, and the initial temperature is 293 Kelvin (about 20°C). Suppose the temperature inside the car increases 15 Kelvin for each hour the car remains in the sun. The car manufacturer has designed the windshield so that it will pop out when the interior pressure is 1.6 times the exterior pressure. How long would these conditions have to be maintained before the windshield pops out?

(b) A car with a sun roof will heat up faster than a car with a solid roof. Suppose the temperature of the interior of a closed car with a sun roof increases at a rate of 20 Kelvin per hour. Suppose the same pressure difference as described in the example will cause the windshield to pop out. How long will it take to reach this pressure?

(c) Older cars do not have windshields that pop out. If an older car has a sun roof and the interior temperature increases by 20 Kelvin per hour, what will the interior temperature be after 1.5 hours in the sun? (Assume the initial temperature is 293 Kelvin.)

(d) A half-full bottle of sunscreen is left in the car. One window is partially open so the air pressure in the car does not build up, but the temperature does. The temperature in the bottle increases from 293

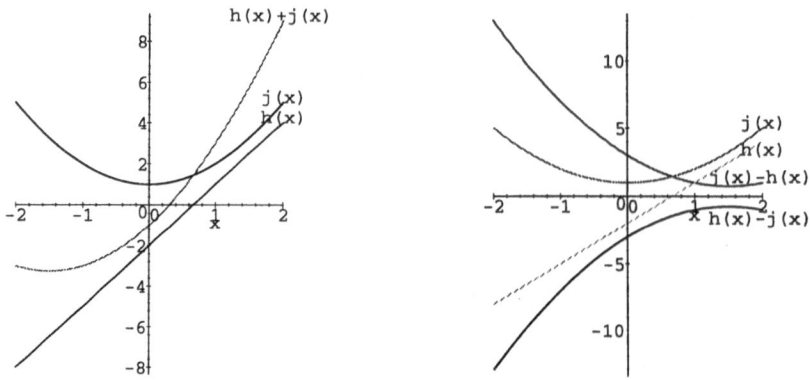

Figure 2.9. (a) Addition of two functions. (b) Subtraction of two functions

Kelvin to 353 Kelvin. Find the pressure within the bottle of sun-screen. Assume there is no change in volume and there is no increase in vapour pressure from the heated lotion.

(e) The plastic bottle described above is empty and expands. The volume is 15% more when heated. What is the new pressure?

2.3 Addition and Subtraction of Functions

Functions add, subtract, multiply and divide through the way those operations act on their output values.

```
> h:=3*x-2;
```
$$h := 3\,x - 2$$

```
> j:=x^2+1;
```
$$j := x^2 + 1$$

```
> 'h+j'=h+j;
```
$$h + j = 3\,x - 1 + x^2$$

```
> 'h+j'=sort(h+j);
```
$$h + j = x^2 + 3\,x - 1$$

Each value of $(h + j)(x)$ is equal to $h(x) + j(x)$. For example, at $x = 1$,

```
> subs(x=1,h);
```
1

```
> subs(x=1,j);
```
$$2$$

```
> subs(x=1,h+j);
```
$$3$$

Graphically, the y−position of a point on the graph of $(h + j)(x)$ is equal to the sum of the distances of the graphs of $h(x)$ and $j(x)$ from the x-axis. See Figure 2.9(a).

```
> t:=textplot([[2,4,'h(x)'],[2,5,'j(x)'],[2,9,'h(x)+j(x)']],
> align=ABOVE):
> p:=plot({h,j,h+j},x=-2..2):
> display({p,t});
```

Unlike adding functions, when subtracting functions, the order in which the operation is performed is important. (Figure 2.9(b)).

```
> 'h-j'=sort(h-j);
```
$$h - j = -x^2 + 3x - 3$$

```
> 'j-h'=sort(j-h);
```
$$j - h = x^2 - 3x + 3$$

```
> t1:=textplot([[2,4,'h(x)'],[2,-1,'h(x)-j(x)']],align=BELOW):
> t2:=textplot([[2,5,'j(x)'],[2,1,'j(x)-h(x)']],align=ABOVE):
> p:=plot({h,j,h-j,j-h},x=-2..2):
> display({p,t1,t2});
```

Notice that the graph of $j(x) - h(x)$ is the reflection in the x-axis of the graph of $h(x) - j(x)$.

Exercise 2.4   Given $f(x) = x^3 - 8$ and $g(x) = x + 4$, plot $f(x)$, $g(x)$, and $f(x) + g(x)$, $-2 \leq x \leq 2$, on the same graph.

Exercise 2.5   A function f is called *even* if $f(-x) = f(x)$ for all x and it is called *odd* if $f(-x) = -f(x)$ for all x. Give examples of even functions and of odd functions. Can you find a function that is odd and even?

Exercise 2.6   Use the identity $f(x) = \frac{1}{2}(f(x) + f(-x)) + \frac{1}{2}(f(x) - f(-x))$ to show that every function is expressible as a sum of an even function and an odd function.

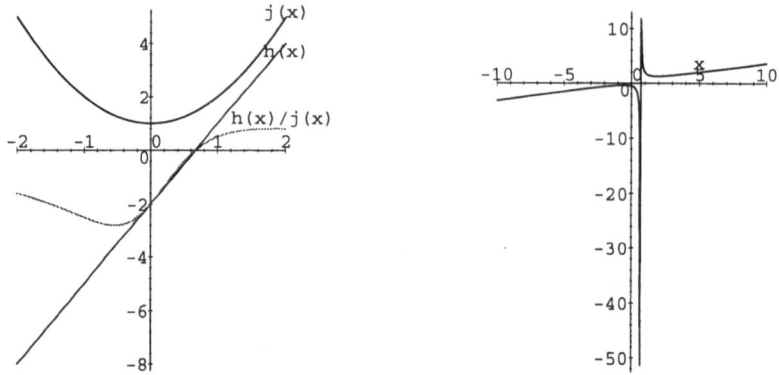

Figure 2.10. (a) Division of functions. (b) The graph of $\dfrac{j(x)}{h(x)}$, on $[-10, 10]$

Exercise 2.7   Let $[[x]]$ be the greatest integer less than or equal to x. Plot the graphs of $[[x]]$, $x - [[x]]$, $\sqrt{x - [[x]]}$, $x + \sqrt{x - [[x]]}$, and $[[x]] + \sqrt{x - [[x]]}$ using *Maple's* `trunc` function to calculate $[[x]]$, ie `trunc(x)` $= [[x]]$.

2.4   Multiplication and Division of Functions

As with addition and subtraction, functions multiply and divide through the way those operations act on their output values. The function $\frac{h}{j}(x)$ is equal to $\frac{h(x)}{j(x)}$ when $j(x) \neq 0$.

```
> 'h/j'=h/j;
```

$$h/j = \frac{3x - 2}{x^2 + 1}$$

Solve for $j(x) = 0$ to determine if the function $\frac{h}{j}(x)$ is undefined at any point.

```
> solve(j,{x});
```

$$\{x = I\}, \{x = -I\}$$

Since there is no real number for which $j(x) = 0$, the function $\frac{h}{j}(x)$ is well-defined for all x values, as seen in Figure 2.10(a).

```
> t1:=textplot([2,4,'h(x)'],align=BELOW):
> t2:=textplot([[2,5,'j(x)'],[2,1,'h(x)/j(x)']],align=ABOVE):
> p:=plot({h,j,h/j},x=-2..2):
> display({p,t1,t2});
```

What happens if we try to divide $j(x)$ by $h(x)$? See Figure 2.10(b).

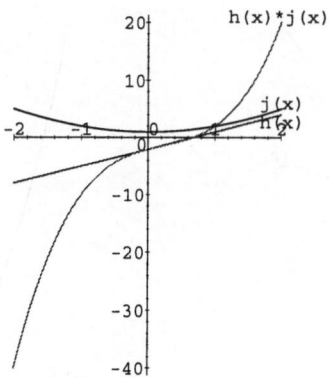

Figure 2.11. Multiplication of functions

```
> 'j/h'=j/h;
```

$$j/h = \frac{x^2 + 1}{3x - 2}$$

```
> solve(h,{x});
```

$$\left\{ x = \frac{2}{3} \right\}$$

```
> plot(j/h,x);
```

This function is undefined at the point at which $h(x) = 0$, namely $x = \frac{2}{3}$; so the graph has a break in it at that point.

When multiplying functions we don't need to worry about breaks, unless the initial functions themselves are undefined at some point. (Figure 2.11).

```
> 'h*j'=sort(expand(h*j));
```

$$h * j = 3x^3 - 2x^2 + 3x - 2$$

```
> t1:=textplot([2,4,'h(x)'],align=BELOW):
> t2:=textplot([[2,5,'j(x)'],[2,20,'h(x)*j(x)']],align=ABOVE):
> p:=plot({h,j,h*j},x=-2..2):
> display({p,t1,t2});
```

Exercise 2.8   Given $f(x) = x^2 - 1$ and $g(x) = x$, plot $f(x)$, $g(x)$, and $f(x) \times g(x)$, $-2 \le x \le 2$, on the same graph.

2.5   Composition of Functions

A function can be thought of as a black box. Put a number in, and a number comes out. Put a function in, and a *composite* function comes out. In this

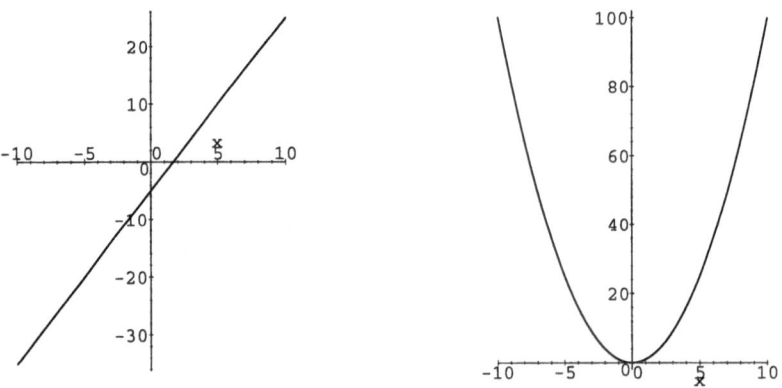

Figure 2.12. (a) The graph of $f(x) = 3x - 5$, on $[-10, 10]$. (b) The graph of $g(x) = x^2$, on $[-10, 10]$

example we shall compose functions from the functions $f(x) = 3x - 5$ and $g(x) = x^2$, both shown in Figure 2.12.

```
> f:=x->3*x-5;
```
$$f := x \to 3x - 5$$

```
> plot(f(x),x=-10..10);
> g:=x->x^2;
```
$$g := x \to x^2$$

```
> plot(g(x),x=-10..10);
> '(f@g)(x)'=(f@g)(x);
```
$$(f@g)(x) = 3x^2 - 5$$

```
> 'f(g(x))'=f(g(x));
```
$$f(g(x)) = 3x^2 - 5$$

The command `(f@g)(x);` gives the same result as `f(g(x));`, literally, 'the value of f at $g(x)$'. These notations are equivalent. In some books $f \circ g$ represents $f@g$. The composite function $f@g(x)$ is graphed in Figure 2.13(a).

```
> t:=textplot([[10,25,'f(x)'],[10,100,'g(x)'],[10,295,'f(g(x))']],
> align=RIGHT):
> p:=plot({f(x),g(x),(f@g)(x)},x=-10..10):
> display({t,p});
> 'g(f(x))'=(g@f)(x);
```
$$g(f(x)) = (3x - 5)^2$$

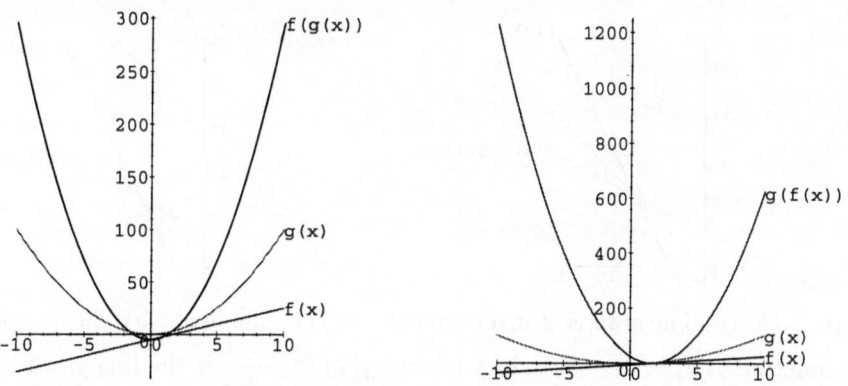

Figure 2.13. (a) The graph of the composite of $f(x) = 3x - 5$ and $g(x) = x^2$. (b) The graph of the composite of $g(x) = x^2$ and $f(x) = 3x - 5$

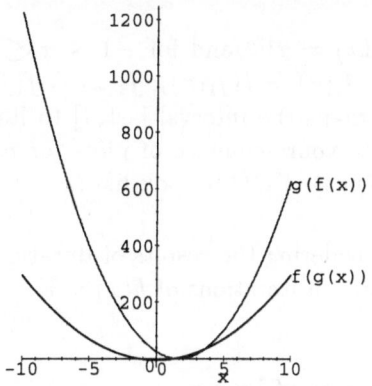

Figure 2.14. The graphs of $(f@g)(x)$ and $(g@f)(x)$

```
> 'g(f(x))'=expand((g@f)(x));
```
$$g(f(x)) = 9\,x^2 - 30\,x + 25$$

The composite function $g@f(x)$ is graphed in Figure 2.13(b).

```
> t:=textplot([[10,25,'f(x)'],[10,100,'g(x)'],[10,620,'g(f(x))']],
> align=RIGHT):
> p:=plot({f(x),g(x),(g@f)(x)},x=-10..10):
> display({t,p});
```

In general, $f(g(x))$ is not equal to $g(f(x))$. Indeed, not often do both exist. The two composite functions are overlayed in Figure 2.14.

```
> p:=plot({(f@g)(x),(g@f)(x)},x):
> t:=textplot([[10,620,'g(f(x))'],[10,295,'f(g(x))']],align=RIGHT):
> display({t,p});
```

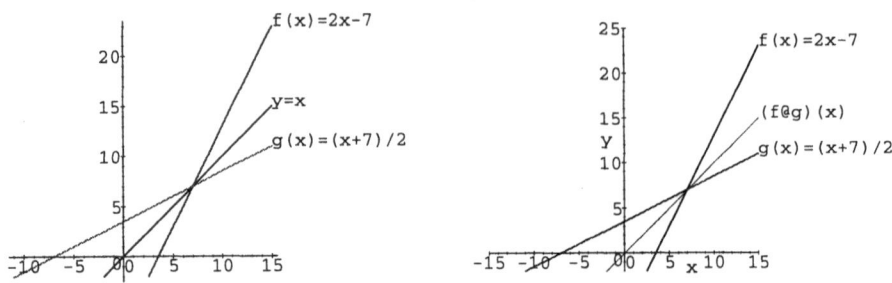

Figure 2.15. (a) The graphs of inverse functions $f(x)$ and $g(x)$. (b) The composite of a function $f(x) = 2x - 7$ and its inverse $g(x) = \dfrac{x+7}{2}$ is the line $y = x$

Exercise 2.9 Given $f(x) = 8x - 7$ and $g(x) = x^2 + x$, find $f(g(x))$, $g(f(x))$, and $f(f(g(x)))$.

Exercise 2.10 Let $f(x) = x^{1/3}$ and for $-1 \le x \le 1$. Consider the iterated functions $f_1(x) = f(x)$, $f_2(x) = f(f_1(x))$, $f_3(x) = f(f_2(x))$, ... Show that each of these new functions maps the interval $[-1, 1]$ to itself and plot their graphs for $n = 1, 2, \ldots$ Animate your sequence of plots for n up to 10. What do you expect the iterations of $g(x) = x^{1/5}$ to look like?

Exercise 2.11 By considering the results of iterations of $f(x) = x^{1/3}$ in Exercise 2.10, deduce the form of iterations of $h(x) = x^3$.

2.6 Inverse Functions

It may happen that we can compose f and g both ways, and that for all x we have $f(g(x)) = g(f(x)) = x$. Then each function undoes what the other does and we say that each is the inverse function of the other. If we write $y = f(x)$ then f has an inverse precisely when this equation has a unique solution for x.

If $y = 2x$, then clearly $x = \frac{1}{2}y$ is an equivalent statement—that is, both x and y have the same meaning in each. Also, if $v = u - 7$, then equivalently, $u = v + 7$. Now, putting these together, if $y = 2x - 7$ then $y + 7 = 2x$ and $x = \frac{1}{2}(y+7)$. We plot both functions $x \mapsto 2x - 7$ and $x \mapsto \frac{1}{2}(y+7)$ in Figure 2.15. Each is the inverse function of the other. We can arrive at this in *Maple* by the following steps:

1. Define *eq1* to be $y = 2x - 7$.

2. Interchange x and y to give *eq2* as $x = 2y - 7$.

3. Plot *eq1* and *eq2*, as well as the line $y = x$, using `implicitplot`.

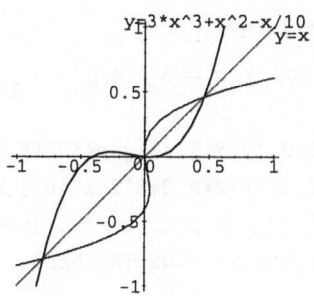

Figure 2.16. The graphs of the function $3x^3 + x^2 - \frac{1}{10}x$ and its inverse.

```
> eq1:=y=2*x-7;
```
$$eq1 := y = 2x - 7$$

```
> eq2:=subs({y=x,x=y},eq1);
```
$$eq2 := x = 2y - 7$$

```
> t:=textplot([[15,23,'f(x)=2x-7'],[15,15,'y=x'],
> [15,11,'g(x)=(x+7)/2']],align={ABOVE,RIGHT}):
> p:=implicitplot({eq1,eq2,y=x},x=-15..15,y=-2..25,
> scaling=CONSTRAINED):
> display({t,p});
```

The graphs of the composite of a function and its inverse are mirror images in the identity function $y = x$. This can be seen in Figure 2.15.

```
> f:=x->2*x-7;
```
$$f := x \rightarrow 2x - 7$$

```
> g:=x->1/2*x+7/2;
```
$$g := x \rightarrow \frac{1}{2}x + \frac{7}{2}$$

```
> '(f@g)(x)'=(f@g)(x);
```
$$(f@g)(x) = x$$

```
> p:=plot({f(x),g(x),(f@g)(x)},x=-15..15,y=-2..25,
> scaling=CONSTRAINED):
> t:=textplot([[15,23,'f(x)=2x-7'],[15,15,'(f@g)(x)'],
> [15,11,'g(x)=(x+7)/2']],align={ABOVE,RIGHT}):
> display({t,p});
```

Here's a more difficult example, plotted in Figure 2.16.

```
> eq3:=y=3*x^3+x^2-x/10;
```
$$eq3 := y = 3x^3 + x^2 - \frac{1}{10}x$$

```
> eq4:=subs({y=x,x=y},eq3);
```

$$eq4 := x = 3\,y^3 + y^2 - \frac{1}{10}\,y$$

```
> t1:=textplot([1,0.9,'y=x'],align={ABOVE,RIGHT}):
> t2:=textplot([0.6,1,'y=3*x^3+x^2-x/10'],align=ABOVE):
> p:=implicitplot({eq3,eq4,y=x},x=-1..1,y=-1..1):
> display({t1,t2,p},scaling=CONSTRAINED);
```

Exercise 2.12 Given $f(x) = 5x - 9$, find the inverse of this function using simultaneous substitutions and plot these relations.

Experiment 2.3 The relation R is defined on the set $1, 2, \ldots, 10$ and has the property that each of these elements is related to every one of the numbers $1, 2, \ldots, 10$. Use **plot** to make a graph of R. It should look like a square array of 100 dots. Let these dots have coordinates (x_i, y_i) for $i = 1, 2, \ldots, 100$. Choose four numbers a, b, c, d and for each construct a function $f : (x_i, y_i) \mapsto (ax_i + by_i, cx_i + dy_i)$. Then f transforms R into a new relation. Use **plot** to investigate the effect of different values for a, b, c, d. What happens when $ad = bc$? What happens when $ad < bc$? Show that the function preserves the area of the dot pattern when $ad - bc = \pm 1$.

Experiment 2.4 Consider the x, y plane and a particle P that begins at time $t = 0$ at $(0,0)$. For each time $t = 1, 2, 3 \ldots$ P moves unit distance along the x−axis or along the y−axis, either forwards or backwards; so before each move it has precisely 4 choices. If it makes this choice completely at random then P illustrates a *random walk*. What is the probability that a random walk consisting of 4 steps actually defines a function? What is the average distance from the origin after 4 steps? What about 10 steps, or n steps? You could also try this on a line, or in space. Try making a *Maple* animation of the steps, or of the collection of all possible random walks consisting of four steps. What bias in movement is needed to ensure that a *function* is generated by a walk in the plane?

Chapter 3

Quadratic Functions

Commands used in this chapter

- `assign`
- `coeff`
- `evalf`
- `expand`
- `plot`
- `plot[options]`
- `plots`
- `plots[display]`

- `plots[textplot]`
- `solve`
- `student`
- `student[completesquare]`
- `student[maximize]`
- `student[minimize]`
- `with`

3.1   Quadratic Functions

Many important functions in mathematics, and especially in its applications, are *continuous functions*. These have the property that small changes in input result in small changes in output. Linear functions have this property. The simplest non-linear continuous functions are the *quadratics*. The general form of a quadratic function is

$$f(x) = ax^2 + bx + c, \text{ where } a \neq 0.$$

Some examples of quadratic functions are:

```
> f:=x->2*x^2;
```

$$f := x \rightarrow 2\,x^2$$

```
> g:=x->4*x^2-13*x+Pi;
```

$$g := x \rightarrow 4\,x^2 - 13\,x + \pi$$

```
> j:=x->-1/7*x^2+2;
```

$$j := x \rightarrow -\frac{1}{7}\,x^2 + 2$$

The following functions are also quadratic, even though they do not appear so at first:

```
> h:=x->(x+1)^2;
```

$$h := x \rightarrow (\,x+1\,)^2$$

```
> expand(h(x));
```
$$x^2 + 2x + 1$$

```
> k:=x->(3*x-2)^2+6;
```
$$k := x \rightarrow (3x - 2)^2 + 6$$

```
> expand(k(x));
```
$$9x^2 - 12x + 10$$

The functions in the next group are not quadratic:
```
> f:=x->(x+3)^2-x^2;
```
$$f := x \rightarrow (x + 3)^2 - x^2$$

```
> expand(f(x));
```
$$6x + 9$$

```
> g:=x->x*(2*x+1)*(x+1);
```
$$g := x \rightarrow x(2x + 1)(x + 1)$$

```
> expand(g(x));
```
$$2x^3 + 3x^2 + x$$

3.2   Parabola

Another form of the general equation of a quadratic function, or *parabola*, is

$$f(x) = a(x - b)^2 + c$$

Each of the coefficients a, b, and c has an effect on the shape and position of the graph of the parabola. The simplest equation of a parabola is $f(x) = x^2$, shown in Figure 3.1(a). All quadratic functions have the characteristic parabolic shape.

The coefficient a determines the steepness and direction of opening of a parabola. The plot that follows (Figure 3.1(b)) illustrates the effect of varying a when $b = c = 0$.

```
> with(plots):
> p:=plot({-2*x^2,x^2,1/2*x^2,2*x^2},x=-2..2,scaling=CONSTRAINED):
> t:=textplot([[2,-8,'a=-2'],[2,2,'a=1/2'],[2,4,'a=1'],
> [2,8,'a=2']],align=RIGHT):
> display({p,t});
```

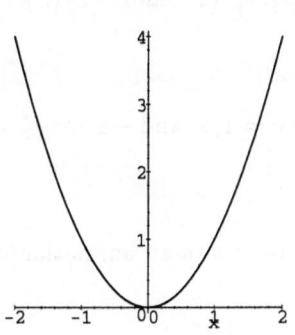 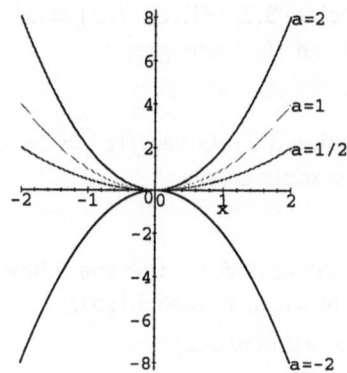

Figure 3.1. (a) Parabola. (b) Effect of varying a on the graph $f(x) = ax^2$

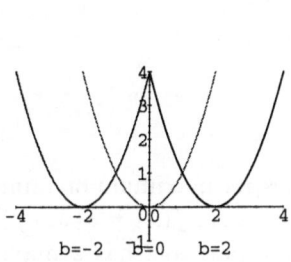 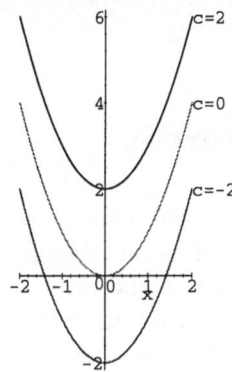

Figure 3.2. (a) Effect of varying b on the graph $f(x) = (x - b)^2$. (b) Effect of varying c on the graph $f(x) = x^2 + c$

The coefficient b determines the position of a parabola in the x-direction. Figure 3.2(a) shows the effect of varying b when $a = 1$ and $c = 0$.

```
> p:=plot({(x+2)^2,x^2,(x-2)^2},x=-4..4,y=-1..4,
> scaling=CONSTRAINED):

> t:=textplot([[-2,-1,'b=-2'],[0,-1,'b=0'],[2,-1,'b=2']],
> align=BELOW):

> display({p,t});
```

The coefficient c determines the position of a parabola in the y-direction. The plot below, shown in Figure 3.2(b), illustrates the effect of varying c when $a = 1$ and $b = 0$.

```
> p:=plot({x^2+2,x^2,x^2-2},x=-2..2,scaling=CONSTRAINED):

> t:=textplot([[2,2,'c=-2'],[2,4,'c=0'],[2,6,'c=2']],align=RIGHT):

> display({p,t});
```

Exercise 3.1   Given $f(x) = x^2$, plot $f(x)$, $2f(x)$, $\frac{1}{2}f(x)$, and $-2f(x)$, $-4 \leq x \leq 4$, on the same graph.

Exercise 3.2   Given $f(x) = 2x^2 + c$, plot $f(x)$ for $c = 1, 2$, and -2, $-4 \leq x \leq 4$, on the same graph.

Experiment 3.1   Use the following *Maple* code to create an animation of the plot shown in Figure 3.1(b).

```
> with(plots):
> animate(a*x^2,x=-4..4,a=-2..2);
```

Alter the `animate` command by replacing `a*x^2` with `(x-a)^2`, and `x^2+a` to animate the plots shown in Figure 3.2. Create your own animations in the same way.

3.3   Vertex

The *vertex* is the point at which a parabola reaches its maximum or minimum value, depending on the sign of a. A parabola of the form $f(x) = a(x - p)^2 + q$ has its vertex at the point (p, q). If $a < 0$, then the parabola has a maximum value at the vertex. Conversely, if $a > 0$, then the parabola has a minimum value at that point. Also, the equation $x - p = 0$ defines the axis of symmetry of the parabola.

Using *Maple*, we can easily estimate the vertex of a parabola $f(x) = ax^2 + bx + c$ by plotting it and using the cursor to point at the vertex.

Example 3.1   What are the coordinates of the vertex of the parabola $f(x) = 3x^2 - 6x - 2$?

```
> f:=x->3*x^2-6*x-2;
```
$$f := x \rightarrow 3x^2 - 6x - 2$$

```
> plot(f(x),x=-2..2);
```

By examining the plot, shown in Figure 3.3(a), the coordinates of the minimum can be estimated to be $(1, -5)$. Knowing this, we can arrange $f(x) = 3x^2 - 6x - 2$ into the form $f(x) = a(x - p)^2 + q$ as follows,

```
> a:=coeff(f(x),x,2);p:=1;q:=-5;
```
$$a := 3$$
$$p := 1$$

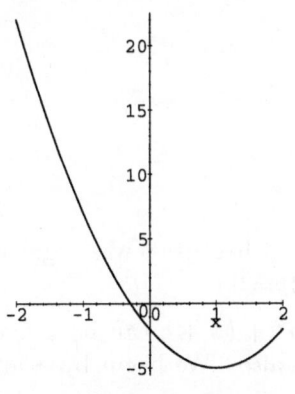 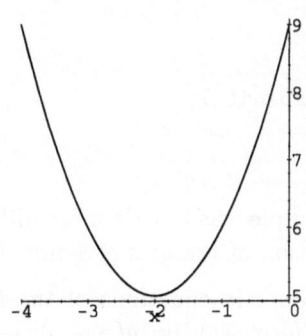

Figure 3.3. (a) The graph of $f(x) = 3x^2 - 6x - 2$, on $[-2, 2]$. (b) The graph of $y = x^2 + 4x + 9$, on $[-4, 0]$

$$q := -5$$

```
> a*(x-p)^2+q;
```
$$3 \left(x - 1 \right)^2 - 5$$

We expand the above equation to be sure that it is the same equation we started with, $f(x) = 3x^2 - 6x - 2$.

```
> expand(");
```
$$3 x^2 - 6 x - 2$$

Example 3.2   Convert the equation $f(x) = x^2 + 4x + 9$ into the form $f(x) = a(x - p)^2 + q$ using the graphical method.

We first plot the parabola (Figure 3.3(b)) and estimate the coordinates of the vertex.

```
> f:=x->x^2+4*x+9;
```
$$f := x \to x^2 + 4 x + 9$$

```
> plot(f(x),x=-4..0,scaling=CONSTRAINED);
> a:=coeff(f(x),x,2);p:=-2;q:=5;
```
$$a := 1$$

$$p := -2$$

$$q := 5$$

```
> a*(x-p)^2+q;
```

$$(x+2)^2 + 5$$

```
> expand(");
```

$$x^2 + 4x + 9$$

This example was a little more difficult than the first because p was negative, so the equation of the axis of symmetry was $x - (-2) = 0$.

Quadratic equations of the form $f(x) = ax^2 + bx + c$ can be converted into the form $f(x) = a(x-p)^2 + q$ algebraically also. We begin by exanding $f(x) = a(x-p)^2 + q$,

```
> expand(a*(x-p)^2+q);
```

$$a x^2 - 2 a x p + a p^2 + q$$

By grouping coefficients of terms of the same order in the equations $f(x) = ax^2 + bx + c$ and $g(x) = ax^2 - 2axp + ap^2 + q$, we arrive at the following set of equations,

```
> eq1:=b=-2*a*p;
```

$$eq1 := b = -2\,a\,p$$

```
> eq2:=c=a*p^2+q;
```

$$eq2 := c = a\,p^2 + q$$

Example 3.3   Express $f(x) = 2x^2 - 8x + 11$ in the form $f(x) = a(x-p)^2 + q$ using the algebraic method above.

```
> f:=x->2*x^2-8*x+11;
```

$$f := x \rightarrow 2\,x^2 - 8\,x + 11$$

```
> a:=coeff(f(x),x,2);b:=coeff(f(x),x,1);c:=coeff(f(x),x,0);
```

$$a := 2$$

$$b := -8$$

$$c := 11$$

```
> solve(eq1,{p});
```

$$\{p = 2\}$$

```
> assign(");
> solve(eq2,{q});
```

$$\{q = 3\}$$

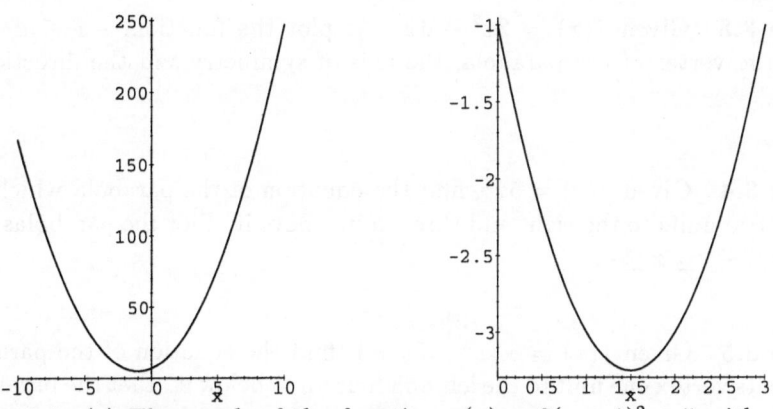

Figure 3.4. (a) The graph of the function $y(x) = 2(x + 1)^2 + 5$ with vertex at $(-1, 5)$. (b) The graph of $y = x^2 - 3x - 1$, on $[0, 3]$

```
> assign(");
> a*(x-p)^2+q;
```

$$2(x - 2)^2 + 3$$

```
> expand(");
```

$$2x^2 - 8x + 11$$

A second and most important algebraic method to convert quadratic equations of the form $f(x) = ax^2 + bx + c$ to the form $f(x) = a(x - p)^2 + q$ is called *completing the square*. This can be done easily using *Maple*.

```
> f:=x->2*x^2+4*x+7;
```

$$f := x \rightarrow 2x^2 + 4x + 7$$

```
> with(student):
> completesquare(f(x));
```

$$2(x + 1)^2 + 5$$

```
> plot(f(x),x);
```

Observe that completing the square does not always result in whole numbers.

```
> f:=x->5*x^2-3*x+2;
```

$$f := x \rightarrow 5x^2 - 3x + 2$$

```
> completesquare(f(x));
```

$$5\left(x - \frac{3}{10}\right)^2 + \frac{31}{20}$$

Exercise 3.3 Given $f(x) = 2x^2 - 6x + 7$, plot the function, $-4 \leq x \leq 4$, and find the vertex of the parabola, the axis of symmetry and the direction of opening.

Exercise 3.4 Given $f(x) = 5x^2$, find the equation of the parabola which has its vertex two units to the right and three units above it. Plot the parabolas over the interval $-4 \leq x \leq 4$.

Exercise 3.5 Given $f(x) = -3x^2 - 7x + 1$, find the equation of the parabola which has its vertex one unit to the left and four units below it. Plot the parabolas over the interval $-4 \leq x \leq 4$.

3.4 Maximum and Minimum Values

It is often necessary to know the maximum or minimum value of a quadratic function to solve problems. There are two commands in the **student** package, **maximize** and **minimize**, which allow us to find them.

```
> f:=x->x^2-3*x-1;
```
$$f := x \rightarrow x^2 - 3x - 1$$

```
> plot(f(x),x=0..3);
```
We can see from the plot (Figure 3.4(b)), that the function $f(x) = x^2 - 3x - 1$ has a minimum value.

```
> minimize(f(x));
```
$$\frac{-13}{4}$$

```
> evalf(minimize(f(x)),4);
```
$$-3.250$$

Notice that the minimum value is found at the vertex. We can get the same result by completing the square or any of the other methods for finding the vertex.

```
> completesquare(f(x));
```
$$\left(x - \frac{3}{2}\right)^2 - \frac{13}{4}$$

Example 3.4 What is the maximum area enclosed by a wall with the dimensions shown in Figure 3.5?

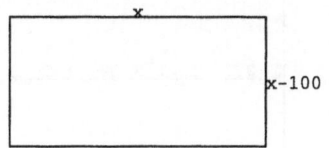

Figure 3.5. Area enclosed by a wall

We begin by calculating the area in terms of x.

> l:=x;
$$l := x$$

> w:=100-x;
$$w := 100 - x$$

> A:=l*w;
$$A := x\,(\,100 - x\,)$$

> maximize(A);
$$2500$$

> solve(A=maximize(A),{x});
$$\{\,x = 50\,\},\{\,x = 50\,\}$$

> assign(");
> w;
$$50$$

So, the maximum area is enclosed by a wall with dimensions 50 unit lengths by 50 unit lengths. Thus, a square is more efficient in capturing area than any other rectangle of the same perimeter.

Experiment 3.2   What area would be enclosed by a regular pentagon of perimeter 50? What about n-sides polygons? Guess the limiting area as the number of sides n approaches infinity.

3.5   Applications of Quadratic Functions

Quadratic functions can be applied to many real world application. For example, the position of an object in free fall, that is, falling under the force of gravity, is

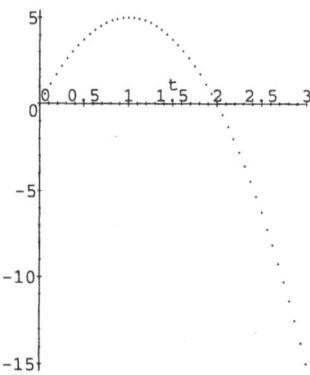

Figure 3.6. Vertical position of a projectile with time

described by the function

$$s(t) = \frac{1}{2}gt^2 + v_0 t + s_0$$

where $s(t)$ is the position of the object at time t, g is the gravitional constant -9.8 m/s², v_0 is the initial velocity of the object, and s_0 is its intial position.

Suppose a projectile is thrown upwards and falls over a cliff with an initial velocity of 10 m/s. If we assume that the projectile is thrown from ground level, then we can set $s_0 = 0$.. The graph of the quadratic function describing the position of the projectile is shown in Figure 3.6.

```
> plot(10*t-5*t^2,t=0..3,style=POINT);
```

Exercise 3.6 A golf ball is hit and in the air follows a parabolic trajectory starting at $x = 0$, $y = 0$ with maximum altitude $y = 20$ m; it strikes the ground at $x = 120$ m. Find an equation for the parabola.

Experiment 3.3 A golf ball is hit horizontally with velocity u m/s off the top of a cliff of height 100 m at time $t = 0$. Its height y reduces with time according to the equation $y = 100 - \frac{1}{2}gt^2$ where $g = 9.8$ m/s² is the acceleration due to gravity. Also, its distance from the cliff at time t is $x = ut$, neglecting air resistance. Hence the trajectory is $y = 100 - \frac{g}{2u^2}x^2$. Plot the unique curve showing the loss of height of the particle. Plot also a family of curves showing the trajectories for a range of initial velocities.

Chapter 4

Solving Quadratic Equations

Commands used in this chapter
- `assign`
- `coeff`
- `evalf`
- `expand`
- `fsolve`
- `op`
- `plot`
- `plot[options]`
- `solve`
- `sqrt`
- `student`
- `student[changevar]`
- `with`

In this chapter, a new form of the `solve` command will be introduced. When we solve equations with more than one solution, we cannot use the `assign` command as we have in the past.

```
> solve(x^2-3*x-4,{x});
```
$$\{ x = 4 \}, \{ x = -1 \}$$

```
> assign(");
> x;
```
$$-1$$

One way is to use the `op` command to extract operands from a solution. Notice that the value to be solved for (x) is not in curly brackets, and that the solution is surrounded by square brackets, this is *Maple's* way of denoting an *ordered* set.

```
> [solve(x^2-3*x-4,x)];
```
$$[4, -1]$$

To extract the first operand from the last output, we type,

```
> op(1,");
```

4

To extract the second operand from the second to last ouput, we type,

> op(2,"");

$$-1$$

The other way is to name the solution, as below.

> root:=[solve(x^2-3*x-4,x)];

$$root := [4, -1]$$

> root[1];

$$4$$

> root[2];

$$-1$$

4.1   Roots of Quadratic Equations

The *roots*, or *zeros*, of the general form of the quadratic equation, $f(x) = ax^2 + bx + c$ are those values of x for which $f(x) = 0$, or $ax^2 + bx + c = 0$.

> root:=[solve(a*x^2+b*x+c=0,x)];

$$root := \left[\frac{1}{2} \frac{-b + \sqrt{b^2 - 4\,a\,c}}{a}, \frac{1}{2} \frac{-b - \sqrt{b^2 - 4\,a\,c}}{a} \right]$$

For example, what are the roots of $f(x) = x^2 - 3x + 2$? As with any new function, the best way to begin investigating its properties is to plot it (Figure 4.1).

> f:=x->x^2-3*x+2;

$$f := x \rightarrow x^2 - 3\,x + 2$$

> plot(f(x),x=-1..3);
> a:=coeff(f(x),x,2);b:=coeff(f(x),x,1);c:=coeff(f(x),x,0);

$$a := 1$$

$$b := -3$$

$$c := 2$$

> root[1];

$$2$$

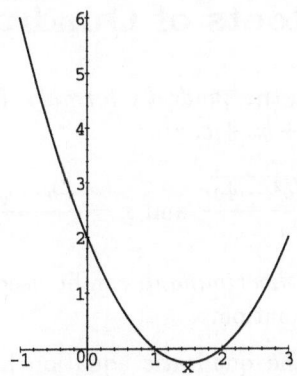

Figure 4.1. The graph of $f(x) = x^2 - 3x + 2$, on $[-1, 3]$

```
> root[2];
```
$$1$$

Of course, the easiest way to find the roots of an equation in *Maple* is to solve it directly.
```
> solve(f(x),{x});
```
$$\{\, x = 2\,\}, \{\, x = 1\,\}$$

Exercise 4.1   Determine the roots of the function $y = 2x^2 - x - 1$ graphically.

Exercise 4.2   Determine the roots of the following functions using any method.

   (a) $y = 2x^2 + 3x + 9$
   (b) $y = x^2 - 9$
   (c) $y = x^2 - 4x + 1$

Exercise 4.3   The roots of $x^2 + bx + c$ are u, v; express $(u^2 + v^2)$ in terms of b, c.

Exercise 4.4   One root of $ax^2 + bx + c = 0$ is twice the other root; prove that $2b^2 = 9ac$.

Experiment 4.1   Completing the square of $x^2 + 2x - 8$ we get $(x + 1)^2 - 9$ so the roots are given by $(x + 1) = 3$ and we find them to be $x = -4$ and $x = 2$. Apply the same procedure to $ax^2 + bx + c$ and deduce the formula for solving quadratics.

4.2   Nature of Roots of Quadratic Equations

As we saw in the last section, the *quadratic formulas* for the roots of the general quadratic equation, $y = ax^2 + bx + c$, are

$$x = \frac{-b + \sqrt{b^2 - 4ac}}{2a} \text{ and } x = \frac{-b - \sqrt{b^2 - 4ac}}{2a}$$

The term $b^2 - 4ac$, called the *discriminant*, can be used to determine the number of real roots of a quadratc equation.

When $b^2 - 4ac = 0$, the quadratic equation has one distinct real root. Consider the function $f(x) = (x - 2)^2$, shown in Figure 4.2(a).

```
> f:=x->(x-2)^2;
```

$$f := x \rightarrow (x - 2)^2$$

```
> plot(f(x),x=-2..6,y=-2..6);
> expand(f(x));
```

$$x^2 - 4x + 4$$

```
> a:=coeff(expand(f(x)),x,2);b:=coeff(expand(f(x)),x,1);
> c:=coeff(expand(f(x)),x,0);
```

$$a := 1$$

$$b := -4$$

$$c := 4$$

```
> discriminant:=(b^2-4*a*c);
```

$$discriminant := 0$$

```
> solve(f(x),{x});
```

$$\{ x = 2 \}, \{ x = 2 \}$$

When $b^2 - 4ac > 0$, the square root of the discriminant is a real number and the quadratic equation, such as $g(x)$ shown in Figure 4.2(b), has two real roots.

```
> g:=x->(x-2)^2-2;
```

$$g := x \rightarrow (x - 2)^2 - 2$$

```
> plot(g(x),x=-1..5,y=-2..4);
> expand(g(x));
```

$$x^2 - 4x + 2$$

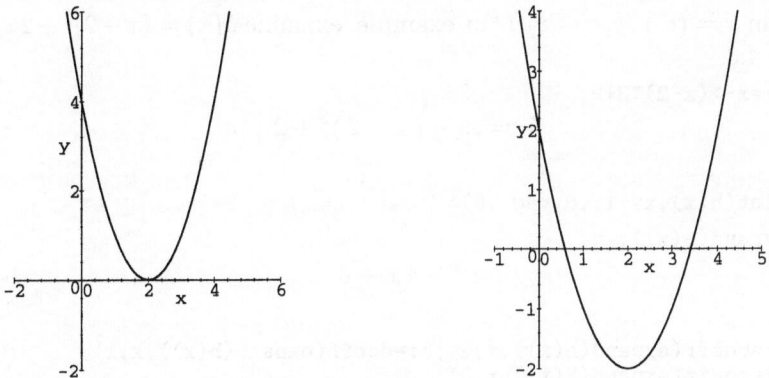

Figure 4.2. (a) Quadratic equation with one real root. (b) Quadratic equation with two real roots

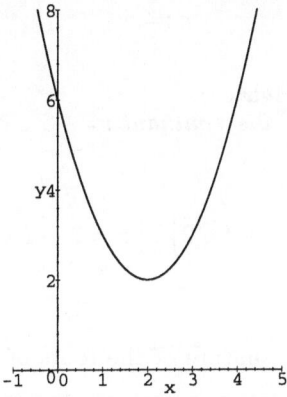

Figure 4.3. Quadratic equation with no real roots

```
> a:=coeff(expand(g(x)),x,2);b:=coeff(expand(g(x)),x,1);
> c:=coeff(expand(g(x)),x,0);
```
$$a := 1$$
$$b := -4$$
$$c := 2$$

```
> discriminant:=(b^2-4*a*c);
```
$$discriminant := 8$$

```
> solve(g(x),{x});
```
$$\left\{ x = 2 + \sqrt{2} \right\}, \left\{ x = 2 - \sqrt{2} \right\}$$

When $b^2 - 4ac < 0$, the quadratic equation has no real roots. The roots are *complex*; in fact they form a *conjugate pair*. (A conjugate pair is a set of solutions of the form $x = (a + ib, a - ib)$.) For example, examine $h(x) = (x-2)^2 + 2$ shown in Figure 4.3.

```
> h:=x->(x-2)^2+2;
```
$$h := x \rightarrow (x-2)^2 + 2$$

```
> plot(h(x),x=-1..5,y=0..8);
> expand(h(x));
```
$$x^2 - 4x + 6$$

```
> a:=coeff(expand(h(x)),x,2);b:=coeff(expand(h(x)),x,1);
> c:=coeff(expand(h(x)),x,0);
```
$$a := 1$$
$$b := -4$$
$$c := 6$$

```
> discriminant:=(b^2-4*a*c);
```
$$discriminant := -8$$

```
> solve(h(x),{x});
```
$$\left\{x = 2 + I\sqrt{2}\right\}, \left\{x = 2 - I\sqrt{2}\right\}$$

Exercise 4.5   Determine the nature of the roots of the following functions.

(a) $y = 3x^2 - 1$

(b) $y = x^2 - 3x + 1$

(c) $y = 8x^2 - x + 1$

Exercise 4.6   For which value of k do the following equations have one distinct real root?

(a) $y = kx^2 - 3x + 2$

(b) $y = x^2 - 2x + k$

(c) $y = x^2 + kx + 5$.

4.3   Quadratic Equations in Other Variables

Not all quadratic equations are defined in terms of x. The functions $f(\cos(\theta)) = 3\cos(\theta)^2 - 7\cos(\theta) + 4$ and $f(2^x) = 2(2^x)^2 - 7(2^x) + 6$ are two such equations.

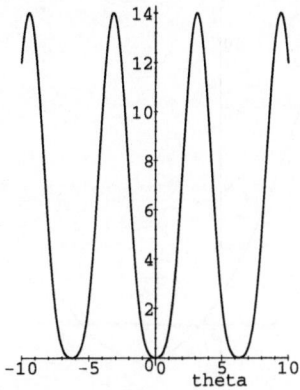
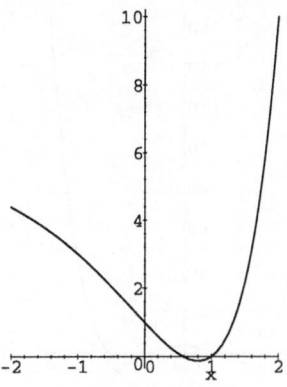

Figure 4.4. (a) The graph of $f(\cos(\theta)) = 3\cos(\theta)^2 - 7\cos(\theta) + 4$, on $[-10, 10]$.
(b) The graph of $f(2^x) = 2(2^x)^2 - 7(2^x) + 6$, on $[-2, 2]$

These plots are shown in Figure 4.4.

```
> f:=cos(theta)->3*cos(theta)^2-7*cos(theta)+4;
```
$$f := \cos(\theta) \rightarrow 3\cos(\theta)^2 - 7\cos(\theta) + 4$$

```
> plot(f(cos(theta)),theta);
> solve(f(cos(theta)),{cos(theta)});
```
$$\left\{\cos(\theta) = \frac{4}{3}\right\}, \{\cos(\theta) = 1\}$$

```
> f:=(2^x)->2*(2^x)^2-7*(2^x)+6;
```
$$f := 2^x \rightarrow 2(2^x)^2 - 7\,2^x + 6$$

```
> plot(f(2^x),x=-2..2);
> solve(f(2^x),{2^x});
```
$$\{2^x = 2\}, \left\{2^x = \frac{3}{2}\right\}$$

Sometimes it is possible to reduce equations into quadratic form using a simple substitution. For example, consider the function $f(x) = x^4 - 6x^2 + 4$ shown in Figure 4.5(a).

```
> with(student):
> f:=x->x^4-6*x^2+4;
```
$$f := x \rightarrow x^4 - 6x^2 + 4$$

```
> plot(f(x),x=-4..4);
```

If we substitute the variable x^2 with w, and y with v, we can reduce this quartic equation into a quadratic equation, $v(w)$, shown in Figure 4.5(b).

```
> v:=changevar(x^2=w,f(x));
```

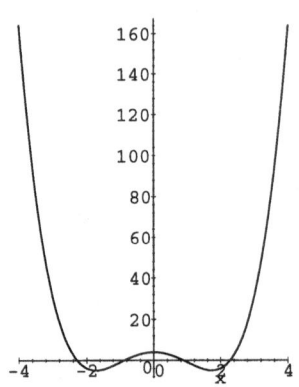 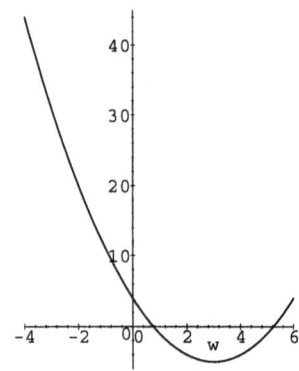

Figure 4.5. (a) The graph of $f(x) = x^4 - 6x^2 + 4$, on $[-4, 4]$. (b) The graph of $v(w) = w^2 - 6w + 4$, on $[-4, 6]$

$$v := w^2 - 6w + 4$$

```
> plot(v,w=-4..6);
> w:=[solve(v)];
```

$$w := \left[3 + \sqrt{5}, 3 - \sqrt{5}\right]$$

```
> x[1]:=sqrt(w[1]);
```

$$x_1 := \frac{1}{2}\sqrt{10} + \frac{1}{2}\sqrt{2}$$

```
> evalf(x[1]);
```

$$2.288245611$$

```
> x[2]:=sqrt(w[2]);
```

$$x_2 := \frac{1}{2}\sqrt{10} - \frac{1}{2}\sqrt{2}$$

```
> evalf(x[2]);
```

$$.8740320490$$

The function $f(x)$ can be solved directly.

```
> solve(f(x));
```

$$\frac{1}{2}\sqrt{10} + \frac{1}{2}\sqrt{2}, \frac{1}{2}\sqrt{10} - \frac{1}{2}\sqrt{2}, -\frac{1}{2}\sqrt{10} + \frac{1}{2}\sqrt{2}, -\frac{1}{2}\sqrt{10} - \frac{1}{2}\sqrt{2}$$

```
> evalf(solve(f(x)),5);
```

$$2.2883, .87410, -.87410, -2.2883$$

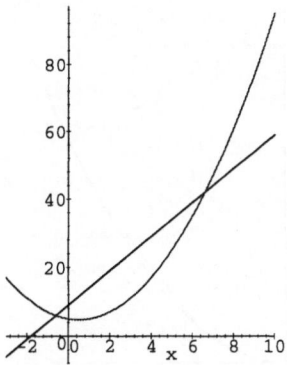

Figure 4.6. Linear-quadratic system $f(x) = (x-2)^2 + 3x + 1$ and $g(x) = 5x + 9$, on $[-3, 10]$

After solving directly we realize that by using a substitution we were not able to find all of the roots.

4.4 Linear-Quadratic Systems

A common problem is to find where a given line, other than the x-axis, and parabola intersect. Consider the intersection of the quadratic $f(x) = (x-2)^2 + 3x + 1$ and line $g(x) = 5x + 9$ as shown in Figure 4.6.

```
> f:=x->(x-2)^2+3*x+1;
```
$$f := x \rightarrow (x-2)^2 + 3x + 1$$

```
> g:=x->5*x+9;
```
$$g := x \rightarrow 5x + 9$$

```
> plot({f(x),g(x)},x=-3..10);
> x:=[solve(f(x)=g(x),x)];
```
$$x := \left[3 + \sqrt{13}, 3 - \sqrt{13}\right]$$

```
> f(x[1]),f(x[2]);
```
$$\left(1 + \sqrt{13}\right)^2 + 10 + 3\sqrt{13}, \left(1 - \sqrt{13}\right)^2 + 10 - 3\sqrt{13}$$

```
> evalf(");
```
$$42.02775638, 5.97224362$$

```
> g(x[1]),g(x[2]);
```
$$24 + 5\sqrt{13}, 24 - 5\sqrt{13}$$

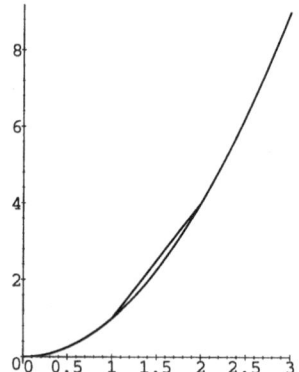

Figure 4.7. The graph of $f(x) = x^2$ and the secant joining $(1, 1^2)$ and $(2, 2^2)$

```
> evalf(");
```
$$42.02775638, 5.97224362$$

4.5   Slope of a Tangent to a Parabola

A *secant* is a line that intersects a curve at two points. For example, consider the secant to the parabola $f(x) = x^2$, that intersects it at $(1, 1^2)$ and $(2, 2^2)$.

```
> f:=x->x^2;
```
$$f := x \to x^2$$

```
> with(plots):
> p1:=plot(f(x),x=0..3):
> p2:=plot([[1,1],[2,4]]):
> display({p1,p2});
```

On a road, the normal way for a steep part to be indicated on a sign is to give a percentage change. Thus, 10% would indicate that the road rises (or falls) by 1 metre for every 10 metres travelled and we say that the *slope* or *gradient* is 'one in ten'. Similarly, a straight line graph has a slope, or gradient, that is simply the ratio of its change in height to change in horizontal distance. Obviously, on *curved* graphs the slope will not be constant and it requires more effort to calculate its value. We return to this in Part II but clearly we can always find the *average* slope of the graph of a function $y = f(x)$ between the points x_1 and x_2 as the ratio

$$\text{average slope } = \frac{f(x_2) - f(x_1)}{x_2 - x_1}$$

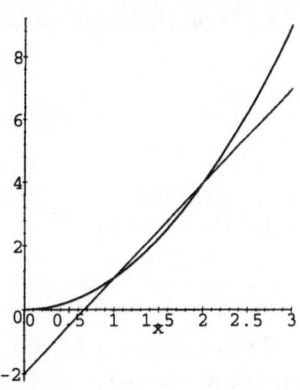 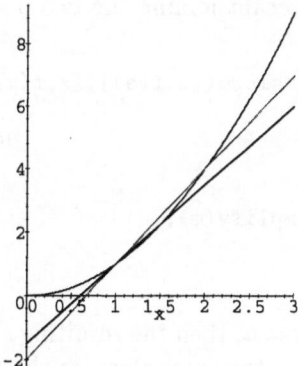

Figure 4.8. (a) The graph of $f(x) = x^2$ and the secant $y = 3x - 2$. (b) The graph of $f(x) = x^2$ and the secants $y = 3x - 2$ and $y = 2.5x - 1.5$

Using this definition for the slope of a line between two points, we can find the equation of the secant, the graph of which is shown in Figure 4.8(a).

```
> m:=(4-1)/(2-1);
```
$$m := 3$$

```
> m=(y-1)/(x-1);
```
$$3 = \frac{y - 1}{x - 1}$$

```
> isolate(",y);
```
$$y = 3\,x - 2$$

```
> plot({f(x),3*x-2},x=0..3);
```

The slope of the secant between two points is an approximation of the slope of the curve in the vicinity of the points. As we move the points closer together, the approximation becomes more accurate. What is the slope of the secant which intersects $f(x)$ at $(1, 1^2)$ and $(1.5, 1.5^2)$? See Figure 4.8(b).

```
> m:=(1.5^2-1)/(1.5-1);
```
$$m := 2.500000000$$

```
> m=(y-1)/(x-1);
```
$$2.500000000 = \frac{y - 1}{x - 1}$$

```
> isolate(",y);
```
$$y = 2.500000000\,x - 1.500000000$$

```
> plot({f(x),3*x-2,2.5*x-1.5},x=0..3);
```

Now, what would happen if we moved the two points on top of one another? Consider the secant joining the two points (a, a^2) and (x, x^2) on the same parabola.

```
> m:=slope([a,f(a)],[x,f(x)]);
```

$$m := \frac{a^2 - x^2}{a - x}$$

```
> simplify(m);
```

$$x + a$$

If we let $x = a$, then the resulting slope is the slope of the *tangent* at (a, a^2). The tangent has the same slope as the parabola at the point of contact.

```
> subs(x=a,");
```

$$2\,a$$

Therefore, the slope of the tangent at any point $(a, f(a))$ on the parabola $f(x) = x^2$ is $2a$.

Example 4.1   What is the slope of the tangent at $x = 2$ to the parabola $y = 2(x - 1)^2 + 3$? See Figure 4.9.

```
> f:=x->2*(x-1)^2+3;
```

$$f := x \rightarrow 2\,(x - 1)^2 + 3$$

```
> slope_of_tangent=slope([a,f(a)],[x,f(x)]);
```

$$slope\_of\_tangent = \frac{2\,(a - 1)^2 - 2\,(x - 1)^2}{a - x}$$

```
> simplify(");
```

$$slope\_of\_tangent = 2\,x - 4 + 2\,a$$

```
> slope_of_tangent=subs(x=a,rhs("));
```

$$slope\_of\_tangent = 4\,a - 4$$

```
> assign(");
> showtangent(2*(x-1)^2+3,x=2,x=-2..3);
> a:=2;
```

$$a := 2$$

```
> slope_of_tangent;
```

$$4$$

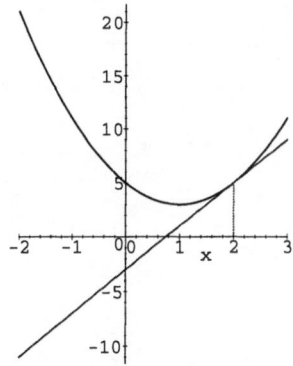

Figure 4.9. The graph $f(x) = 2(x-1)^2 + 3$ and the tangent at $x = 2$

Exercise 4.7 Find the equation of the tangent at the point $(-1, 0)$ on the parabola $y = (x+1)^2$.

Exercise 4.8 Find the equation of the tangent at $x = 1$ on the parabola $f(x) = x^2 + 3x - 1$.

Exercise 4.9 A line perpendicular to a tangent is called a *normal* line; its slope is equal to the negative inverse of the slope of the tangent. Find the equations of the tangent and normal lines to the parabola $y = 4x^2$ at the points $x = 0, \pm 0.2, \pm 0.4, \dots, \pm 1$. Construct two plots of the parabola, showing on one the tangent lines and on the other the normal lines. Overlay these plots using the `display` command.

Experiment 4.2 A snowboard halfpipe has cross section given approximately by $y = (1 - (1 - (\frac{x}{2})^4)^{1/4}$ metres. Ignoring friction of the board on the snow, the height of a snowboarder who dropped into the pipe at time $t = 0$ and $x = 0$ is estimated to be $h(t) = -4t^2$ after t seconds. Plot the cross sectional shape of the halfpipe and estimate the maximum speed of the snowboarder.

Chapter 5

Polynomial Functions

Commands used in this chapter

- assign
- divide
- evalf
- expand
- plot
- plot[options]
- quo
- rem

- rhs
- simplify
- solve
- student
- student[showtangent]
- student[slope]
- with

5.1 Polynomial Functions

The general form of a *polynomial* function is

$$f(x) = ax^n + bx^{n-1} + cx^{n-2} + \ldots + k$$

where $a, b, c, \ldots, k$ are constants. The *order*, or *degree*, of a polynomial is the value of the highest exponent on the independent variable, x. We say that the function above is 'nth order', or 'of degree n'. The higher the order of a polynomial function, the greater the *curvature* and the more curvature changes in general. We shall study curvature changes in more detail in Chapter 18.

5.2 Division of Polynomials

The *Maple* divide command is used to detemine if a given dividend and divisor will divide exactly. Only if the remainder is zero will it perform the division.

> divide(x^2-4,x+2,'quotient1');

$$true$$

> quotient1;

$$x - 2$$

> dividend:=x^3+2*x-1;

$$dividend := x^3 + 2x - 1$$

```
> divisor:=x+1;
```
$$divisor := x + 1$$

```
> divide(dividend,divisor,'quotient2');
```
$$false$$

```
> quotient2;
```
$$quotient2$$

If the remainder is non-zero, then the **divide** command returns 'false' and the value of the quotient (*quotient2*) remains unassigned.

If the dividend and divisor do not divide exactly, then we must use a different method to determine the quotient and remainder. The **quo** command calculates the quotient of a division by polynomials and assigns the value of the remainder to a variable name, in this case *remainder*. The **rem** command calculates the remainder and assigns the value of the quotient to a variable name, in this case *quotient3*. It is necessary to use only one of these commands, but both are shown below as a demonstration.

```
> quo(dividend,divisor,x,'remainder');
```
$$x^2 - x + 3$$

```
> remainder;
```
$$-4$$

```
> rem(dividend,divisor,x,'quotient3');
```
$$-4$$

```
> quotient3;
```
$$x^2 - x + 3$$

```
> (divisor)*(quotient3)+remainder;
```
$$(x+1)(x^2 - x + 3) - 4$$

It's always a good idea to check that the result is correct.

```
> expand(");
```
$$x^3 + 2x - 1$$

Exercise 5.1   Divide the polynomial $y = 2x^3 + x^2 - 5x + 2$ by $x - 1$.

Exercise 5.2   Is the polynomial $y = x^3 - 4x^2 + 2x - 1$ divisible by $x - 3$? If not, find the quotient and remainder.

Exercise 5.3   Find the remainders on dividing $x^5 - 5x^4 + x^3 - 7$ by $(x-1)$ and by $(x-1)^2$.

Factor Theorem

The Factor Theorem states that $x - a$ is a factor of $f(x)$ if and only if $f(a) = 0$. We first illustrate that if $f(x) = 0$ has solution $x = a$ then $x - a$ is a factor of $f(x)$.

```
> f:=x->x^2-5*x+6;
```
$$f := x \rightarrow x^2 - 5x + 6$$

```
> solve(f(x),{x});
```
$$\{x = 3\}, \{x = 2\}$$

```
> divide(f(x),x-3,'quotient4');
```
$$true$$

```
> quotient4;
```
$$x - 2$$

```
> divide(f(x),x-2,'quotient5');
```
$$true$$

```
> quotient5;
```
$$x - 3$$

The converse is also true. If $x - a$ is a factor of $f(x)$, then $f(a) = 0$.

```
> f:=x->2*x^2-x-3;
```
$$f := x \rightarrow 2x^2 - x - 3$$

```
> divide(f(x),x+1,'quotient6');
```
$$true$$

```
> quotient6;
```
$$2x - 3$$

```
> f(-1);
```
$$0$$

Exercise 5.4   Find quadratic equations with the following roots.

    (a) (3,6)

    (b) (2,-1)

    (c) (9,0)

Exercise 5.5   Show that if $y = f(x) = \frac{x+b}{ax-1}$ then $x = f(y)$. Can you find other functions with this property?

Exercise 5.6   Show that $x^2 + 5x + 6 = (x+2)(x+3)$ and so

$$\frac{2x^2 + 15x + 25}{x^2 + 5x + 6} = 2 + \frac{3}{x+2} + \frac{2}{x+3}$$

Exercise 5.7   Use the Binomial Theorem in the form

$$(1-x)^{-1} = 1 - x + x^2 + \ldots$$

to express $\frac{1}{1-x}$ in terms of such a series of powers of x, for $\mid x \mid < 1$. Hence deduce that $2 = 1 + \frac{1}{2} + \frac{1}{2^2} + \frac{1}{2^3} + \ldots$

Exercise 5.8   Find a when $(x-1)$ is a factor of $7x^3 + ax^2 - 2x + 1$.

5.3   Product and Sum of Roots

The roots of the general quadratic function, $ax^2 + bx + c = 0$, are related in such a way that the equation can be rewritten in the form $x^2 -$ sum of roots$x +$ product of roots. You can easily show why this is so by solving $ax^2 + bx + c = 0$ and calculating the sum and product of its roots.

```
> f:=x->a*x^2+b*x+c;
```
$$f := x \rightarrow a\,x^2 + b\,x + c$$

```
> root:=[solve(f(x),x)];
```
$$root := \left[\frac{1}{2}\frac{-b + \sqrt{b^2 - 4\,a\,c}}{a}, \frac{1}{2}\frac{-b - \sqrt{b^2 - 4\,a\,c}}{a}\right]$$

```
> 'Product of roots'=root[1]*root[2];
```
$$Product\ of\ roots = \frac{1}{4}\frac{\left(-b + \sqrt{b^2 - 4\,a\,c}\right)\left(-b - \sqrt{b^2 - 4\,a\,c}\right)}{a^2}$$

```
> expand(");
```
$$Product\ of\ roots = \frac{c}{a}$$

```
> 'Sum of Roots'=root[1]+root[2];
```
$$Sum\ of\ Roots = \frac{1}{2}\frac{-b+\sqrt{b^2-4\,a\,c}}{a} + \frac{1}{2}\frac{-b-\sqrt{b^2-4\,a\,c}}{a}$$

```
> expand(");
```
$$Sum\ of\ Roots = -\frac{b}{a}$$

To see how this relationship is useful, we must rearrange the form of the general quadratic so that the coefficient of x^2 is 1. The resulting equation is $x^2+\frac{b}{a}x+\frac{c}{a} = 0$, or $x^2-(-\frac{b}{a})x+\frac{c}{a}=0$. Using this form of the general quadratic we can determine the equation of a parabola from its roots.

Example 5.1   Find the equation of a quadratic with roots $x = -3$ and $x = 5$.

```
> root1:=-3;
```
$$root1 := -3$$

```
> root2:=5;
```
$$root2 := 5$$

```
> 'Product of roots':=root1*root2;
```
$$Product\ of\ roots := -15$$

```
> 'Sum of Roots':=root1+root2;
```
$$Sum\ of\ Roots := 2$$

```
> y:=x^2-'Sum of Roots'*x+'Product of roots';
```
$$y := x^2 - 2\,x - 15$$

```
> solve(y,{x});
```
$$\{\,x = 5\,\},\{\,x = -3\,\}$$

Exercise 5.9   Find the sum and product of the roots of the following equations.

(a) $y = x^2 + 4x + 9$

(b) $y = 2x^2 - 5x + 9$

Exercise 5.10 Find the second root for each of the following equations.

(a) $y = x^2 + 4x - c$, root $= 1$

(b) $y = x^2 - bx + 3$, root $= 2$

5.4 Related Roots

Using the above method, it is possible to determine the equations of quadratics having roots related to those of a known equation. For example, what is the quadratic equation with roots twice those of $y = x^2 + x - 12$?

```
> f:=x->x^2+1*x-12;
```
$$f := x \rightarrow x^2 + x - 12$$

```
> root:=[solve(f(x),x)];
```
$$root := [3, -4]$$

```
> newr1:=root[1]*2;
```
$$newr1 := 6$$

```
> newr2:=root[2]*2;
```
$$newr2 := -8$$

```
> 'Sum of Roots':=newr1+newr2;
```
$$Sum\ of\ Roots := -2$$

```
> 'Product of roots':=newr1*newr2;
```
$$Product\ of\ roots := -48$$

```
> y:=x^2-'Sum of Roots'*x+'Product of roots';
```
$$y := x^2 + 2x - 48$$

Exercise 5.11 What is the equation of a polynomial whose roots are two less than those of $y = x^2 - x - 3$?

5.5 Roots of Higher Order Polynomials

There are formulas to determine the roots of polynomials of degree three, but they are cumbersome. However, the roots of polynomial functions of higher order

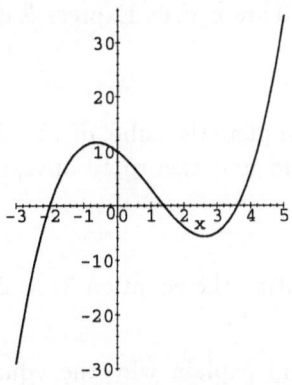 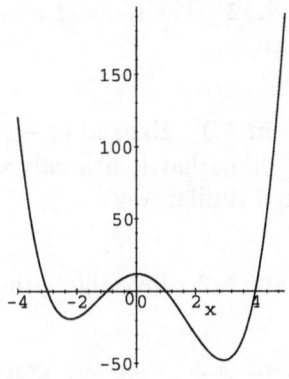

Figure 5.1. (a) The graph of $f(x) = x^3 - 3x^2 - 5x + 10$, on $[-3, 5]$. (b) The graph of $f(x) = x^4 - x^3 - 13x^2 + x + 12$, on $[-4, 5]$

than 2 can be easily estimated graphically. Consider the third order function $f(x) = x^3 - 3x^2 - 5x + 10$ shown in Figure 5.1(a).

```
> f:=x->x^3-3*x^2-5*x+10;
```
$$f := x \rightarrow x^3 - 3x^2 - 5x + 10$$

```
> plot(f(x),x=-3..5);
```
They can also be determined numerically using solve.

```
> solve(f(x),x);
```
$$-2, \frac{5}{2} + \frac{1}{2}\sqrt{5}, \frac{5}{2} - \frac{1}{2}\sqrt{5}$$

```
> evalf(solve(f(x),x),3);
```
$$-2., 3.62, 1.38$$

To find the roots of the *quartic*, or fourth order, function $f(x) = x^4 - x^3 - 13x^2 + x + 12$ we first examine Figure 5.1(b) and estimate the roots using the cursor to point at them.

```
> f:=x->x^4-x^3-13*x^2+x+12;
```
$$f := x \rightarrow x^4 - x^3 - 13x^2 + x + 12$$

```
> plot(f(x),x=-4..5);
```
Then we solve the function to determine the roots exactly.

```
> evalf(solve(f(x),{x}));
```
$$\{x = 1.\}, \{x = 4.\}, \{x = -3.\}, \{x = -1.\}$$

Exercise 5.12   The roots of $x^3 + bx^2 + cx + d = 0$ are u, v, w. Express b, c, d in terms of u, v, w.

Experiment 5.1   Expand $(x-1)^3$ and hence 'complete the cube' of $x^3 - 3x^2 + 3x - 9$ to show that it has only one real root, and find this root. Investigate $(x - a)^3$ in a similar way.

Experiment 5.2   Investigate the curves representing the equation $5x = 2y^3$.

Experiment 5.3   Plot the graph of $x^5 + x^3$ and explain why the equation $y = x^5 + x^3$ has one and only one solution for each value of y. By successively plotting subintervals find the solution of $4 = x^5 + x^3$ to 3 significant figures.

Chapter 6

Exponential Functions

Commands used in this chapter
- evalf
- exp
- expand
- fsolve
- plot
- plot[options]
- solve

6.1   Properties of Exponentials

Some of the properties of exponentials are demonstrated below:
```
> x*x*x*x;
```
$$x^4$$

```
> 1/y/y/y/y/y;
```
$$\frac{1}{y^5}$$

```
> x^(a+b)=expand(x^(a+b));
```
$$x^{(a+b)} = x^a \, x^b$$

```
> x^(a-b)=expand(x^(a-b));
```
$$x^{(a-b)} = \frac{x^a}{x^b}$$

```
> (x^a)^b=expand((x^a)^b);
```
$$\left(x^a \right)^b = x^{(a\,b)}$$

```
> (x*y)^a=expand((x*y)^a);
```
$$\left(x \, y \right)^a = x^a \, y^a$$

```
> (x/y)^a=expand((x/y)^a);
```
$$\left(\frac{x}{y} \right)^a = \frac{x^a}{y^a}$$

```
> x^0;
```
$$1$$

```
> x^1;
```
$$x$$

```
> expand(x^(-a));
```
$$\frac{1}{x^a}$$

An exponential expression is an expression which contains an exponent. For example,

```
> eq:=2^(2*x+1)=8^3;
```
$$eq := 2^{(2x+1)} = 512$$

We can solve this expression using the `solve` or `fsolve` command as we would any other expression in x.

```
> fsolve(eq,{x});
```
$$\{\, x = 4.000000000 \,\}$$

Exercise 6.1 Solve the following equations for x.

(a) $2^x = 32$

(b) $5^{3x-1} = 7^x$

(c) $7^x = 49^4$

6.2   Scientific Notation

Scientific, or *standard*, *notation* is the convention used in science of placing a decimal after the first non-zero digit of a number, and then multiplying it by the appropriate power of 10. In this way, large numbers such as 2,589,000,000,000 are more conveniently represented as 2.589×10^{12}. Likewise, very small numbers such as 0.00000000928 are reported as 9.28×10^{-9}. In this example, 9.28 is the *standard factor* of 0.00000000928, and the -9 is the *standard exponent*. *Maple* does not always report numbers in standard notation, in fact, it uses standard form only for numbers less than 10^-6. Also, *Maple* places the decimal point so that the standard factor is less than one.

```
> 0.00009999;
```
$$.00009999$$

```
> 0.000001;
```
$$.1\,10^{-5}$$

```
> 1000000000000000000000000000000000;
```
$$1000000000000000000000000000000000$$

```
> 1.2*10^10;
```
$$.1200000000\,10^{11}$$

6.3   Table of Values

The natural exponential function, given by $exp(x) = e^x$, is the most important
function in all of mathematics. The value $e = exp(1)$ is an *irrational* number
(so not any exact ratio of whole numbers), of value about 2.72. The values in
Table 6.1 should give some feeling for the expression 'exponentially increasing'.

n	e^n	e^{-n}
0	1.000	1.000
1	2.718	0.368
2	7.389	0.135
3	20.086	0.050
4	54.598	0.018
5	148.413	0.008
6	403.429	0.002

Table 6.1. Some approximate numerical values of the exponential function

6.4   Exponential Growth and Decay

Two natural phenomena which can be described by exponential functions are
simple population growth, such as of bacterial cultures, and the decomposition
of radioactive elements. Using measured characteristics of a culture or sample,
we can create a *model* which we can then use to predict future populations or
concentrations. We shall study exponential growth and decay in more depth in
Chapter 22.

The general form the exponential growth and decay expression that we shall be using in this chapter is

$$y(t) = y_0 2^{\frac{t}{k}}$$

where y_0 is the value of y at time $t = 0$, and k is the *half-life* if $k < 0$, or the *doubling period* if $k > 0$. The half-life is the time required for half of a sample of radioactive material to decay into another element; the doubling period is the time required for a population to double itself.

Example 6.1   Consider a colony of bacteria with a doubling period of one day, and an initial population of 100 bacteria. What will be the number of bacteria in 10 days?

```
> population:=initial_population*2^(t/doubling_period);
```

$$population := initial\_population \, 2^{\left(\frac{t}{doubling\_period}\right)}$$

```
> initial_population:=100;
```

$$initial\_population := 100$$

```
> doubling_period:=1;
```

$$doubling\_period := 1$$

```
> t:=10;
```

$$t := 10$$

```
> population;
```

$$102400$$

The growth of the population with time is graphed in Figure 6.1(a).

```
> t:='t':plot(100*2^(t),t=0..10);
```

Exercise 6.2   A certain colony of bacteria has an initial population of 2500. If the population after 3 days is 8000, what is the doubling period? In how many days will the population be 20,000?

Example 6.2   Estimate the half-life of carbon-14 if 29.8% of the original amount remains radioactive after 10,000 years.

```
> amount:=initial_amount*2^(-t/half_life);
```

$$amount := initial\_amount \, 2^{\left(-\frac{t}{half\_life}\right)}$$

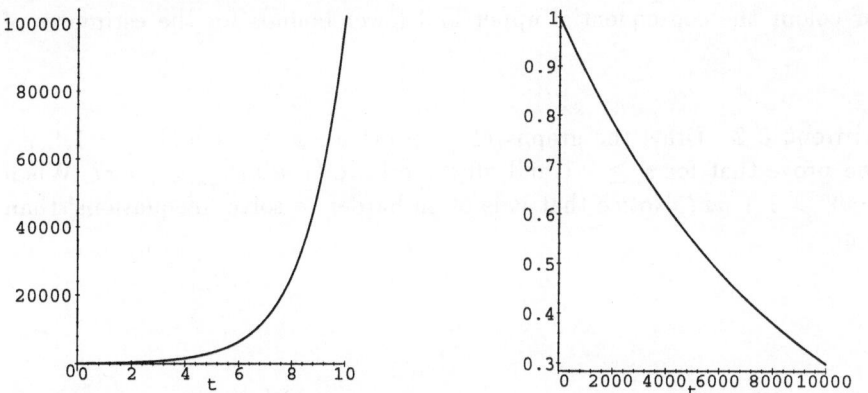

Figure 6.1. (a) Bacterial population growth example. (b) Radioactive decay example

```
> initial_amount:=1;
```
$$initial\_amount := 1$$

```
> amount:=.298*initial_amount;
```
$$amount := .298$$

```
> t:=10000;
```
$$t := 10000$$

```
> evalf(solve(amount=initial_amount*2^(-t/half_life),half_life),3);
```
$$5730.$$

Carbon-14 decays according to the graph shown in Figure 6.1(b).
```
> t:='t':plot(1*2^(-t/""),t=0..10000);
```

Exercise 6.3 The half-life of one of the isotopes of polonium, Po^{216}, is 0.15 s. How long would it take for a sample of Po^{216} to decay to 1% of its original size?

Experiment 6.1 The so called 'iceman', recently found in the Alps, was carrying a bow. Suppose that a sample of the wood from this bow was analyzed and found to have $r\%$ of its carbon in the form of carbon-14. Since the fraction of carbon-14 began to decay from the time the bow ceased to be living wood, prepare a table to show the values of $r\%$ if the bow is $1000, 2000, \ldots 10,000$ years old. Try to express $r\%$ as a function of the age of the bow and plot it on a graph. The half-life of carbon-14 may be taken to be 5730 years. Try first to guess the values in the table. Suppose that the experimental determination of the value of the fraction of carbon-14 has a possible error of $\pm 10\%$. Show on your plot in another colour the consequential upper and lower bounds for the estimates of age.

Experiment 6.2   Draw the graphs of $(1 + x)^n$ for $x \geq -1$ and $n = 1, 2, \ldots$ Can you prove that for $x \geq -1$ and all n we have $(1 + x)^n \geq 1 + nx$? When is $(1 + x)^n > 1 + nx$? Notice that it is often harder to solve 'inequations' than equations.

Chapter 7

Logarithmic Functions

Commands used in this chapter
- combine
- evalf
- expand
- log
- plot
- plot[options]
- plots
- plots[logplot]
- solve
- with

7.1 Properties of Logarithms

The natural logarithm function, represented by $\ln(x)$ or $\log_e(x)$ or just $\log(x)$, is the inverse of the natural exponential function so $\exp(\ln(x)) = x = \ln(\exp(x))$. In *Maple,* the base for the log function is e, however, you can ask for base 10 in *Maple* with log10(x), as we shall see later. (Be careful when using e in *Maple.* You must input a capital E, but *Maple* will return a lower case e.) When mathematicians say log they always mean to base e.

```
> x=a^y;
```
$$x = a^y$$

```
> combine(solve(x=a^y,{y}));
```
$$\left\{ y = \frac{\ln(x)}{\ln(a)} \right\}$$

```
> log(p*q)=expand(log(p*q));
```
$$\ln(pq) = \ln(p) + \ln(q)$$

```
> log(p/q)=expand(log(p/q));
```
$$\ln\left(\frac{p}{q}\right) = \ln(p) - \ln(q)$$

```
> log(p^b)=expand(log(p^b));
```
$$\ln(p^b) = b\ln(p)$$

Logarithmic expressions, like exponential expressions, can be solved using the `solve` command.

```
> solve(log10(y)=3.5,{y});
```
$$\{y = 3162.277660\}$$

Exercise 7.1   Solve the following logarithmic equations.

(a) $\log(x) = 9$, base 3

(b) $7\log(4) = 2x$, base 16

(c) $\log(10) = 5$, base x

Exercise 7.2   Solve the following equalities.

(a) $\log(x) + \log(2x) = 1$, base 3

(b) $\log(x^2) - \log(x) = \log(3)$, base 10

(c) $\log(x + 7) + \log(x^2) = \log(9)$, base 5

7.2   Values of Logarithmic Functions

Table 7.1 lists some of the values of the base 10 and base e logarithmic functions. These values were calculated using the following *Maple* commands.

n	$\log_{10}(n)$	$\log_e(n)$
1	0	0
2	0.30	0.69
3	0.48	1.10
4	0.60	1.39
5	0.70	1.61
6	0.78	1.79

Table 7.1. Some approximate numerical values of the logarthmic function.

```
> evalf(log10(1)),evalf(log(1));
```
$$0, 0$$

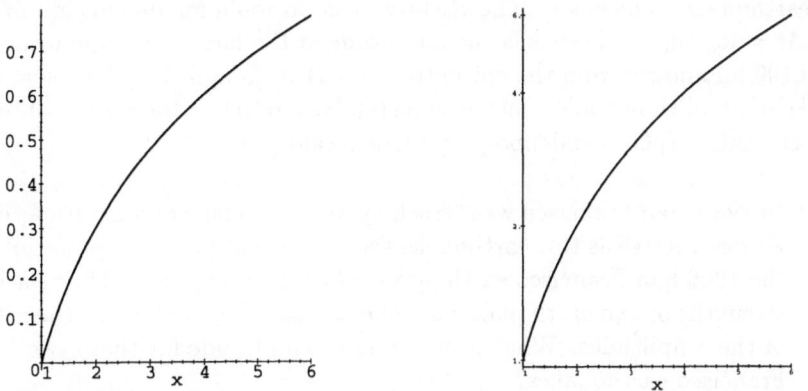

Figure 7.1. (a) Base 10 logarithmic function. (b) Log-linear plot of $y = x$

```
> evalf(log10(2)),evalf(log(2));
                .3010299957, .6931471806

> evalf(log10(3)),evalf(log(3));
                .4771212547, 1.098612289

> evalf(log10(4)),evalf(log(4));
                .6020599913, 1.386294361

> evalf(log10(5)),evalf(log(5));
                .6989700043, 1.609437912

> evalf(log10(6)),evalf(log(6));
                .7781512504, 1.791759469
```

The plot of the function $y = \log_{10}(x)$ on a linear scale has the same shape as the plot of $y = x$ on a log-linear scale. Compare the graphs in Figure 7.1; apparently, `logplot` does not respond to our request for a larger font.

```
> plot(log10(x),x=1..6);
> with(plots):
> logplot(x,x=1..6);
```

Exercise 7.3 A scale for measuring the magnitude of earthquakes was developed in 1935 by Charles F. Richter of the California Institute of Technology. The so-called Richter Scale allows the 'size' of earthquakes to be compared. The Richter formula computes the magnitude of a quake from the logarithm of the amplitude of waves recorded by seismographs. Included in the formula is an adjustment to compensate for the variation in the distance of the seismograph and the earthquake's epicentre. The Richter Scale formula for magnitude M is given by $M = \log 10(\frac{x}{c})$ where x is the amplitude of the largest seismic wave as measured 100 kilometres from the epicentre and c is the amplitude of a reference earthquake of amplitude 1 micrometre on a standard graph at the same distance from the epicentre. (p63, Goldenberg and Greenwald [7])

(a) In 1989, San Francisco was struck by an earthquake of magnitude 7.1. As destructive as this earthquake was, it was not nearly as powerful as the 1906 San Francisco earthquake, which measured 8.3. The relative strengths of two earthquakes may be compared by looking at the ratio of the amplitudes. What is the ratio of amplitiude for these two San Francisco earthquakes?

(b) The largest earthquake magnitude ever measured was 8.9 for an earthquake in Japan in 1933. The largest earthquake magnitude ever recorded in the United States was 8.5 for the Alaskan earthquake of 1964. Determine the ratio of amplitude for these earthquakes.

(c) When the amplitude of an earthquake is tripled, by how much does the magnitude increase?

(d) Assume that there are two earthquakes of unequal strength. If the ratio of amplitudes is 1.5 and the weaker earthquake has magnitude 5.6, determine the magnitude of the stronger earthquake.

(e) If the difference in magnitudes of two earthquakes is 0.5, determine the ratio of amplitudes for the two earthquakes.

Experiment 7.1   Stirling's formula gives an approximation to the factorials of large numbers using

$$\log(n!) \doteqdot \left(n + \frac{1}{2}\right)\log(n) - n + \log(\sqrt{2n})$$

Investigate how accurate this is by plotting the difference between the two expressions for $n = 10, 10^2, 10^3, \ldots, 10^{100}$.

Experiment 7.2   Plot the function $p(x) = e^{-x^2}$ on the interval $[-3, 3]$. It has the familiar bell shape that often arises in distributions of marks on examinations, or heights of people, errors in experiments. Plot also the function $q(x) = \log(p(x))$ for the same interval. Clearly $q(x) = -x^2$, so we expect an inverted parabola. Now consider $l(x) = \log |q(x)|$, the log of the absolute value of q(x). We expect a log-linear plot, using the command `logplot`, to be a straight line, $l(x) = 2\log(x)$. So, by applying the log function twice, we can 'straighten out' the original bell curve. Now investigate the function $h(x) = e^{p(x)}$ by similar methods. What about $g(x) = e^{h(x)}$, etc?

Experiment 7.3 Consider the family of polynomials of degree n

$$f_n(x) = 1 - x + \frac{x^2}{2!} - \frac{x^3}{3!} + \ldots + \frac{(-x)^n}{n!}$$

Construct an animation of the plots $f_0(x), f_1(x), f_2(x), \ldots, f_{100}(x)$ for $-1 \leq x \leq 1$. Construct also an animation of the differences $(e^{-x} - f_n(x))$ for $n = 1, 2, \ldots, 100$. For $-1 < x \leq 1$ investigate the function

$$g_n(x) = x - \frac{x^2}{2} + \frac{x^3}{3} - \frac{x^4}{4} + \ldots + (-1)^{n+1}\frac{x^n}{n}$$

and construct an animation of the differences $(\log(1 + x) - g_n(x))$ for $n = 1, 2, \ldots, 100$.

7.3   pH Scale

From chemistry, you may know that the *pH* scale for acidity is a base 10 logarithmic function of the form pH $= -\log_{10}[H]$, or, the pH of a solution is equal to the negative base 10 logarithm of the concentration of hydrogen ions.

```
> pH:=-log[10](H);
```

$$pH := -\frac{\ln(H)}{\ln(10)}$$

A neutral solution has a pH of 7, corresponding to a hydrogen ion concentration $[H] = 10^{-7}$.

```
> solve(pH=7,{H});
```

$$\left\{ H = \frac{1}{10000000} \right\}$$

```
> evalf(");
```

$$\{ H = .1000000000\, 10^{-6} \}$$

Acids have a high hydrogen ion concentration but a low pH (between 1 and 7),

```
> solve(pH=1,{H});
```

$$\left\{ H = \frac{1}{10} \right\}$$

whereas bases have a low hydrogen ion concentration but a high pH (between 7 and 14).

```
> solve(pH=14,{H});
```

$$\left\{ H = \frac{1}{100000000000000} \right\}$$

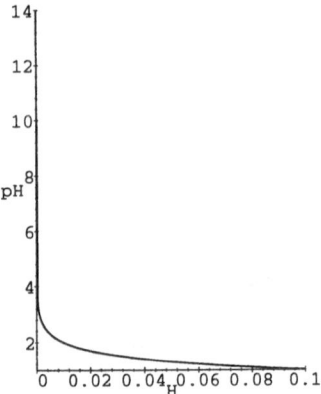

Figure 7.2. pH scale

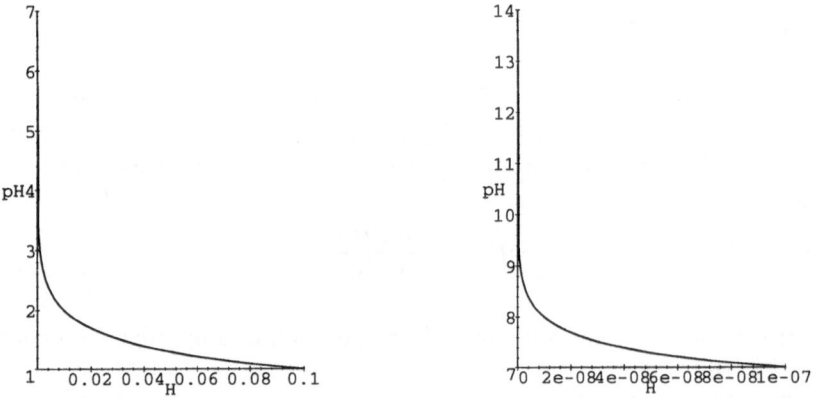

Figure 7.3. (a) Acidic range of pH scale. (b) Basic range of pH scale

The next three `plot` commands graph the relationship between hydrogen ion concentration and pH for the entire pH scale (Figure 7.2), the acidic range (Figure 7.3(a)), and the basic range (Figure 7.3(b)).

```
> plot(pH,H=10^(-14)..10^(-1),'pH'=1..14);
> plot(pH,H=10^(-7)..10^(-1),'pH'=1..7);
> plot(pH,H=10^(-14)..10^(-7),'pH'=7..14);
```

Exercise 7.4 What is the concentration of H^+ ions in a cola, the pH of which is 3.4?

7.4   Simple Interest

Simple interest is calculated as follows,

$$A = P(1 + in)$$

where A is the accumulated amount, P is the principal, i is the interest rate for a given period of time, and n is the number of time units. Normally, interest is calculated as a percentage per annum, that means per year, denoted % pa, so n is measured in years.

Example 7.1 If \$500 is invested at 4% pa simple interest, what will be the amount after 6 years?

```
> A:=P*(1+i*n);
```
$$A := P(1+in)$$

```
> P:=500;
```
$$P := 500$$

```
> i:=0.04;
```
$$i := .04$$

```
> n:=6;
```
$$n := 6$$

```
> A;
```
$$620.00$$

Exercise 7.5 Find the simple interest of each of the following.

(a) \$500 for 2 years at 4% pa

(b) \$800 for 4 years at 3% pa

7.5 Compound Interest

Compound interest differs from simple interest in that the interest earned is added to the principal. For each preceding time interval, the principal increases and the interest earned increases accordingly. It is calculated as follows,

$$A = P(1+i)^n$$

where A is the accumulated amount, P is the principal, i is the interest rate for a given period of time, and n is the number of time units as before.

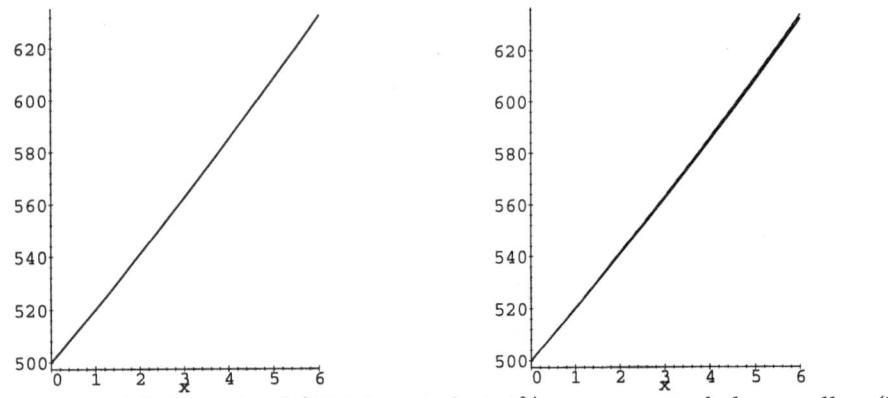

Figure 7.4. (a) Growth of $500 invested at 4% pa compounded annually. (b) Growth of $500 invested at 4% pa compounded semi-annually

Example 7.2   If $500 is invested at 4% pa compounded annually, what will be the amount after 6 years?

```
> A:=P*(1+i)^n;
```
$$A := 632.6595095$$

```
> P:=500;
```
$$P := 500$$

```
> n:=6;
```
$$n := 6$$

```
> i:=0.04;
```
$$i := .04$$

```
> evalf(A,5);
```
$$632.66$$

Figure 7.4(a) shows how the amount increases with time.

```
> plot(500*(1.04)^x,x=0..6);
```

If the same amount is invested for the same time period but the interest is compounded semi-annually, what will be the final amount?

```
> i:=0.04/2;
```
$$i := .02000000000$$

```
> n:=6*2;
```
$$n := 12$$

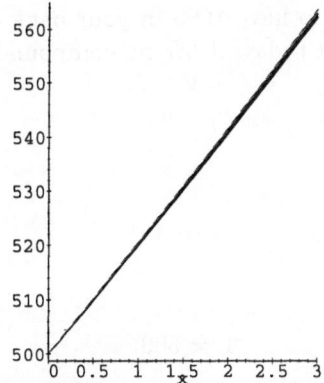

Figure 7.5. Growth of $500 invested at 4% pa compounded daily

```
> A:=P*(1+i)^n:
```
```
> evalf(A,5);
```
$$634.12$$

Figure 7.4(b) overlays the growth curve of $500 compounded annually and semi-annually.

```
> plot({500*(1.04)^x,500*(1.02)^(2*x)},x=0..6);
```

And if the interest is compounded daily, what will be the final amount?

```
> i:=0.04/365;
```
$$i := .0001095890411$$

```
> n:=6*365;
```
$$n := 2190$$

```
> A:=P*(1+i)^n:
```
```
> evalf(A,5);
```
$$635.62$$

See Figure 7.5 and compare the growth rate of $500 compounded annually, semi-anually, and daily.

```
> plot({500*(1.04)^x,500*(1.02)^(2*x),500*(1.00010959)^(365*x)},
> x=0..3 );
```

Banks often lend large sums as 'bridging loans' to customers who wish to buy a house before their own is sold; then they use daily interest. What would you expect a bank to say if you asked for your savings account to have interest compounded every second?

Example 7.3   If you wish to have $600 in your bank account in 3 years, how much money must you invest today at 6% pa compounded annually?

```
> i:=0.06;
```
$$i := .06$$

```
> n:=3;
```
$$n := 3$$

```
> A:=600;
```
$$A := 600$$

```
> evalf(solve(A=P*(1+i)^n,P),5);
```
$$503.78$$

Exercise 7.6   Calculate the amount for

  (a) $500 for 2 years at 4% pa compounded annually

  (b) $500 for 2 years at 4% pa compounded daily

  (c) $800 for 4 years at 3% pa compounded semi-annually.

Experiment 7.4   Two friends are considering whether to buy a house or rent an apartment. They have £50,000 in the bank, gaining interest at 4.5% pa and they have to pay income tax at 25% on this interest. The house in which they are interested costs £100,000 and they can obtain a mortgage of value in the range of 8.2% pa. Over 5 years they expect that the cost of maintaining the house would amount to about £4000, but the capital value of the house would increase to £110,000. Apartments with suitable accommodation cost from £1000 to £1800 per month. Create a worksheet with tables, graphs, and animations to help decide on a best strategy. Since interest rates on bank accounts fluctuate, arrange for graphs to show the consequences of changing from 4.5% to 4% and 5%.

7.6   Equivalent Rates

The *equivalent rate* is the interest rate compounded annually which is equivalent to a given interest rate compounded over a shorter time period.

Example 7.4 Find the equivalent rate for 6% pa compounded semi-annually.

```
> P:=1;
```
$$P := 1$$

```
> n:=1;
```
$$n := 1$$

```
> i1:=0.06/2;
```
$$i1 := .03000000000$$

```
> n1:=2;
```
$$n1 := 2$$

```
> A:=P*(1+i1)^n1;
```
$$A := 1.060900000$$

```
> solve(A=P*(1+i)^n,i);
```
$$.06090000000$$

An interest rate of 6.09% pa compounded annually is equivalent to 6% pa compounded semi-annually.

Chapter 8

Circular Functions

Commands used in this chapter

- arccos
- arcsin
- arctan
- assign
- convert
- cos
- csc
- plot
- plot[options]
- plots

- plots[display]
- plots[implicitplot]
- plots[textplot]
- sec
- sin
- solve
- subs
- tan
- trig
- with

8.1 Primary Trigonometric Functions

Consider a circle of radius r, and a point $P(x, y)$ on its circumference at angle t *radians* (recall that π radians is 180°) above the x-axis, then the primary trigonometric functions are defined as:

$$\sin(t) = \frac{y}{r}, \cos(t) = \frac{x}{r}, \text{ and } \tan(t) = \frac{y}{x}.$$

The graphs of the primary trigonometric functions $y = \sin(x)$, $y = \cos(x)$, and $y = \tan(x)$ are shown in Figure 8.1(a), Figure 8.1(b), and Figure 8.2, respectively.

```
> plot(sin(x),x=-2*Pi..2*Pi);

> plot(cos(x),x=-2*Pi..2*Pi);

> plot(tan(x),x=-2*Pi..2*Pi,y=-10..10);
```

The tangent function is defined by $\tan(t) = \frac{\sin(t)}{\cos(t)}$ and is not defined when $\cos(t) = 0$. The discontinuity is shown in the plot as a vertical line but, in fact, the value of $\tan(t)$ is undefined at those points; $\tan(t)$ has a vertical asymptote there.

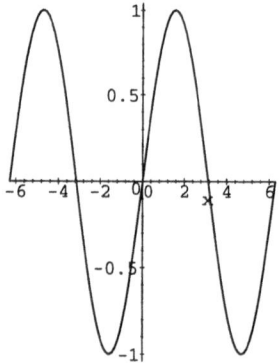 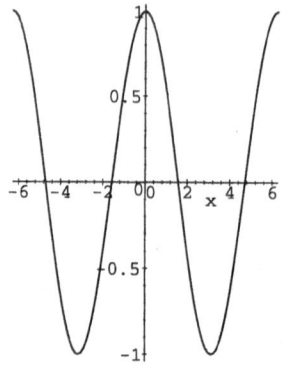

Figure 8.1. (a) Sine function. (b) Cosine function

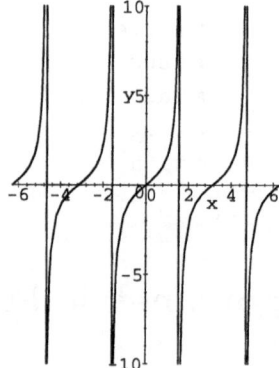

Figure 8.2. Tangent function

8.2 Reciprocal Trigonometric Functions

The reciprocal trigonometric functions are defined as the reciprocals of primary trigonometric functions:

$$\csc(t) = \frac{r}{y}, \sec(t) = \frac{r}{x}, \text{ and } \cot(t) = \frac{x}{y}.$$

The graphs of the reciprocal trigonometric functions $y = \csc(x)$, $y = \sec(x)$, and $y = \cot(x)$ are shown in Figure 8.3(a), Figure 8.3(b), and Figure 8.4, respectively. Once again, these are defined only when the denominator is nonzero.

```
> csc(t);
```
$$\csc(t)$$

```
> convert(",sincos);
```
$$\frac{1}{\sin(t)}$$

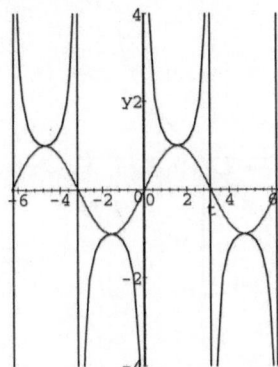 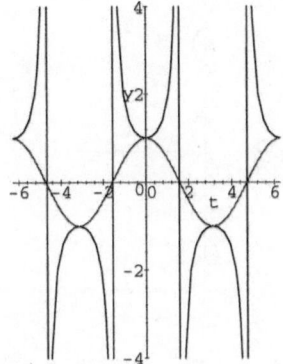

Figure 8.3. (a) Sine and cosecant functions. (b) Cosine and secant functions

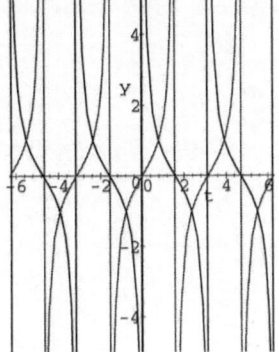

Figure 8.4. Tangent and cotangent functions

The reciprocal trigonometric functions have asymptotes at the zeros of the circular functions.

```
> plot({sin(t),csc(t)},t=-2*Pi..2*Pi,y=-4..4);
> plot({cos(t),sec(t)},t=-2*Pi..2*Pi,y=-4..4);
```

Similarly, the reciprocal trigonometric functions have zeros at the asymptotes of the circular function.

```
> plot({tan(t),cot(t)},t=-2*Pi..2*Pi,y=-5..5);
```

8.3 Inverse Circular Functions

The equation $y = \sin(t)$ has, for given y in $[-1, 1]$ a unique solution for t in $[-\frac{\pi}{2}, \frac{\pi}{2}]$ because over this range $\sin(t)$ *increases* with t. So there is a partial inverse to the sine function.

The inverse circular functions are determined by simultaneously substitut-

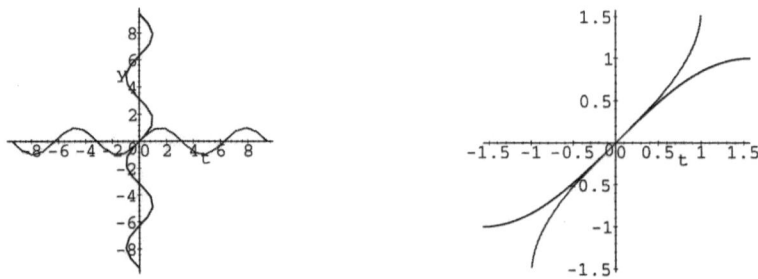

Figure 8.5. (a) Sine function and its inverse. (b) Sine and arcsine functions

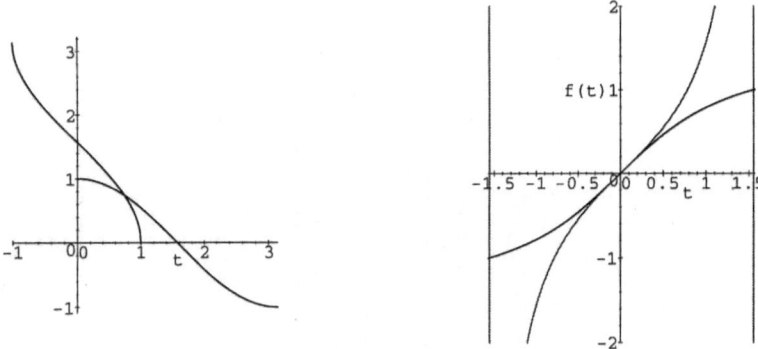

Figure 8.6. (a) Cosine and arccosine functions. (b) Tangent and arctangent functions

ing y for t, and t for y into the primary trigonometric functions.

```
> y=sin(t);
```

$$y = \sin(t)$$

```
> subs({t=y,y=t},");
```

$$t = \sin(y)$$

```
> solve(",{y});
```

$$\{y = \arcsin(t)\}$$

```
> with(plots):
> implicitplot({y=sin(t),t=sin(y)},t=-3*Pi..3*Pi,y=-3*Pi..3*Pi);
```

From the graph, shown in Figure 8.5(a), it is obvious that the inverse of the whole sine function is not a function, only a relation. For each value of

t there are an infinite number of values of y. Therefore, the domains of the inverse sine and cosine functions are restricted to $[-\frac{\pi}{2}, \frac{\pi}{2}]$ and $[0, \pi]$ respectively. The graphs of the inverse trigonometric functions $y = \arcsin(x)$, $y = \arccos(x)$, and $y = \arctan(x)$ are shown in Figure 8.5(b), Figure 8.6(a), and Figure 8.6(b), respectively, over *just* the intervals on which the inverse functions exist.

```
> plot({sin(t),arcsin(t)},t=-Pi/2..Pi/2,scaling=CONSTRAINED);
> display({plot(cos(t),t=0..Pi),plot(arccos(t),t=-1..1)},
> scaling=CONSTRAINED);
> plot({tan(t),arctan(t)},t=-Pi/2..Pi/2,f(t)=-2..2,
> scaling=CONSTRAINED);
```

8.4   Transformations

Each of the constants a, b, c, and d has a different effect on the function $y = a\sin(bt - c) + d$. The constant a is the *amplitude* of the sine function. So if a is negative, then the function will be reflected in the y-axis. If $-1 < a < 1$, the function will be compressed . If $a < -1$ or $a > 1$, the function will be stretched. See Figure 8.7(a).

```
> p:=plot({sin(t),2*sin(t),-1/2*sin(t)},t=-3/2*Pi..3/2*Pi):
> te:=textplot([[3/2*Pi,-2,'a=2'],[3/2*Pi,-1,'a=1'],
> [3/2*Pi,1/2,'a=-1/2']],align=RIGHT):
> display({p,te});
```

The *period* of the sine function $y = \sin(bt - c)$ is inversely proportional to b and is equal to $\frac{2\pi}{|b|}$. You can understand the effect of increasing b by noting that it multiplies the variable t, so the same point on the wave is reached sooner, at smaller t. See Figure 8.7(b).

```
> p1:=plot({sin(t),sin(2*t),sin(-1/2*t)},t=-3/2*Pi..3/2*Pi):
> plot({sin(t),sin(2*t),sin(-1/2*t)},t=-3/2*Pi..3/2*Pi);
```

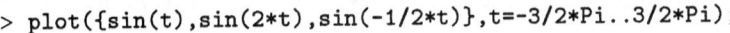

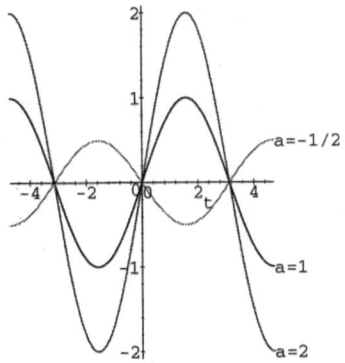
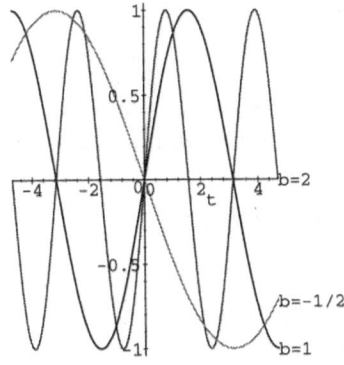

Figure 8.7. (a) Effect of varying a on the graph $f(x) = a\sin(t)$. (b) Effect of varying b on the graph $f(x) = \sin(bt)$

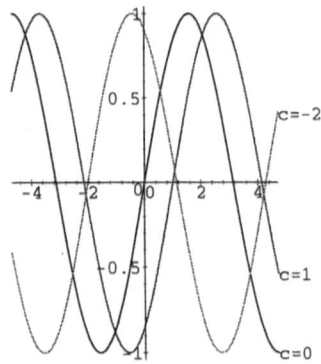   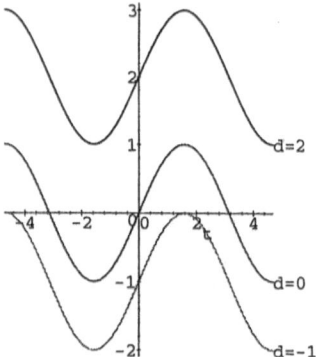

Figure 8.8. (a) Effect of varying c on the graph $f(x) = \sin(t - c)$. (b) Effect of varying d on the graph $f(x) = \sin(t) + d$

```
> te:=textplot([[3/2*Pi,0,'b=2'],[3/2*Pi,-1,'b=1'],
> [3/2*Pi,-0.72,'b=-1/2']],align=RIGHT):

> display({p1,te});
```

If $c \neq 0$, the function will be translated c units in the x-direction. See Figure 8.8(a).

```
> p2:=plot({sin(t),sin(t+2),sin(t-1)},t=-3/2*Pi..3/2*Pi):

> te:=textplot([[3/2*Pi,0.41,'c=-2'],[3/2*Pi,-1,'c=0'],
> [3/2*Pi,-0.54,'c=1']],align=RIGHT):

> display({p2,te});
```

If $d \neq 0$, the function will be translated d units in the y-direction. See Figure 8.8(a).

```
> p3:=plot({sin(t),sin(t)+2,sin(t)-1},t=-3/2*Pi..3/2*Pi):

> te:=textplot([[3/2*Pi,1,'d=2'],[3/2*Pi,-1,'d=0'],[3/2*Pi,-2,'d=-1']],
> align=RIGHT):

> display({p3,te});
```

Exercise 8.1 Plot the following circular functions from -4π to 4π and determine the period and amplitude of each.

(a) $y = \sin(x)$

(b) $y = \cos(x)$

(c) $y = 3\cos(2x)$

Exercise 8.2 Compare the plots of the functions $f(x) = |\sin(x)|$ and $g(x) = \sin(|x|)$.

Exercise 8.3   In a certain city the average temperature $f(t)$ on the tth day of the year (with $t = 1$ on January 1) is given by a formula of the form $f(t) = A + B\sin(k(t - \alpha))$. (The phase shift, α, can be thought of as the distance between the y-axis and the closest point at which the function intersects the x-axis.) (p438, Edwards and Penney [6])

(a) Plot the function using the values $A = 15$, $B = -7$, $k = 1/50$, and $\alpha = -\pi/4$ to get an idea of the shape of the curve. What must k be for the period of $f(t)$ to be 365 days (where $period = \frac{2\pi}{k}$)?

(b) The minimum daily average temperature of 10°C occurs on January 20 and the maximum of 25°C occurs half a year later. What are the values of A, B, and α?

(c) On what days of the year is the average temperature 15°C? Repeat for 20°C?

Experiment 8.1   Create an animation of the plot shown in Figure 8.7(a) using the following *Maple* commands.

```
> with(plots):
> animate(a*sin(t),t=-2*Pi..2*Pi,a=-2..2);
```

Create other animations by replacing a*sin(t) with sin(a*t), sin(t-a), and sin(t)+a. What happens when you vary two coefficients at once, ie a*sin(a*t)?

8.5   Addition of Circular Functions

A common application of the addition of circular functions is in representing sound waves. Suppose we have two sound waves, $y = \cos(2t)$ and $y = \sin(t)$. The graph of the sound wave that results, shown in Figure 8.9, is equal to the sum of the graphs of the waves. Notice that at $t = -\frac{\pi}{2}$, the value of the resultant wave

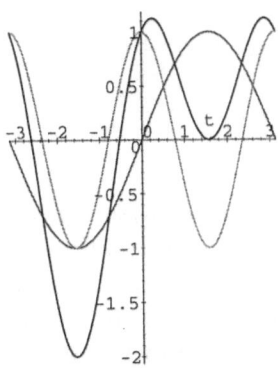

Figure 8.9. Addition of circular functions

is -2, twice the amplitude of either sound wave. This is known as *constructive interference*. At $t = \frac{\pi}{2}$, the value of one sound wave is 1, and the other is -1. The two waves cancel each other out here, so that the value of the resultant wave is 0. This is an example of *destructive interference*. Where are the other points of such constructive and destructive interference?

```
> plot({cos(2*t),sin(t),cos(2*t)+sin(t)},t=-Pi..Pi);
```

Exercise 8.4   One French horn emits a 200 Hz note and a second horn emits a 400 Hz note. If the amplitudes of the wave forms that represent each note are equal, show that the resultant of the combined sounds differs from each original sound and has a frequency of 200 Hz.

8.6   Simple Harmonic Motion

Periodic phenomena, such as a reciprocating piston or a pendulum, are examples of *simple harmonic motion* and can be described by circular functions. Imagine a pendulum swinging back and forth under the force of gravity, with negligible friction and air resistance. The speed of the pendulum is greatest when the pendulum is vertical and gradually decreases as it swings up until it stops and changes direction. The horizontal displacement x of the pendulum with time t is described by the function

$$x = a \sin(2\pi ft)$$

where f is the frequency, or number of cycles per unit time and $p = \frac{1}{f}$ is the period.

Suppose our swinging pendulum has a frequency of 2 cycles per second, or 2 Hz, and amplitude of 1 cm. Then the expression that describes its position is, in centimetres of displacement is,

```
> x:=sin(4*Pi*t);
```
$$x := \sin(4\pi t)$$

```
> plot(x,t=0..5);
```

Experiment 8.2   From mechanics or physics, a pendulum of length l undergoing small oscillations has period given approximately by $p = 2\pi\sqrt{\frac{l}{g}}$ where $g = 9.8$ ms^{-2} is the acceleration due to gravity. Suppose that you market pendulum clock kits with pendulums consisting of a small metal disk, which can be moved and locked in position, on a thin metal rod. When delivered, the pendulums are set to have the correct period with an amplitude $a = 0.05$ m when their length is $l = 0.2$ m. However, the pendulum expands in warm weather and the

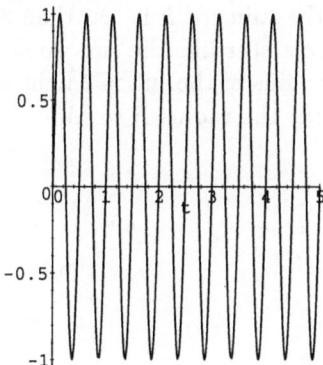

Figure 8.10. Simple harmonic motion example

position of the metal disk must be moved to restore the correct period. Plot a graph of period against length for the range of l from 0.18 m to 0.22 m. Also, construct a table to show the approximate adjustments needed in the position of the disk to compensate for changes in l of $\pm 1\%, \pm 2\%, \ldots, \pm 10\%$. You sell also some clocks to a physics laboratory and they need to use the next order approximation $p = 2\pi\sqrt{\frac{l}{g}}(1 + \frac{a^2}{16})$. Use plot3d to show how p depends on l in the same range as before, and on a in the range 0.04 to 0.1 m. Provide the modified table of adjustments for the case $a = 0.05$ m.

Experiment 8.3   A vibrating strip of metal has amplitude of vibration $a(t) = 3\cos(2\pi t)$ in air, where t is measured in seconds. At time $t = 1$, the strip is immersed in a viscous fluid and the amplitude diminishes according to the function $a(t) = 3e^{-t}\cos(2\pi t)$. This function is represented in *Maple* by,

```
> readlib(piecewise):
> a:=t->piecewise(1,t<1,3*cos(2*Pi*t),t>=1,3*exp(-t)*cos(2*Pi*t));
```

Plot this function for $t \geq 0$ and find the time taken for the amplitude of vibrations to fall below 1.

Experiment 8.4   In relativity theory it is common to use time units for everything. This is achieved by setting the velocity of light and the gravitational constant equal to one. Then we find that the mass of the sun is about 0.5×10^{-5} s. From Einstein's theory it is possible to approximate the period T of a circular orbit of radius R around a mass M in time units by

$$T = 2\pi R\sqrt{\frac{R}{M} - 3} \quad \text{for } R > 3M$$

Plot a family of curves for T against R, for a range of values of M. What is the estimate of the radius of Earth's orbit in time units (ie light-seconds)? Clearly, something goes wrong when the value of R is less than $3M$—this is a black hole situation. Find out how dense a star like the sun would need to be if it had its radius $R < 3M$. The normal radius of the sun is 2 light-seconds. What would be the period of orbits just outside the radius $R = 3M$?

Chapter 9

Trigonometry

Commands used in this chapter

- assign
- convert
- cos
- csc
- eval
- evalf
- geometry
- geometry[bisector]
- geometry[coordinates]
- geometry[inter]
- geometry[is_right]
- geometry[point]
- geometry[sides]
- geometry[triangle]
- interface
- linalg
- linalg[add]
- linalg[angle]
- linalg[dotprod]
- linalg[vector]

- op
- plot
- plot[options]
- plots
- plots[display]
- plots[polygonplot]
- plots[textplot]
- read
- readlib
- sec
- simplify
- sin
- solve
- sqrt
- student
- student[isolate]
- tan
- trigsubs
- verboseproc
- with

9.1   Basic Trigonometry

You should commit to memory the sides and angles of two simple triangles:

1. half of a unit square, formed by a diagonal line, gives a $45°, 45°, 90°$ triangle with sides $1, 1, \sqrt{2}$

2. half of an *equilateral triangle* with side length 2 gives a $30°, 60°, 90°$ triangle with sides $1, 2, \sqrt{3}$

Exercise 9.1   Using this information and definitions of the circular functions, make a table of $\cos(x)$, $\sin(x)$, and $\tan(x)$ for $x = 30°, 45°, 60°$.

The basic circular functions of sine and cosine are covered in Chapter 8.

```
> 2*Pi*rad=360;
```

$$2\,\pi\,rad = 360$$

```
> with(student):
> isolate(2*Pi*rad=360,rad);
```

$$rad = 180\,\frac{1}{\pi}$$

```
> convert(60*degrees,radians);
```

$$\frac{1}{3}\,\pi$$

Arguments in *Maple's* trigonometric functions must be given in radians.

```
> evalf(sin(Pi/3),5);
```

$$.86605$$

```
> evalf(sin(60),5);
```

$$-.30481$$

Exercise 9.2   Express the following in radians.

(a) 30°

(b) 270°

(c) −210°

Exercise 9.3   Express the following in degrees.

(a) $\frac{\pi}{4}$ rad

(b) $-\frac{9\pi}{4}$ rad

(c) 4π rad

9.2   Trigonometric Identities

Maple contains a library of trigonometric identities, some of which you may recognise, others you may not meet until later courses.

```
> readlib(trigsubs):
```

We can obtain a list of equivalent expressions to $\tan(x)$ as follows:

```
> trigsubs(tan(x));
```

$$\left[\tan(x), \tan(x), 2\,\frac{\tan\left(\dfrac{1}{2}x\right)}{1-\tan\left(\dfrac{1}{2}x\right)^2}, \frac{\sin(x)}{\cos(x)}, \frac{\sin(2x)}{1+\cos(2x)}, \frac{1-\cos(2x)}{\sin(2x)},\right.$$

$$\frac{1}{\cot(x)}, \frac{1}{\cot(x)}, 2\,\frac{\cot\left(\dfrac{1}{2}x\right)}{\cot\left(\dfrac{1}{2}x\right)^2-1}, 2\,\frac{1}{\cot\left(\dfrac{1}{2}x\right)-\tan\left(\dfrac{1}{2}x\right)},$$

$$\left.-\frac{I\left(e^{(Ix)}-e^{(-Ix)}\right)}{e^{(Ix)}+e^{(-Ix)}}\right]$$

Maple will accept only certain functions into its library search command. Even combinations of trigonometric functions found using the `trigsubs` command may not be accepted as search parameters. Here are some equivalents for $\sin(x)^2$.

```
> trigsubs(sin(x)^2);
```

$$\left[\sin(x)^2, 1-\cos(x)^2, \frac{1}{2}-\frac{1}{2}\cos(2x), \sin(x)^2, \sin(x)^2,\right.$$

$$4\sin\left(\frac{1}{2}x\right)^2\cos\left(\frac{1}{2}x\right)^2, \frac{1}{\csc(x)^2}, \frac{1}{\csc(x)^2}, 4\,\frac{\tan\left(\dfrac{1}{2}x\right)^2}{\left(1+\tan\left(\dfrac{1}{2}x\right)^2\right)^2},$$

$$\left.-\frac{1}{4}\left(e^{(Ix)}-e^{(-Ix)}\right)^2\right]$$

Maple does perform trigonometric substitutions. Here are some examples:

```
> simplify(sin(x)^2+cos(x)^2);
```
$$1$$

```
> simplify(sin(x)*cot(x),trig);
```
$$\cos(x)$$

```
> simplify(sin(x)^2,trig);
```
$$1-\cos(x)^2$$

```
> simplify(sec(x)^2,trig);
```

$$\frac{1}{\cos(x)^2}$$

```
> f:=convert(1+tan(x)^2,sincos);
```
$$f := 1 + \frac{\sin(x)^2}{\cos(x)^2}$$

```
> g:=convert(f,tan);
```
$$g := 1 + 4\,\frac{\tan\left(\frac{1}{2}x\right)^2}{\left(1 - \tan\left(\frac{1}{2}x\right)^2\right)^2}$$

```
> h:=simplify(g,trig);
```
$$h := \frac{1}{\cos(x)^2}$$

Maple does not always return the answer you expect, but it is an equally valid answer!

Maple's `simplify` command is not very useful for proving trigonometric identities because it does not show the intermediate steps leading to the final solution. Its `trigsubs` command can be helpful as a list of intermediate identities to be used when proving indentities by hand.

Exercise 9.4   Express the following in terms of sine and cosine.

(a) $\csc(x)$

(b) $\sec(x)^2$

(c) $\tan(x)\sin(x)$

Exercise 9.5   Find an equivalent form of the following.

(a) $\csc(x)^2$

(b) $-\tan(x)^2$

(c) $1 + \cot(x)^2$

9.3   Trigonometric Equations

A common last step in the solution of a problem is the need to solve a trigonometric equation for a particular numerical value. From your knowledge of the

graphs of trigonometric functions, you can usually make a guess of the answer
to check against your working.

```
> sin(x)=0.9;
```
$$\sin(x) = .9$$

```
> solve(sin(x)=0.9,{x});
```
$$\{ x = 1.119769515 \}$$

```
> assign(");
> evalf(convert(x,degrees),5);
```
$$64.159 \ degrees$$

```
> sin(y)^2-2*cos(y)=0.5;
```
$$\sin(y)^2 - 2\cos(y) = .5$$

```
> solve(sin(y)^2-2*sin(y)=0.5,{y});
```
$$\{ y = 1.570796327 - 1.437955921 \, I \}, \{ y = -.2266812021 \}$$

Exercise 9.6   Solve the following trigonometric equations.

(a) $\sin(x) = \frac{1}{2}$

(b) $\sin(x)^2 + 2\sin(x) - 3 = 0$

(c) $2\sin(x) = \cos(x)$

Exercise 9.7   Show that the general solution of $\cos(4x) = 0$ is $x = (2n+1)\frac{\pi}{8}$,
and the general solution of $\sin(x) = 0$ is $x = n\pi$, for $n = 0, 1, 2, \ldots$ Hence deduce
the general solution of $\sin(5x) - \sin(3x) = 0$.

Exercise 9.8   Show that

$$3\cos x + \sin x = b(\cos x \cos a + \sin x \sin a) = b\cos(x - a)$$

where $b\cos a = 3$ and $b\sin a = 1$, and so $\tan a = \frac{1}{3}$ and $b = \sqrt{10}$. Hence solve the
equation $3\cos x + \sin x = 1$ for values of x between 0 and 4π.

Exercise 9.9   Use the approximations $\sin\theta \doteq \theta$ and $\cos\theta \doteq 1 - \frac{\theta^2}{2}$ to obtain
an approximation to $\frac{\sin 2\theta}{4 - \cos 3\theta}$. Over what range is this approximation accurate to
three significant figures?

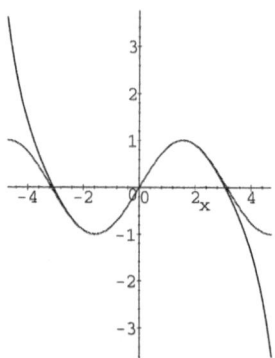

Figure 9.1. The graphs of the sine function and its power series expansion about $x = 0$ to the 7th power, on $[-\frac{3}{2}\pi, \frac{3}{2}\pi]$

Experiment 9.1   Investigate graphically the range of x values for which

(a) $\sin x \le x$

(b) $\tan x \ge x$

Experiment 9.2   By plotting on successively smaller intervals, convince yourself that there is an x with $0 < x < \frac{\pi}{2}$, such that $\sin x = \frac{1}{x}$, and find it to 3 decimal places.

9.4   Power Series Expansions

Maple can give us *power series expansions* for trigonometric functions about any given point. For example, near $x = 0$,

```
> series(sin(x),x,9);
```

$$x - \frac{1}{6}\,x^3 + \frac{1}{120}\,x^5 - \frac{1}{5040}\,x^7 + \mathrm{O}(\,x^9\,)$$

If we overlay the graphs of the sine function and its power series expansion about $x = 0$, as shown in Figure 9.1, we can see that the farther away from $x = 0$, the poorer the approximation becomes.

```
> plot({sin(x),x-1/6*x^3+1/120*x^5-1/5040*x^7},x=-3/2*Pi..3/2*Pi);
> series(cos(x),x,9);
```

$$1 - \frac{1}{2}\,x^2 + \frac{1}{24}\,x^4 - \frac{1}{720}\,x^6 + \frac{1}{40320}\,x^8 + \mathrm{O}(\,x^9\,)$$

```
> series(tan(x),x,9);
```

$$x + \frac{1}{3}x^3 + \frac{2}{15}x^5 + \frac{17}{315}x^7 + O(x^9)$$

We can, of course, improve the approximation by including higher order terms.

To obtain a power series expansion in terms of distance x from, say the point $\frac{\pi}{9}$, we use,

```
> series(sin(Pi/9+x),x,9);
```

$$\sin\left(\frac{1}{9}\pi\right) + \cos\left(\frac{1}{9}\pi\right)x - \frac{1}{2}\sin\left(\frac{1}{9}\pi\right)x^2 - \frac{1}{6}\cos\left(\frac{1}{9}\pi\right)x^3 +$$
$$\frac{1}{24}\sin\left(\frac{1}{9}\pi\right)x^4 + \frac{1}{120}\cos\left(\frac{1}{9}\pi\right)x^5 - \frac{1}{720}\sin\left(\frac{1}{9}\pi\right)x^6 -$$
$$\frac{1}{5040}\cos\left(\frac{1}{9}\pi\right)x^7 + \frac{1}{40320}\sin\left(\frac{1}{9}\pi\right)x^8 + O(x^9)$$

```
> evalf(");
```

$$.3420201433 + .9396926208\,x - .1710100717\,x^2 - .1566154368\,x^3 +$$
$$.01425083931\,x^4 + .007830771840\,x^5 - .0004750279768\,x^6 -$$
$$.0001864469486\,x^7 + .8482642442\,10^{-5}\,x^8 + O(x^9)$$

9.5 Right-Angled Triangles

A *right-angled triangle* is any triangle with an interior angle of 90°. The side opposite the right angle ($\overline{AC}$ in Figure 9.2) is called the *hypotenuse*.

Suppose we are interested in determining $\angle C$. Then the side $\overline{AB}$ is called the *opposite* side and the side $\overline{BC}$ is the *adjacent* side, and

$$\sin(C) = \text{opposite/hypotenuse},$$
$$\cos(C) = \text{adjacent/hypotenuse, and}$$
$$\tan(C) = \text{opposite/adjacent}.$$

Recall the two right-angled triangles from the first exercise in this chapter, with angles 45°, 45°, 90° and 30°, 60°, 90° respectively.

Triangles can be defined in a number of ways. They can be defined by three vertices, the lengths of three sides, the lengths of two sides and the angle between them, or by three lines. The package geometry contains this information.

We define a triangle $\triangle T1$ to have two sides of length 2 and 3, and included angle $\frac{\pi}{2}$, using the geometry package:

```
> with(geometry):
```

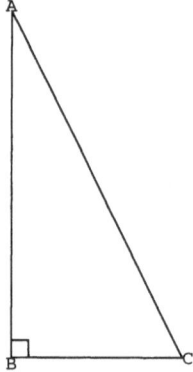

Figure 9.2. Right-angled triangle $\triangle ABC$

```
> triangle(T1,[2,angle=Pi/2,3]);
                          T1
```

```
> is_right(T1);
```
$$true$$

Obviously, $\triangle T1$ is a right-angled triangle because we specified the angle between the lines as 90°. The **op** command will now summarize the properties of $\triangle T1$ for us.

```
> op(T1);
```
$$table([$$
$$form = triangle$$
$$given = \left[2, angle = \frac{1}{2}\pi, 3\right]$$
$$sides = \left[2, 3, \sqrt{13}\right]$$
$$])$$

Pythagoras' Theorem states that

$$\text{hypotenuse}^2 = \text{opposite}^2 + \text{adjacent}^2$$

(Commit to memory the prototype of all right-angled triangles, with sides 3, 4, and 5. Obviously it satisfies Pythagoras' Theorem since $9 + 16 = 25$.) Using this theorem, we can calculate the length of the third side of $\triangle T1$.

```
> opp:=2;
```
$$opp := 2$$

```
> adj:=3;
```
$$adj := 3$$

```
> hyp:=sqrt(opp^2+adj^2);
```
$$hyp := \sqrt{13}$$

So, we have just confirmed *Maple's* result.
```
> sin(z)=opp/hyp;
```
$$\sin(z) = \frac{2}{13}\sqrt{13}$$

```
> solve(sin(z)=opp/hyp,z);
```
$$\arcsin\left(\frac{2}{13}\sqrt{13}\right)$$

```
> convert(solve(sin(z)=opp/hyp,z),degrees);
```
$$180\,\frac{\arcsin\left(\frac{2}{13}\sqrt{13}\right)\,degrees}{\pi}$$

```
> evalf(convert(solve(sin(z)=opp/hyp,z),degrees));
```
$$33.69006751\ degrees$$

```
> sin(w)=adj/hyp;
```
$$\sin(w) = \frac{3}{13}\sqrt{13}$$

```
> evalf(convert(solve(sin(w)=adj/hyp,w),degrees));
```
$$56.30993246\ degrees$$

Now we sum the angles to be sure that they add up to 180°.
```
> sum_of_angles:=90+33.7+56.3;
```
$$sum\_of\_angles := 180.0$$

The opposite and adjacent sides were assigned arbitrarily. It doesn't matter which is which, as long as you are consistent.

Exercise 9.10 Solve the right-angled triangle $\triangle QRS$ given that $\angle R = 90°$, $s = 5$, and $q = 7$.

Exercise 9.11 On a clear day in a flat desert how far away is the horizon seen by a person whose eyes are 1.5 m above the ground? You may take the radius of the Earth to be 6400 km, and you will need to know Pythagoras' theorem for a right-angled triangle.

Experiment 9.3 Let the diameter of a sphere be D. Suppose that a telescope is mounted at height h above the surface. At what distance is the horizon seen by the telescope? In the case that $h \ll D$ show that the distance to the horizon is approximately given by the geometric mean of h and D. Plot a graph to show the error in this approximation for the cases $h/D = 10^0, 10^{-5}, 10^{-10}, 10^{-15}, 10^{-20}, 10^{-25}$.

9.6   Law of Sines

We declare a new triangle $\triangle T2$ with sides 3 and 4, and included angle $\pi/6$ ...

```
> triangle(T2,[3,angle=Pi/6,4]);
```
$$T\mathit{2}$$

... and then catalogue its properties.

```
> op(T2);
```
$$\text{table([}$$
$$\quad form = triangle$$
$$\quad given = \left[3, angle = \frac{1}{6}\pi, 4\right]$$
$$\text{])}$$

The `sides` command calculates the lengths of the sides of $\triangle T2$.

```
> sides(T2);
```
$$\left[3, 4, \sqrt{25 - 12\sqrt{3}}\right]$$

The Law of Sines states that, for any triangle $\triangle ABC$, with sides of lengths a, b, and c, and opposite angles A, B, and C, respectively,

$$\frac{a}{\sin(A)} = \frac{b}{\sin(B)} = \frac{c}{\sin(C)}.$$

This relationship is, in fact, two independent equations. *Maple* can calculate a, b, and, c no matter how the triangle is defined, but we must know one of the angles to solve for the other two.

```
> a:=3;
```
$$a := 3$$

```
> B:=Pi/6;
```
$$B := \frac{1}{6}\pi$$

```
> c:=4;
```
$$c := 4$$

```
> b:=sqrt(25-12*sqrt(3));
```
$$b := \sqrt{25 - 12\sqrt{3}}$$

```
> A:=evalf(convert(solve(a/sin(A)=b/sin(B),A),degrees));
```
$$A := 46.93568648 \; degrees$$

```
> C:=evalf(convert(solve(b/sin(B)=c/sin(C),C),degrees));
```
$$C := 76.93568627 \; degrees$$

```
> sum_angles:=A+evalf(convert(Pi/6,degrees))+C;
```
$$sum\_angles := 153.8713728 \; degrees$$

Why do the angles not sum to 180°?

```
> sine_of_calculated_angle:=evalf(sin(solve(b/sin(B)=c/sin(C),C)));
```
$$sine\_of\_calculated\_angle := .9741169491$$

```
> correct_angle:=evalf(180*degrees-convert(B,degrees)-A);
```
$$correct\_angle := 103.0643135 \; degrees$$

```
> sine_of_correct_angle:=evalf(sin(convert(correct_angle,
> radians)));
```
$$sine\_of\_correct\_angle := .9741169474$$

It would seem that the angles 76.94° and 103.06° have the same sine. One way to avoid this problem would be to calculate either A or C using the Sine Law and then solve for the third by subtracting from the total. But how do we know that the angle we calculate using the Sine Law is correct? For this reason, it is better to use the Law of Cosines to solve triangles.

Exercise 9.12   Solve the following triangles using the Sine Law.

(a) $\triangle ABC$, given $a = 10$, $b = 15$, $\angle C = \frac{3\pi}{4}$

(b) $\triangle XYZ$, given $\angle X = 50°$, $z = 5$, $y = 3.5$

9.7   Law of Cosines

The Law of Cosines states that, for any triangle $\triangle ABC$:

$$a^2 = b^2 + c^2 - 2bc\cos(A),$$

$$b^2 = c^2 + a^2 - 2ca\cos(B), \text{ and}$$

$$c^2 = a^2 + b^2 - 2ab\cos(C).$$

This is a set of three equations with six unknowns. Unless the triangle is defined in terms of the coordinates of the vertices, three of the parameters must be known in order to define the triangle using `triangle`. But we need know only three of the parameters to solve the triangle using the Cosine Law. Therefore, it is not necessary to define the triangle in order to solve it.

```
> a:=3;b:=4;c:=6;
```
$$a := 3$$
$$b := 4$$
$$c := 6$$

```
> A:=evalf(convert(solve(a^2=b^2+c^2-2*b*c*cos(A),A),degrees),5);
```
$$A := 26.384 \; degrees$$

```
> B:=evalf(convert(solve(b^2=c^2+a^2-2*c*a*cos(B),B),degrees),5);
```
$$B := 36.337 \; degrees$$

```
> C:=evalf(convert(solve(c^2=a^2+b^2-2*a*b*cos(C),C),degrees),5);
```
$$C := 117.28 \; degrees$$

```
> sum_angles:=A+B+C;
```
$$sum\_angles := 180.001 \; degrees$$

So we do indeed believe that we have a triangle defined by three lengths—you could construct it with a compasses by drawing arcs.

Example 9.1   Solve $\triangle ABC$ given that $a = 2$, $b = 4$, and $\angle C = \frac{\pi}{4}$.

```
> a:=2;b:=4;C:=Pi/4;
```
$$a := 2$$
$$b := 4$$
$$C := \frac{1}{4}\pi$$

```
> c:=evalf(solve(c=sqrt(a^2+b^2-2*a*b*cos(C))),5);
```
$$c := 2.9472$$

```
> A:=evalf(convert(solve(a^2=b^2+c^2-2*b*c*cos(A),A),degrees),5);
```
$$A := 28.677 \ degrees$$

```
> B:=evalf(convert(solve(b^2=c^2+a^2-2*c*a*cos(B),B),degrees),5);
```
$$B := 106.33 \ degrees$$

```
> sum_angles:=convert(C,degrees)+A+B;
```
$$sum\_angles := 180.007 \ degrees$$

Maple allows the user to write and save procedures. For long calculations that you perform frequently, you can save quite a bit of time by writing a program. An example of a user written program is 'angles'. It can solve any type of triangle that is accepted by the **triangle** function.

```
> read 'angles.m';
> triangle(T,[2,angle=Pi/4,4]);
```
$$T$$

```
> angles(T);
```

$$\left[180 \frac{\arccos\left(\frac{8}{17}\sqrt{5-2\sqrt{2}} + \frac{3}{34}\sqrt{5-2\sqrt{2}}\sqrt{2}\right) \ degrees}{\pi}, 45 \ degrees, \right.$$
$$\left. 180 \frac{\arccos\left(\frac{1}{17}\sqrt{5-2\sqrt{2}} - \frac{3}{17}\sqrt{5-2\sqrt{2}}\sqrt{2}\right) \ degrees}{\pi} \right]$$

```
> evalf(");
```
$$[\, 28.67505008 \ degrees, 45. \ degrees, 106.3249499 \ degrees \,]$$

To view that series of commands that make up 'angles', enter the following commands. (Any built-in or external *Maple* procedure can be listed this way.)

```
> interface(verboseproc=2):
> eval(angles):
```

The output is omitted here, but it can be found in Appendix C.1.

Exercise 9.13   Solve the following triangles.

(a) $\triangle JKL$, given $\angle J = 130°$, $\angle L = 30°$, $k = 10$

(b) $\triangle PQR$, given $p = 4$, $q = 6$, $\angle R = \frac{\pi}{4}$

Exercise 9.14   If a 50 m tall structure casts an 80 m long shadow, what is the angle of the sun's elevation?

Experiment 9.4   Design a sundial to use near the equator; there the sun always rises in the east and sets in the west, passing vertically overhead at noon. Design also a moondial.

9.8   Vectors

Let's begin by calling in a new package called `linalg`. This package contains commands used in linear algebra.

> `with(linalg):`

To define a 2-dimensional *vector*, $\vec{a}$, from (0,0) to (3,4), we use the `vector` command as follows.

> `a:=vector(2,[3,4]);`

$$a := [\,3\ 4\,]$$

By projecting the $x-$ and $y-$components of this vector onto their axes, we create a right-angled triangle. We can now calculate the length of the vector using Pythagoras' Theorem.

> `length_a:=sqrt(a[1]^2+a[2]^2);`

$$length\_a := 5$$

Similarly, we can define a vector, $\vec{b}$, joining the points (0,0) and (2,1).

> `b:=vector(2,[2,1]);`

$$b := [\,2\ 1\,]$$

> `length_b:=sqrt(b[1]^2+b[2]^2);`

$$length\_b := \sqrt{5}$$

To determine the angle between the vectors $\vec{a}$ and $\vec{b}$, we use the `angle` command. First, however, imagine the two vectors drawn on graph paper and make a guess at the angle.

> `angle(a,b);`

$$\arccos\left(\frac{2}{25}\sqrt{25}\sqrt{5}\right)$$

> `convert(",degrees);`

$$180\,\frac{\arccos\left(\frac{2}{25}\sqrt{25}\sqrt{5}\right)\ degrees}{\pi}$$

```
> evalf(");
```
$$26.56505115 \; degrees$$

Suppose we wish to know the x−component of the vector $\vec{c}(x,2)$ which is 30° from vector $\vec{b}(2,1)$.

```
> convert(30*degrees,radians);
```
$$\frac{1}{6}\pi$$

```
> evalf(1/6*Pi);
```
$$.5235987758$$

```
> b:=vector(2,[2,1]);
```
$$b := [\, 2 \; 1 \,]$$

```
> c:=vector(2,[x,2]);
```
$$c := [\, x \; 2 \,]$$

```
> angle(b,c);
```
$$\mathrm{arccos}\left(\frac{1}{5}\frac{(2x+2)\sqrt{5}}{\sqrt{x^2+4}}\right)$$

```
> solve(angle(b,c)=0.524,{x});
```
$$\{\, x = 1.319356115 \,\}$$

Now we wish to determine both coordinates of a vector $\vec{b}(x,y)$ knowing that the length of $\vec{b}$ is 7 units, the coordinates of a second vector $\vec{a}(1,0)$, and that the angle between them is 0.8 radians. (This is a very tricky method. It is much easier to use the dot product.)

```
> length_hyp_squared:=(x^2+y^2);
```
$$length\_hyp\_squared := x^2 + y^2$$

```
> a:=vector(2,[1,0]);
```
$$a := [\, 1 \; 0 \,]$$

```
> b:=vector(2,[x,y]);
```
$$b := [\, x \; y \,]$$

```
> c:=angle(a,b);
```
$$c := \mathrm{arccos}\left(\frac{x}{\sqrt{x^2+y^2}}\right)$$

Normally we would solve this system of equations simultaneously, but *Maple* will not accept the command in this case. We can find an expression for y in terms of x two ways. The first is to solve the equation $x^2 + y^2 = 7$.

```
> s1:=[solve(length_hyp_squared=7,y)];
```
$$s1 := \left[\sqrt{-x^2 + 7}, -\sqrt{-x^2 + 7} \right]$$

```
> y1:=op(1,s1);
```
$$y1 := \sqrt{-x^2 + 7}$$

The second is to solve the equation $\arccos(\frac{x}{\sqrt{x^2+y^2}}) = 0.8$.

```
> s2:=[solve(c=0.8,y)];
```
$$s2 := \left[1.029638557\sqrt{x^2}, -1.029638557\sqrt{x^2} \right]$$

```
> y2:=op(1,s2);
```
$$y2 := 1.029638557\sqrt{x^2}$$

With these two expressions for y, we can solve for x.

```
> solution1:=[solve(y1^2=y2^2,x)];
```
$$solution1 := [-1.843312690, 1.843312690]$$

Now we shall substitute each value of x into the expressions $y1$ and $y2$. If after the substitution the values of $y1$ and $y2$ are equal, we shall know these solutions are valid.

```
> x:=solution1[1];
```
$$x := -1.843312690$$

```
> y1;
```
$$1.897945818$$

```
> y2;
```
$$1.897945818$$

```
> x:=solution1[2];
```
$$x := 1.843312690$$

```
> y1;
```
$$1.897945818$$

```
> y2;
```
$$1.897945818$$

One valid set of coordinates for vector $\vec{b}$ is (1.843,1.898), but we still have to consider the other solutions for $s1$ and $s2$.

```
> y1:=op(2,s1);
```
$$y1 := -\sqrt{-x^2 + 7}$$

```
> y2:=op(2,s2);
```
$$y2 := -1.029638557\sqrt{x^2}$$

```
> solution2:=[solve(y1^2=y2^2,x)];
```
$$solution2 := [-1.843312690, 1.843312690]$$

```
> x:=solution2[1];
```
$$x := -1.843312690$$

```
> y1;
```
$$-1.897945818$$

```
> y2;
```
$$-1.897945818$$

```
> x:=solution2[2];
```
$$x := 1.843312690$$

```
> y1;
```
$$-1.897945818$$

```
> y2;
```
$$-1.897945818$$

Another possible set of coordinates for vector $\vec{b}$ is $(1.843, -1.898)$.

To add vectors, we simply add the $x-$ and $y-$components to form a new vector.

```
> d:=vector(2,[1,3]);
```
$$d := [1\,3]$$

```
> e:=vector(2,[3,4]);
```
$$e := [3\,4]$$

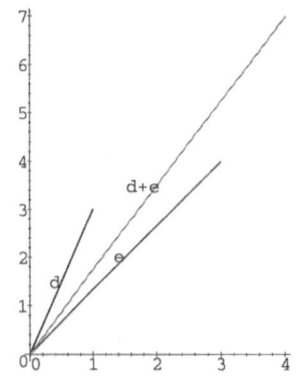

Figure 9.3. Addition of vectors

```
> f:=add(d,e);
```
$$f := [4\,7]$$

Figure 9.3 is a plot of vectors $\vec{d}$ and $\vec{e}$, and the resulting vector $\vec{f}$.
```
> with(plots):
> p:=plot({[0,0,1,3],[0,0,3,4],[0,0,4,7]}):
> t:=textplot([[0.4,1.5,'d'],[1.8,3.5,'d+e'],[1.4,2,'e']]):
> display({p,t});
```
The magnitude of $\vec{f}$ is calculated using Pythagoras' Theorem.
```
> magnitude_of_f:=sqrt(f[1]^2+f[2]^2);
```
$$magnitude\_of\_f := \sqrt{65}$$

Exercise 9.15   Determine the angle enclosed by two vectors, $\vec{a}(6,7)$ and $\vec{b}(4,3)$.

Exercise 9.16   The angle between vectors $\vec{c}(2,3)$ and $\vec{d}(x,1)$ is 35°. Find x.

9.8.1   Dot Product

The *dot product* of two vectors is equal to the product of the magnitudes of the vectors multiplied by the cosine of the angle between them, so,

$$\vec{a} \cdot \vec{b} = \mid \vec{a} \mid\mid \vec{b} \mid \cos \theta$$

It is also equal to the sum of the products of the $x-$ and $y-$components,

$$\vec{a} \cdot \vec{b} = a_x b_x + a_y b_y$$

Let's calculate the dot product of $\vec{d}(1,3)$ and $\vec{e}(3,4)$ defined in the last section using the two expressions described above and the `dotprod` command.

```
> length_d:=sqrt(d[1]^2+d[2]^2);
```
$$length\_d := \sqrt{10}$$

```
> length_e:=sqrt(e[1]^2+e[2]^2);
```
$$length\_e := 5$$

```
> angle(d,e);
```
$$\arccos\left(\frac{3}{50}\sqrt{10}\sqrt{25}\right)$$

```
> ang:=evalf(");
```
$$ang := .3217505546$$

```
> dot_product1:=length_d*length_e*cos(ang);
```
$$dot\_product1 := 4.743416490\sqrt{10}$$

```
> evalf(");
```
$$15.00000000$$

```
> dot_product2:=d[1]*e[1]+d[2]*e[2];
```
$$dot\_product2 := 15$$

```
> dot_product3:=dotprod(d,e);
```
$$dot\_product3 := 15$$

Example 9.2 Find the vector $\vec{h}$ knowing that it is three units long and the angle between it and vector $\vec{g}(3,1)$ is 35°.

```
> g:=vector(2,[3,1]);
```
$$g := [\,3\,1\,]$$

```
> ang:=convert(35*degrees,radians);
```
$$ang := \frac{7}{36}\pi$$

```
> evalf(ang);
```
$$.6108652381$$

```
> h:=vector(2,[x,y]);
```
$$h := [\,x\,y\,]$$

```
> dotprod(g,h);
```
$$3\,x + y$$

```
> length_of_g:=sqrt(g[1]^2+g[2]^2);
```
$$length\_of\_g := \sqrt{10}$$

```
> length_of_h:=3;
```
$$length\_of\_h := 3$$

By equating one form of the dot product to another, we can find an expression for y in terms of x.

```
> dot_product:=length_of_g*length_of_h*cos(ang)=dotprod(g,h);
```
$$dot\_product := 3\,\sqrt{10}\cos\left(\frac{7}{36}\,\pi\right) = 3\,x + y$$

```
> solve(",{y});
```
$$\left\{ y = 3\,\sqrt{10}\cos\left(\frac{7}{36}\,\pi\right) - 3\,x \right\}$$

```
> assign(");
```

Using the expression for the length of $\vec{h}$, we can now solve for x knowing y in terms of x.

```
> length_of_h:=sqrt(h[1]^2+h[2]^2)=3;
```
$$length\_of\_h := \sqrt{x^2 + \left(3\,\sqrt{10}\cos\left(\frac{7}{36}\,\pi\right) - 3\,x\right)^2} = 3$$

```
> s:=evalf(");
```
$$s := \sqrt{x^2 + (\,7.771158630 - 3.\,x\,)^2} = 3.$$

```
> [solve(s,x)];
```
$$[\,2.875489974, 1.787205204\,]$$

Now we shall substitute each value of x into our expression for y.

```
> x1:=op(1,");
```
$$x1 := 2.875489974$$

```
> x2:=op(2,"");
```
$$x2 := 1.787205204$$

```
> y1:=evalf(subs(x=x1,y));
```
$$y1 := -.855311292$$

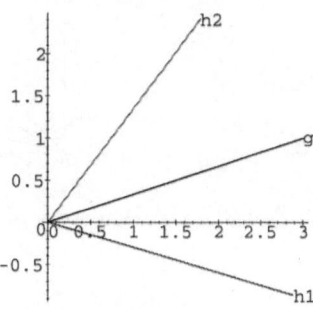

Figure 9.4. Dot product example

> y2:=evalf(subs(x=x2,y));
$$y2 := 2.409543018$$

> h1:=vector(2,[x1,y1]);
$$h1 := [\,2.875489974 - .855311292\,]$$

> evalf(angle(g,h1));
$$.6108652382$$

> h2:=vector(2,[x2,y2]);
$$h2 := [\,1.787205204\ 2.409543018\,]$$

> evalf(angle(g,h2));
$$.6108652382$$

There are two vectors, $\vec{h1}$ and $\vec{h2}$, which are three units long and 35° from vector $\vec{g}(3,1)$. The three vectors are graphed in Figure 9.4.

```
> display({plot({[[0,0],[3,1]],[[0,0],[h1[1],h1[2]]],[[0,0],
> [h2[1],h2[2]]]}),textplot([[3,1,'g'],[2.89,-0.86,'h1'],
> [1.79,2.41,'h2']],align=RIGHT)},scaling=CONSTRAINED);
```

Exercise 9.17   Use the dot product to determine the angle between the vectors $\vec{a}(5,3)$ and $\vec{b}(7,-4)$.

Exercise 9.18   What is the x-coordinate of the vector $\vec{a}(x,-1)$ so that it forms a 40° angle with the vector $\vec{g}(3,3)$.

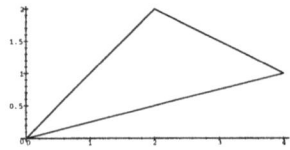
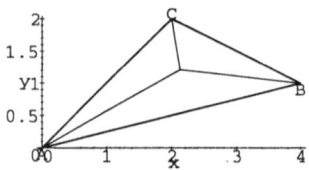

Figure 9.5. (a) $\triangle ABC$. (b) Bisectors of $\triangle ABC$

9.9 Bisector of a Triangle

The line through an interior angle of a triangle which is equidistant from the sides of the triangle which enclose the angle is called a *bisector*. There is a bisector for each of the three interior angles of a triangle.

We begin by defining $\triangle ABC$ with vertices at $A(0,0)$, $B(2,2)$, and $C(4,1)$.

```
> triangle(ABC,[point(A,0,0),point(B,2,2),point(C,4,1)]);
```
$$ABC$$

The following plot is shown in Figure 9.5(a).

```
> polygonplot([coordinates(A),coordinates(B),coordinates(C)]);
> bisector(ABC,A,bisector_of_A);
```
$$bisector\_of\_A$$

```
> bisector_of_A[equation];
```
$$\frac{\left(-2\sqrt{2}-2\sqrt{17}\right)x}{\sqrt{17}+2\sqrt{2}}+\frac{\left(8\sqrt{2}+2\sqrt{17}\right)y}{\sqrt{17}+2\sqrt{2}}=0$$

```
> evalf(bisector_of_A[equation],4);
```
$$-1.593\,x+2.814\,y=0$$

```
> bisector(ABC,B,bisector_of_B);
```
$$bisector\_of\_B$$

```
> bisector_of_B[equation];
```
$$\frac{\left(2\sqrt{5}+2\sqrt{2}\right)x}{\sqrt{5}+2\sqrt{2}}+\frac{\left(4\sqrt{2}-2\sqrt{5}\right)y}{\sqrt{5}+2\sqrt{2}}-12\,\frac{\sqrt{2}}{\sqrt{5}+2\sqrt{2}}=0$$

```
> evalf(bisector_of_B[equation],4);
```
$$1.442\,x+.2338\,y-3.350=0$$

```
> bisector(ABC,C,bisector_of_C);
```
$$bisector\_of\_C$$

```
> bisector_of_C[equation];
```
$$\frac{\left(\sqrt{5}-\sqrt{17}\right)x}{\sqrt{5}+\sqrt{17}}+\frac{\left(-2\sqrt{17}-4\sqrt{5}\right)y}{\sqrt{5}+\sqrt{17}}+6\frac{\sqrt{17}}{\sqrt{5}+\sqrt{17}}=0$$

```
> evalf(bisector_of_C[equation],4);
```
$$-.2967\,x-2.703\,y+3.890=0$$

All of the bisectors of a triangle intersect at the same point. There is no command which allows us to determine where three lines intersect, but using the inter command we can pair together different bisectors and see if the points are the same. The bisectors of $\triangle ABC$ are shown in Figure 9.5(b).

```
> inter(bisector_of_A,bisector_of_B);
```
$$bisector\_of\_A\_intersect\_bisector\_of\_B$$

```
> evalf(coordinates(inter(bisector_of_A,bisector_of_B)));
```
$$[\,2.128947516, 1.205389601\,]$$

```
> inter(bisector_of_B,bisector_of_C);
```
$$bisector\_of\_B\_intersect\_bisector\_of\_C$$

```
> evalf(coordinates(inter(bisector_of_B,bisector_of_C)));
```
$$[\,2.128947516, 1.205389601\,]$$

```
> inter(bisector_of_A,bisector_of_C);
```
$$bisector\_of\_A\_intersect\_bisector\_of\_C$$

```
> evalf(coordinates(inter(bisector_of_A,bisector_of_C)));
```
$$[\,2.128947516, 1.205389601\,]$$

```
> display({plot([coordinates(A),coordinates(B),coordinates(C),
> coordinates(A)]),textplot([[0,0,'A'],[4,1,'B']],align=BELOW),
> textplot([2,2,'C'],align=ABOVE),implicitplot
> (bisector_of_A[equation],x=0..2.129,y=0..1.205),
> implicitplot(bisector_of_B[equation],x=0..4,y=1.205..2),
> implicitplot(bisector_of_C[equation],x=2.129..4,y=0..1.205)});
```

Experiment 9.5 Using the method presented above, produce a plot, like that in Figure 9.5(b), of the bisectors of $\triangle FGH$ with vertices $F(0,0)$, $G(2,-1)$, and $H(5,3)$.

Experiment 9.6   A spider and a fly live in a rectangular room which is 3 m high, 4 m wide and 5 m long. The spider can only walk on walls, floor, and ceiling. Now, the fly can fly between any two points and normally waits until the spider is within 1 m before flying to a point as far away as possible from the spider. If they start off in diagonally opposite corners and the spider always walks the shortest route towards the fly, stopping while the fly is in the air, what do you expect will be the first few moves? Hint: It may help to imagine the walls, floor and ceiling as a box made from one peice of flat material by folding. Then shortest distances for the spider are straight lines on this 'plan' of the room. You need to mark the edges that are 'identified' together to make a real room.

Chapter 10

Similar Figures

Commands used in this chapter

- geometry
- geometry[are_similar]
- geometry[area]
- geometry[coordinates]
- geometry[distance]
- geometry[inter]
- geometry[line]
- geometry[perpendicular]
- geometry[point]
- geometry[triangle]

- plot
- plot[options]
- plots
- plots[display]
- plots[implicitplot]
- plots[polygonplot]
- plots[textplot]
- solve
- sqrt
- with

10.1 Similar Figures

Two polygons are said to be *similar* if corresponding angles are the same and all of the ratios of the lengths of the corresponding sides are equal.

10.1.1 Similar Triangles

The criteria for similar polygons apply to similar triangles, but, due to the nature of triangles, we can make three additional statements which apply only to similar triangles. If two triangles have all corresponding angles equal, or if the lengths of the corresponding sides are proportional, then the triangles are similar. Also, if two triangles have one corresponding angle, and the lengths of the sides that enclose the angle are proportional, then the triangles are similar.

Here is a series of triangles to examine. Looking at Figures 10.1 to 10.3, which would you guess are similar?

```
> with(geometry):

> with(plots):

> triangle(ABC,[point(A,0,0),point(B,2,3),point(C,4,1)]);
```
$$ABC$$

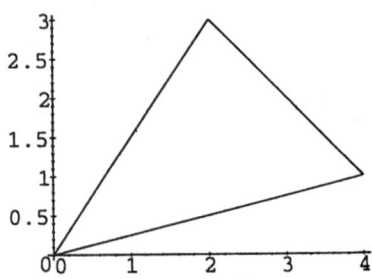

Figure 10.1. $\triangle ABC$

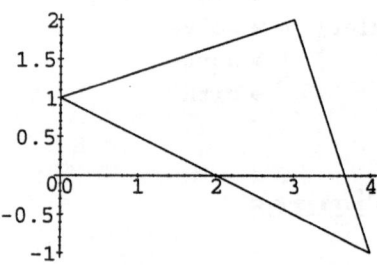

Figure 10.2. $\triangle FGH$

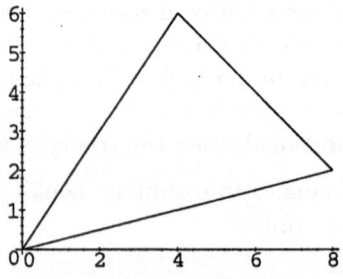

Figure 10.3. $\triangle JKL$

```
> polygonplot([coordinates(A),coordinates(B),coordinates(C)]);
> triangle(FGH,[point(F,0,1),point(G,3,2),point(H,4,-1)]);
```
$$FGH$$

```
> polygonplot([coordinates(F),coordinates(G),coordinates(H)]);
> triangle(JKL,[point(J,0,0),point(K,4,6),point(L,8,2)]);
```
$$JKL$$

```
> polygonplot([coordinates(J),coordinates(K),coordinates(L)]);
> triangle(M,[2*sqrt(2),sqrt(17),sqrt(13)]);
```
$$M$$

Using the **are_similar** command, we can determine which triangles are indeed similar. First, let's compare $\triangle ABC$ and $\triangle FGH$.
```
> are_similar(ABC,FGH);
```
$$false$$

$\triangle ABC$ and $\triangle FGH$ are not similar. How about $\triangle ABC$ and $\triangle M$?
```
> are_similar(ABC,M);
```
$$true$$

These triangles are similar. ($\triangle M$ was defined by the length of its sides which are the same as $\triangle ABC$.)

Example 10.1   Find x and y in Figure 10.4 that will make $\triangle T1$ and $\triangle T2$ similar triangles.

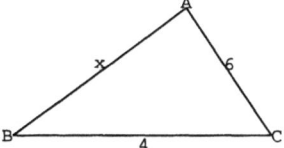   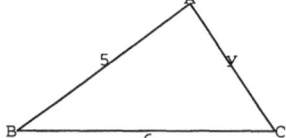

Figure 10.4. (a) $\triangle T1$. (b) $\triangle T2$

Without using any geometry, we can solve for the missing sides by splitting the proportion $x : 6 : 4 = 5 : y : 9$ into two equations and solving for x and y with the **solve** command.
```
> solve({x/5=4/9,6/y=4/9},{y,x});
```
$$\left\{ x = \frac{20}{9}, y = \frac{27}{2} \right\}$$

Another method is to define both triangles and use the `are_similar` command.

```
> triangle(T1,[x,6,4]);
```
$$T1$$

```
> triangle(T2,[5,y,9]);
```
$$T2$$

```
> are_similar(T1,T2);
```

THE TRIANGLES ARE SIMILAR IF ONE OF THE FOLLOWIN
G IS TRUE

$$
\left\{ \left[\frac{1}{81} \frac{x^2 y^2 - 2916}{y^2} = 0, \frac{1}{81} x^2 - \frac{16}{25} = 0 \right], \right.
$$
$$
\left[\frac{1}{81} x^2 - \frac{36}{25} = 0, \frac{1}{81} \frac{x^2 y^2 - 1296}{y^2} = 0 \right],
$$
$$
\left[\frac{1}{25} \frac{25 x^2 - 36 y^2}{y^2} = 0, \frac{1}{81} \frac{81 x^2 - 16 y^2}{y^2} = 0 \right],
$$
$$
\left[\frac{1}{9} \frac{9 x^2 - 4 y^2}{y^2} = 0, \frac{1}{25} \frac{25 x^2 - 16 y^2}{y^2} = 0 \right],
$$
$$
\left[\frac{1}{25} \frac{x^2 y^2 - 900}{y^2} = 0, \frac{1}{25} x^2 - \frac{16}{81} = 0 \right],
$$
$$
\left. \left[\frac{1}{25} x^2 - \frac{4}{9} = 0, \frac{1}{25} \frac{x^2 y^2 - 400}{y^2} = 0 \right] \right\}
$$

Let's solve one of the sets of equations listed above.

```
> solve({1/25*(x^2*y^2-900)/y^2,1/25*x^2-16/81},{y,x});
```

$$
\left\{ x = \frac{20}{9}, y = \frac{27}{2} \right\}, \left\{ x = \frac{20}{9}, y = \frac{-27}{2} \right\}, \left\{ y = \frac{27}{2}, x = \frac{-20}{9} \right\},
$$
$$
\left\{ y = \frac{-27}{2}, x = \frac{-20}{9} \right\}
$$

All but the first solution are extraneous because the length of a side must be positive.

Exercise 10.1   If a 50 m tall structure casts an 80 m shadow at a given time of day, what would be the length of the shadow cast by a 1.8 m tall man?

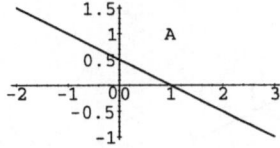 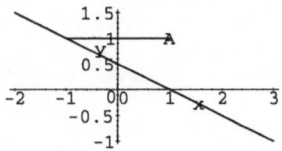

Figure 10.5. (a) Line $x + 2y = 1$ and point $A(1, 1)$. (b) Horizontal line connecting $A(1, 1)$ to $x + 2y = 1$

10.2   Length of a Perpendicular

The shortest distance from a given point, P, to a given line, l, is a line passing through P and *perpendicular* to line l. This perpendicular line is called a normal line. For example, what is the length of the normal line through the point $A(1, 1)$ and intersecting the line $x + 2y = 1$?

We wish to solve this problem using similar figures. Let's begin by plotting the line $x + 2y = 1$ and the point $A(1, 1)$, shown in Figure 10.5(a).

```
> point(A,1,1);
```
$$A$$

```
> line(l,[x+2*y=1]);
```
$$l$$

```
> t1:=textplot([1,1,'A']):
> p1:=implicitplot(l[equation],x=-2..3,y=-3..3):
> display({t1,p1},scaling=CONSTRAINED);
```

Next, draw a line parallel to the x−axis joining A to the point at which the two lines intersect. See Figure 10.5(b).

```
> solve({y=1,x+2*y=1},{x,y});
```
$$\{ x = -1, y = 1 \}$$

```
> p2:=implicitplot(y=1,x=-1..1,y=-3..3):
> display({t1,p1,p2},scaling=CONSTRAINED);
```

Finally, we must find the equation of the line perpendicular to $x + 2y = 1$ and passing through A, the length of which we are trying to find. We know that the slope of a normal line to a given line is the negative inverse of the slope of the given line. By inspection, the slope of $x + 2y = 1$, or $y = \frac{1-x}{2}$, is $-\frac{1}{2}$, therefore the slope of the normal line is 2. Knowing the slope and a point, we can determine the equation of the normal line. The complete triangle is shown in Figure 10.6.

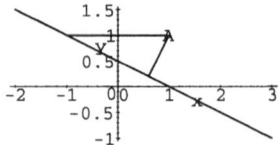

Figure 10.6. Right-angled triangle line formed by $x + 2y = 1$ and $A(1,1)$

```
> 2=(y-1)/(x-1);
```
$$2 = \frac{y-1}{x-1}$$

```
> with(student):
> isolate("",y);
```
$$y = 2x - 1$$

```
> solve({",y=(1-x)/2},{x,y});
```
$$\left\{ x = \frac{3}{5}, y = \frac{1}{5} \right\}$$

```
> p3:=implicitplot(y=2*x-1,x=3/5..1,y=-3..3):
> display({t1,p1,p2,p3},scaling=CONSTRAINED);
```

The triangle we have just drawn is similar to the triangle formed by the line $x + 2y = 1$ and the two axes. Therefore, the length of the normal line is proportional to the length of the corresponding line in the smaller triangle, which is the portion of the $y-$axis which forms the triangle.

```
> hyp_large:=1-(-1);
```
$$hyp\_large := 2$$

```
> hyp_small:=sqrt(0.5^2+1^2);
```
$$hyp\_small := 1.118033989$$

```
> height_small:=0.5;
```
$$height\_small := .5$$

```
> normal/height_small=hyp_large/hyp_small;
```
$$2.000000000\ normal = 1.788854382$$

```
> isolate(",normal);
```
$$normal = .8944271910$$

To check this result, we use *Maple's* distance command which finds the shortest distance between a point and a line, that is, the length of the normal line.

> 'length of perpendicular':=distance(A,l);

$$length\ of\ perpendicular := \frac{2}{5}\sqrt{5}$$

> evalf(");

$$.8944271912$$

We can find the equation of the normal line, using the perpendicular function.

> perpendicular(A,l,perp);

$$perp$$

> perp[equation];

$$-y + 2x - 1 = 0$$

The general equation for the distance from a point $P(x_1, y_1)$ to the line $ax + by + c = 0$ is $\dfrac{|ax_1 + by_1 + c|}{\sqrt{a^2 + b^2}}$.

> a:=1;b:=2;c:=-1;

$$a := 1$$
$$b := 2$$
$$c := -1$$

> p:=[1,1];

$$p := [1,1]$$

> abs(a*p[1]+b*p[2]+c)/sqrt(a^2+b^2);

$$\frac{2}{5}\sqrt{5}$$

> evalf(");

$$.8944271912$$

Exercise 10.2 What is the shortest distance between the point $F(3,3)$ and the line $y = 2x + 6$ and what is the equation of the line joining them?

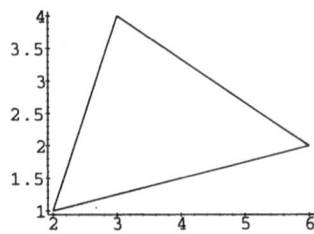

Figure 10.7. △KLM

10.3   Areas of Similar Triangles

As you know from previous work, the area of a triangle is equal to one-half of the
base times the perpendicular height. With a simple equation like this, you may
wonder why we need *Maple* to calculate the area of a triangle. In this example,
we know the vertices of the △KLM (Figure 10.7) but not the length of the base
or the height, so we must use the built-in command area.

```
> triangle(KLM,[point(K,2,1),point(L,3,4),point(M,6,2)]);
```
$$KLM$$

```
> polygonplot([coordinates(K),coordinates(L),coordinates(M)]);
> area(KLM);
```
$$\frac{11}{2}$$

The areas of similar triangles are related in the following way.

```
> triangle(ABC,[point(A,0,0),point(B,2,3),point(C,4,1)]);
```
$$ABC$$

```
> triangle(GHJ,[point(G,0,0),point(H,4,6),point(J,8,2)]);
```
$$GHJ$$

```
> area(ABC);
```
$$5$$

```
> AB2:=(distance(A,B))^2;
```
$$AB2 := 13$$

```
> area(GHJ);
```
$$20$$

```
> GH2:=(distance(G,H))^2;
```
$$GH2 := 52$$

The ratio of the sides of the triangles ...

```
> AB2/GH2;
```
$$\frac{1}{4}$$

... is equal to the ratio of the areas of the triangles.

```
> area(ABC)/area(GHJ);
```
$$\frac{1}{4}$$

10.4   Dilatations and Similar Figures

A *dilatation* of factor k is a mapping of the form $(x, y) \mapsto (kx, ky)$. For example, $\triangle FGH$ is a dilatation of factor 2 of $\triangle ABC$, both shown in Figure 10.8. The two triangles are overlayed in Figure 10.9.

```
> triangle(ABC,[point(B,0,0),point(C,3,2),point(A,2,5)]);
```
$$ABC$$

```
> triangle(FGH,[point(B,0,0),point(F,6,4),point(H,4,10)]);
```
$$FGH$$

```
> tria:=plot([0,0,3,2,2,5,0,0],scaling=CONSTRAINED):
> text1:=textplot([0,0,'B'],align={ABOVE,RIGHT}):
> text2:=textplot([3,2,'C'],align=BELOW):
> text3:=textplot([2,5,'A'],align=ABOVE):
> display({tria,text1,text2,text3});
> trib:=plot([0,0,6,4,4,10,0,0],scaling=CONSTRAINED):
> t1:=textplot([0,0,'B'],align={ABOVE,RIGHT}):
> t2:=textplot([6,4,'F'],align=BELOW):
> t3:=textplot([4,10,'H'],align=ABOVE):
> display({trib,t1,t2,t3});
> display({tria,trib,text2,text3,t1,t2,t3},scaling=CONSTRAINED);
```

Any two triangles that are dilatations of each other are similar.

```
> are_similar(ABC,FGH);
```
$$true$$

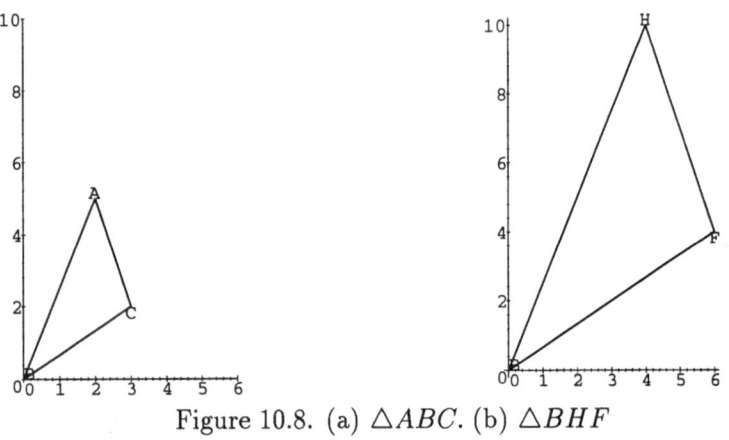

Figure 10.8. (a) $\triangle ABC$. (b) $\triangle BHF$

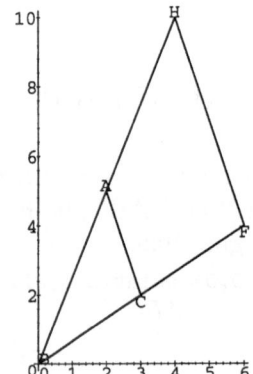

Figure 10.9. Dilatation of 2 from $\triangle ABC$ to $\triangle BHF$

But the inverse is not necessarily true. The triangles $\triangle KLM$ and $\triangle QRS$ are similar, but are not dilatations of each other.

```
> triangle(KLM,[point(K,0,0),point(L,2,2),point(M,0,4)]);
> triangle(QRS,[point(B,0,0),point(F,2,-2),point(H,0,-4)]);
> are_similar(KLM,QRS);
```
$$true$$

Experiment 10.1 Start with a unit square, which has area 1, and make it into two triangles, each of area $\frac{1}{2}$, by drawing a diagonal. Draw a parallel line joining mid points on the two adjacent sides of the triangle, hence forming a smaller, similar triangle. What is the area of this new triangle? Repeat the process indefinitely to fractalize the corners of the square, and find the sequence of areas of triangles formed.

Experiment 10.2 Show that equilateral triangles can fill the plane, and that they can be organized to do so in groups of six. So the plane can be filled with hexagons and this turns out to be the shape which most efficiently acts as a 'service area' around its centre point, for a service offered in a plane (think of delivering pizzas). The solution to the problem in the one-dimensional case is obviously to use intervals as service areas. In three-dimensions, the solution turns out to consist of truncated octahedra; that is, octahedra with the vertices cut off, yielding a 14-sided figure. Try to construct the faces of the truncated octahedra and make them from paper. An even more *practical* experiment is to put dried peas in a strong box, filling it tightly and closing it but for two small holes. Introduce water into one hole, allowing air to escape from the other until the box is full. The peas swell and fill the box after a few hours. If you take them out and examine them you will find that they have on average 14 sides and resemble truncated octahedra.

Chapter 11

Circles and Spheres

Commands used in this chapter

- cos
- evalf
- geom3d
- geom3d[area]
- geom3d[center]
- geom3d[coordinates]
- geom3d[radius]
- geom3d[sphere]
- geom3d[volume]
- geometry
- geometry[are_tangent]
- geometry[area]
- geometry[circle]
- geometry[detailf]
- geometry[point]
- geometry[tangentpc]

- map
- op
- plot
- plot[options]
- plots
- plots[implicitplot]
- plots[parametric]
- plots[polarplot]
- plots[sphereplot]
- sin
- solve
- sqrt
- student
- student[isolate]
- student[showtangent]
- with

11.1   Circle

The equation of a circle of radius r, centred at $x = y = 0$, is $x^2 + y^2 = r^2$. The coordinates of a point $P(x, y)$ on the circle $x^2 + y^2 = r^2$ and the radius of the

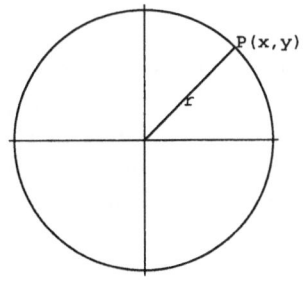
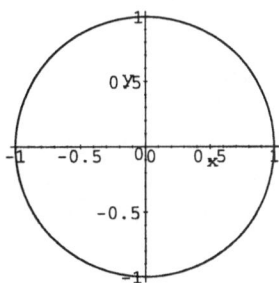

Figure 11.1. (a) Unit circle. (b) The graph of $x^2 + y^2 = 1$, on $[-1, 1]$

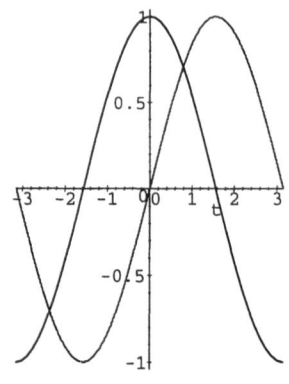
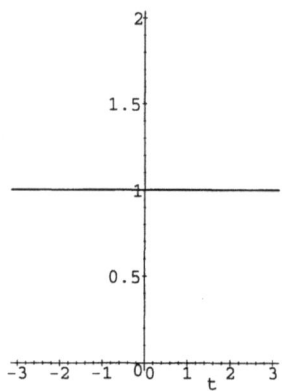

Figure 11.2.   (a) Sine and cosine functions.   (b) The graph of $\sin^2 + \cos^2$, on $[-\pi, \pi]$

the circle are related by Pythagoras' Theorem. See Figure 11.1(a).

```
> origin:=[0,0];
```

$$origin := [0,0]$$

```
> P:=[x,y];
```

$$P := [x,y]$$

```
> sqrt((origin[1]-P[1])^2+(origin[2]-P[2])^2)=r;
```

$$\sqrt{x^2 + y^2} = r$$

```
> map(x -> x^2,");
```

$$x^2 + y^2 = r^2$$

For simplicity, let's assume a radius of $r = 1$ for the time being. The graph of this circle is shown in Figure 11.1(b).

```
> with(plots):
> implicitplot(x^2+y^2=1,x=-1..1,y=-1..1);
```

There are several ways to plot a circle. The implicit plot you just saw is probably the most straightforward way to do it. Another method is the parametric plot in which the variables x and y are expressed in terms of an independent variable, t, called a *parameter*. What are two functions whose sum of squares is equal to one? The sine and cosine functions are two such functions. (Figure 11.2).

```
> plot({sin(t),cos(t)},t=-Pi..Pi);
> plot(sin(t)^2+cos(t)^2,t=-Pi..Pi);
```

We shall see in Chapter 23 that $x = r\cos(t)$ and $y = r\sin(t)$. The resulting parametric plot is shown in Figure 11.3(a).

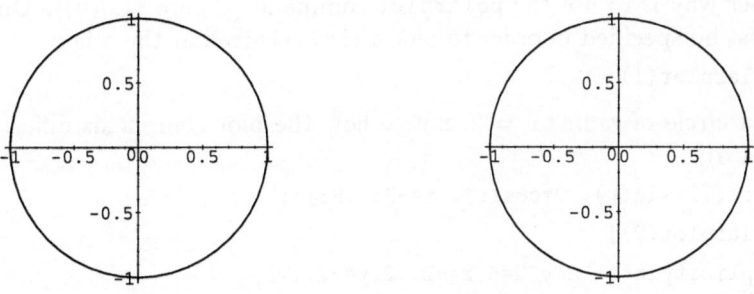

Figure 11.3. (a) Parametric plot. (b) Polar plot of circle of radius 1

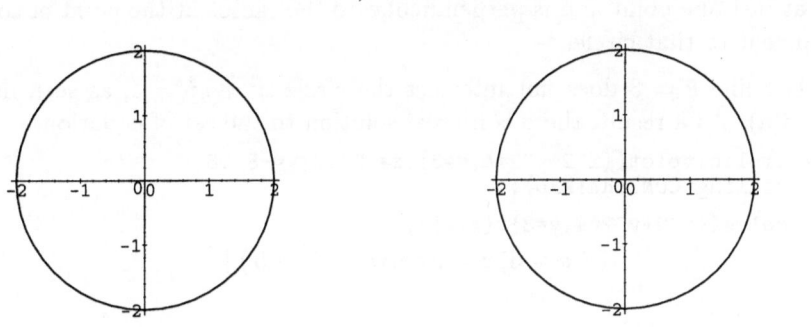

Figure 11.4. (a) Parametric plot of circle of radius 2. (b) Polar plot of circle of radius 2

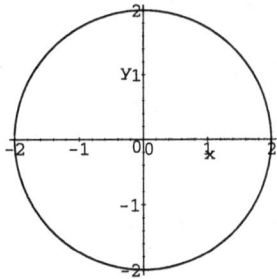

Figure 11.5. Implicit plot of circle of radius 2

```
> plot([cos(t),sin(t),t=-Pi..Pi]);
```

Yet another way is to use the `polarplot` command (Figure 11.3(b)). Only the radius must be specified in order to plot a circle centred at the origin.

```
> polarplot(1);
```

For a circle of radius $r = 2$, notice how the plot commands differ. (Figure 11.4-11.5)

```
> plot([2*sin(t), 2*cos(t), t=-Pi..Pi]);
> polarplot(2);
> implicitplot(x^2+y^2=4,x=-2..2,y=-2..2);
```

11.1.1   Intersection of a Line and a Circle

A line and a circle may intersect at one or two points. Any line that intersects a circle at just one point and is perpendicular to the radius at the point of contact is a tangent to that circle.

The line $y = 3$ does not intersect the circle $x^2 + y^2 = 4$, as seen in Figure 11.6(a). As a result, there is no real solution to the set of equations.

```
> implicitplot({x^2+y^2=4,y=3},x=-5..5,y=-5..5,
> scaling=CONSTRAINED);
> solve({x^2+y^2=4,y=3},{x,y});
```
$$\{\, y = 3, x = \mathrm{RootOf}(\_Z^2 + 5)\,\}$$

```
> allvalues(");
```
$$\left\{ y = 3, x = I\sqrt{5}\right\}, \left\{ y = 3, x = -I\sqrt{5}\right\}$$

The line $y = 2$ intersects the circle $x^2 + y^2 = 4$ (see Figure 11.6(b)) in one point. To prove that the line is a tangent to the circle, use the **are_tangent** command.

```
> implicitplot({x^2+y^2=4,y=2},x=-5..5,y=-5..5,
> scaling=CONSTRAINED);
> solve({x^2+y^2=4,y=2},{x,y});
```
$$\{\, y = 2, x = 0\,\}$$

```
> with(geometry):
> line(l,[y=2]),circle(c,[x^2+y^2=4]);
```
$$l, c$$

```
> are_tangent(c,l);
```
$$true$$

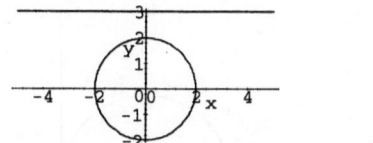 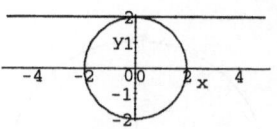

Figure 11.6. (a) Circle $x^2 + y^2 = 4$ and line $y = 3$. (b) Circle $x^2 + y^2 = 4$ and line $y = 2$

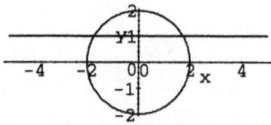

Figure 11.7. Circle $x^2 + y^2 = 4$ and line $y = 1$

Since the line $y = 1$ passes through the circle $x^2 + y^2 = 4$ (Figure 11.7), there are two real solutions to the set of equations.

```
> implicitplot({x^2+y^2=4,y=1},x=-5..5,y=-5..5,
> scaling=CONSTRAINED);
> solve({x^2+y^2=4,y=1},{x,y});
```
$$\{ x = \text{RootOf}(\_Z^2 - 3), y = 1 \}$$

```
> allvalues(");
```
$$\{ x = \sqrt{3}, y = 1 \}, \{ x = -\sqrt{3}, y = 1 \}$$

Exercise 11.1 Plot the relations $y - x + 1 = 0$ and $(x - 1)^2 + y^2 = 1$ to determine if they intersect. If so, calculate their point(s) of intersection.

11.1.2   Tangent to a Circle

We start with an easy case. Find the tangent to the circle $x^2 + y^2 = 1$ shown in Figure 11.8(a) at the point $x = 1$.

```
> implicitplot(x^2+y^2=1,x=-1..1,y=-1..1);
```

Simply by examining the plot, we know that the tangent is a vertical line passing through the point $(1, 0)$, the equation of which is $x = 1$. The first step is to define the circle and the point on the circle in which we are interested.

```
> point(a,1,0),circle(cir,[x^2+y^2=1]);
```
$$a, cir$$

```
> tangentpc(a,cir,tan1);
```
$$tan1$$

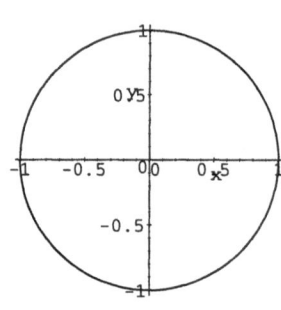
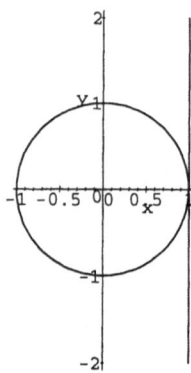

Figure 11.8. (a) The graph of $x^2 + y^2 = 1$, on $[-1, 1]$. (b) Circle of radius 1 and tangent $y = 1$

```
> tan1[equation];
```
$$x - 1 = 0$$

```
> cir[equation];
```
$$x^2 + y^2 - 1 = 0$$

```
> are_tangent(cir,tan1);
```
$$true$$

The following plot, Figure 11.8(b), shows the tangent to the circle at $y = 1$.
```
> implicitplot({cir[equation],tan1[equation]},x=-2..2,y=-2..2,
> scaling=CONSTRAINED);
```
Now find the tangents to the circle at $x = 0.5$.
```
> solve({cir[equation],x=0.5},{x,y});
```
$$\{\, y = -.8660254038,\, x = .5000000000\,\},$$
$$\{\, y = .8660254038,\, x = .5000000000\,\}$$

First, we shall examine the tangent at the point $(0.5, 0.8660254038)$.
```
> point(b,0.5,0.8660254038);
```
$$b$$

```
> tangentpc(b,cir,tan2);
```
$$\text{tan2}$$

```
> tan2[equation];
```
$$.5\,x + .8660254038\,y - 1 = 0$$

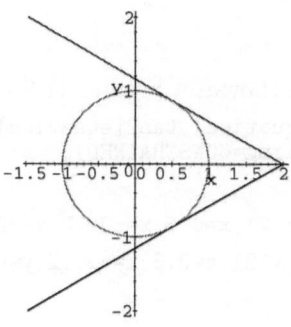

Figure 11.9. Tangents to circle of radius 1 at $x = 0.5$

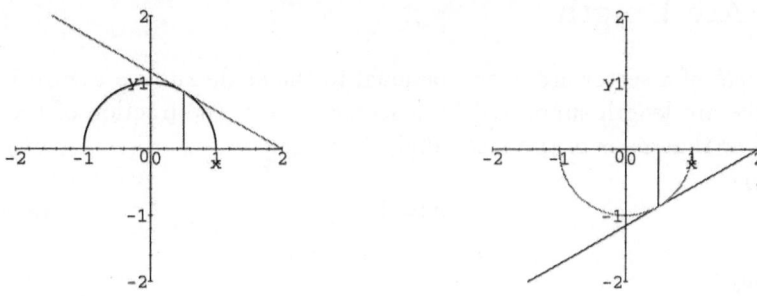

Figure 11.10. (a) Tangent to circle of radius 1 at $x = 0.5$. (b) Tangent to circle of radius 1 at $x = 0.5$

```
> are_tangent(cir,tan2);
```
$$true$$

Now, find the tangent at the point $(0.5, -0.8660254038)$.
```
> point(c,0.5,-0.8660254038);
```
$$c$$

```
> tangentpc(c,cir,tan3);
```
$$tan3$$

```
> tan2[equation];
```
$$.5\,x + .8660254038\,y - 1 = 0$$

```
> are_tangent(cir,tan2);
```

true

The tangents to the circle are shown in Figures 11.9 and 11.10.

```
> implicitplot({cir[equation],tan2[equation],tan3[equation]},
> x=-2..2,y=-2..2,scaling=CONSTRAINED);
> with(student):
> showtangent(sqrt(1-x^2),x=0.5,x=-2..2,y=-2..2);
> showtangent(-sqrt(1-x^2),x=0.5,x=-2..2,y=-2..2);
```

Experiment 11.1   Determine the equations of the tangents to the circle $(x - 2)^2 + (y + 1)^2 = 5$ at $x = 3$ using the method introduced in Chapter 4.

11.1.3   Arc Length

The *arc length* of a sector arc is proportional to the angle $\angle a$ (see Figure 11.11); precisely, the arc length supported by a sector of $\angle a$ is the fraction of the total circumference that $\angle a$ is of the total angle 2π radians.

```
> r:=1;
```
$$r := 1$$

```
> a:=Pi/4;
```
$$a := \frac{1}{4}\pi$$

```
> arc_length:=a/(2*Pi)*2*Pi*r;
```
$$arc\_length := \frac{10}{3}\pi$$

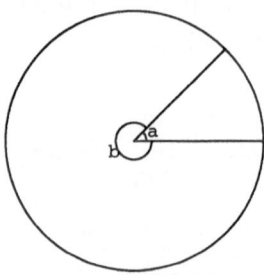

Figure 11.11. Arc length

Example 11.1   A 20 cm diameter pie is cut into sixths. What is the arc length of each piece?

```
> r:=10;a:=1/6*2*Pi;
```

$$r := 10$$

$$a := \frac{1}{3}\pi$$

```
> arc_length;
```

$$\frac{10}{3}\pi$$

```
> evalf(");
```

$$10.47197551$$

11.1.4   Area Bounded by a Circle

Find the area of the circle *c2*. Strictly speaking, we find the area of the *disk* bounded by the circle.

```
> circle(c2,[(x+2)^2+(y-sqrt(3))^2=4]);
```

$$c2$$

```
> detailf(c2);
```

$$[\, center = [-2., 1.732050808\,], radius = 2.\,]$$

We know from previous courses that the area of a circle is equal to π times the radius squared.

```
> ar:=Pi*radius(c2)^2;
```

$$ar := 4\,\pi$$

But, we can also calculate the area of *c2* using the built-in *Maple* function `area`.

```
> area(c2);
```

$$4\,\pi$$

```
> evalf(area(c2));
```

$$12.56637062$$

The area of a sector is also proportional to the angle $\angle a$.

```
> a:=Pi/4;
```

$$a := \frac{1}{4}\pi$$

```
> r:=1;
```
$$r := 1$$

```
> sector_area:=a/(2*Pi)*Pi*r^2;
```
$$sector\_area := \frac{1}{8}\pi$$

Example 11.2 What is the area of each piece of pie in Example 11.1?

```
> r:=10;a:=1/6*2*Pi;
```
$$r := 10$$

$$a := \frac{1}{3}\pi$$

```
> sector_area;
```
$$\frac{50}{3}\pi$$

```
> evalf(");
```
$$52.35987758$$

Exercise 11.2 The height of a circular cylinder is equal to its radius. Express its total surface area A (including both ends) as a function of its volume. (p38, Edwards and Penney [6])

Exercise 11.3 A piece of wire 100 cm long is cut into two pieces of lengths x and $100 - x$. The first piece is bent into the shape of a square and the second is bent into the shape of a circle. Express as a function of x the sum A of the area of the square and the circle. (p38, Edwards and Penney [6])

11.2 Sphere

The sphere is the 2-dimensional surface which is the counterpart to the circle: it consists of the set of points at fixed distance r from a given point. Its equation is $x^2 + y^2 + z^2 = r^2$ and its graph is shown in Figure 11.12. Of course, if we choose the plane $z = 0$ we have the circle $x^2 + y^2 = r^2$.

```
> sphereplot(2,theta=0..2*Pi,phi=0..2*Pi,scaling=CONSTRAINED,
> axes=NONE);
```

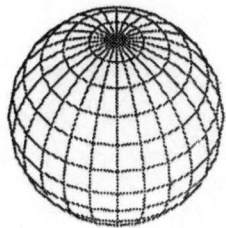

Figure 11.12. Sphere of radius 1

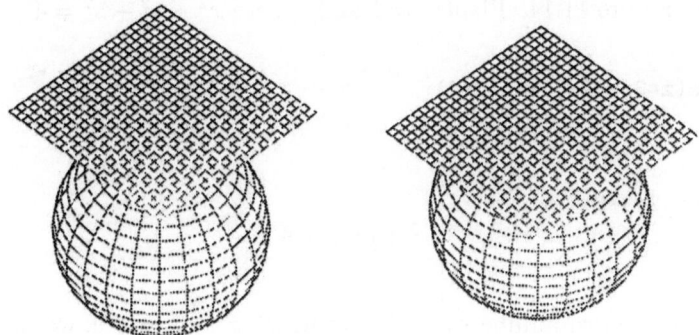

Figure 11.13. (a) Plane $z = 3$ and sphere $x^2 + y^2 + z^2 = 4$. (b) Plane $z = 2$ and sphere $x^2 + y^2 + z^2 = 4$

11.2.1   Intersection of a Plane and a Sphere

Whereas the intersection of a line and a circle may result in one or two points, the intersection of a plane and a sphere may result in a single point, or a circle. In this section we shall be considering a sphere of radius $r = 2$, centred at the origin, $(0, 0, 0)$, the equation of which is $x^2 + y^2 + z^2 = 4$, and its intersection with horizontal planes only, ie $x = y = 0$.

We begin by studying the intersection of the sphere with the plane $z = 3$, shown in Figure 11.13(a). We know that this plane cannot intersect the sphere because it lies 3 units away from the $x-$ and $y-$axes, while the radius of the sphere extends only 2 units from the origin.

```
> display({sphereplot(2,theta=0..2*Pi,phi=0..Pi),plot3d(3,-2..2,
> -2..2)},scaling=CONSTRAINED,axes=NONE);
```

Substitute the value $z = 3$ into the equation for the sphere to be sure that we are correct.

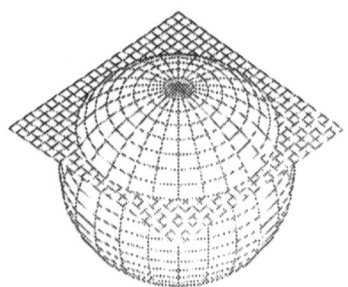

Figure 11.14. Plane $z = 1$ and sphere $x^2 + y^2 + z^2 = 4$

```
> subs(z=3,x^2+y^2+z^2=4);
```
$$x^2 + y^2 + 9 = 4$$

```
> x^2+y^2=4-9;
```
$$x^2 + y^2 = -5$$

Since there are no real numbers x and y for which $x^2 + y^2 = -5$, we can conclude that the plane and the sphere do not intersect, and that the solution is complex.

Now, let's determine whether or not the plane $z = 2$ intersects the sphere. We can see in Figure 11.13(b) that the plane appears to rest on top of the sphere at the point $(0, 0, 2)$.

```
> display({sphereplot(2,theta=0..2*Pi,phi=0..Pi),plot3d(2,-2..2,
> -2..2)},scaling=CONSTRAINED,axes=NONE);
```

Once again, we substitute the value $z = 2$ into our equation.

```
> subs(z=2,x^2+y^2+z^2=4);
```
$$x^2 + y^2 + 4 = 4$$

```
> x^2+y^2=4-4;
```
$$x^2 + y^2 = 0$$

In order for this equation to balance, x and y must both equal zero. Therefore, the plane and sphere intersect at a single point.

Finally, how does the plane $z = 1$ intersect the sphere? In Figure 11.14(a), we can see that the plane slices through the sphere. When viewed from above we can see that the relation that results is a circle.

```
> display({sphereplot(2,theta=0..2*Pi,phi=0..Pi),plot3d(1,-2..2,
> -2..2)},scaling=CONSTRAINED,axes=NONE);
```

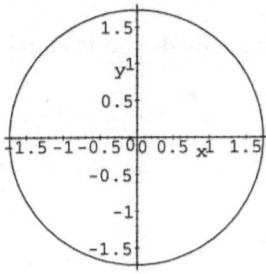 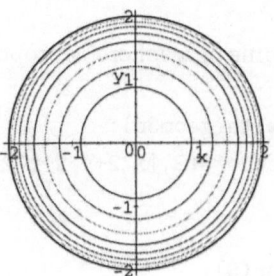

Figure 11.15. (a) Circle formed by the intersection of plane $z = 1$ and sphere $x^2 + y^2 + z^2 = 4$. (b) Circles formed by the interestions of various horizontal planes with sphere $x^2 + y^2 + z^2 = 4$

```
> subs(z=1,x^2+y^2+z^2=4);
```
$$x^2 + y^2 + 1 = 4$$

```
> x^2+y^2=4-1;
```
$$x^2 + y^2 = 3$$

```
> implicitplot(",x=-2..2,y=-2..2,scaling=CONSTRAINED);
```

The radius of the resulting circle depends on the position of the horizontal plane. A plane intersecting the sphere at its widest point, namely $z = 0$, will result in a circle with radius $r = 2$, equal to that of the sphere. As the plane moves away from $z = 0$, the radius of the circle decreases until at $z = 2$, the plane and sphere intersect at a point. The following table lists 10 such circles, which are plotted in Figure 11.15(b). Although the spacing between the planes is constant, the spacing between the circles is not. Why?

```
> seq(subs(z=t/5,x^2+y^2+z^2=4),t=0..10);
```
$$x^2 + y^2 = 4, x^2 + y^2 + \frac{1}{25} = 4, x^2 + y^2 + \frac{4}{25} = 4, x^2 + y^2 + \frac{9}{25} = 4,$$
$$x^2 + y^2 + \frac{16}{25} = 4, x^2 + y^2 + 1 = 4, x^2 + y^2 + \frac{36}{25} = 4, x^2 + y^2 + \frac{49}{25} = 4,$$
$$x^2 + y^2 + \frac{64}{25} = 4, x^2 + y^2 + \frac{81}{25} = 4, x^2 + y^2 + 4 = 4$$

```
> implicitplot({"},x=-2..2,y=-2..2,scaling=CONSTRAINED);
```

Experiment 11.2 Investigate the intersection of the sphere $x^2 + y^2 + z^2 = 4$ with the plane $x + y + z = 1$.

11.2.2   Volume and Surface Area

The volume of a sphere is proportional to the radius cubed, and is equal to $\frac{4}{3}\pi r^3$.

```
> with(geom3d):
> sphere(S,[x^2+y^2+z^2=4]);
```
$$S$$

```
> op(S);
```
$$
\text{table}([
$$
$$
\begin{aligned}
radius &= \sqrt{4} \\
given &= [\, x^2 + y^2 + z^2 = 4\,] \\
form &= sphere \\
equation &= (\, x^2 + y^2 + z^2 - 4 = 0\,) \\
center &= center\_of\_S
\end{aligned}
$$
$$
])
$$

```
> coordinates(center_of_S);
```
$$[\,0,0,0\,]$$

Using the equation for volume of a sphere,
```
> vol:=4/3*radius(S)^3*Pi;
```
$$vol := \frac{32}{3}\,\pi$$

```
> evalf(vol);
```
$$33.51032165$$

Using the built-in *Maple* command to find the volume of a sphere,
```
> volume(S);
```
$$\frac{16}{3}\,\pi\,\sqrt{4}$$

```
> evalf(volume(S));
```
$$33.51032164$$

And for surface area,
```
> sa:=4*Pi*radius(S)^2;
```
$$sa := 16\,\pi$$

```
> area(S);
```
$$16\,\pi$$

```
> evalf(area(S));
```
$$50.26548246$$

Exercise 11.4   A spherical cloud of toxic powder forms over a factory following a fire. The radius of the sphere is 100 m and contains a uniform dispersion of powder at a density of $0.001 \ kg \cdot m^{-3}$. This cloud subsequently is deposited on the ground over a circular region of radius 100 m. Find the total mass of powder in the cloud and find also the average areal density in $kg \cdot m^{-2}$ of powder deposited on the ground.

Exercise 11.5   A giant Sequoia tree, perhaps the largest tree in the world, stands 82.9 m tall and is about 10.4 m in diameter at its base. Its habitat is the central California mountain range, the Sierra Nevada, in the Sequoia National Park. The great age, size, and rapid growth of this tree have contributed to its fame. It has the heroic name General Sherman. (p4, Goldenberg and Greenwald [7])

(a) We can approximate the volume of a tree by disregarding the branch structure and considering the trunk as a cylinder. If the General Sherman has an average radius of 2.3 m, what is its volume?

(b) Assuming the General Sherman produces an annual growth ring of 0.00009 m, how much new wood is added to the tree (trunk) each year?

(c) Suppose the paper industry cultivated a fast-growing tree that could grow from a seedling to a height of 15.2 m in one year. If the increase in volume of this tree were to match that of the General Sherman, what would be the average radius of the tree?

(d) The baobab tree, native to Australia and Africa, has a trunk that measures as much as 18.3 m in diameter. However, this tree only grows to a height of 12.2 m. Approximate the amount of new wood this tree would produce each year. Use a cylindrical model with a radius of 12.2 m, a height of 18.3 m, and an annual growth ring of 0.0009 m.

(e) Model the baobab tree as a cone. Approximate the amount of new wood it would produce in one year if the increase in thickness is 0.0009 m.

Experiment 11.3   A solid sphere of radius 1 is sliced like a loaf of bread into n disks with curved edges, each disk having thickness $\frac{1}{n}$. Compute a table of radii of the two faces of the disks for $n = 2$. Estimate the volume of each disk by assuming that it is part of a cone, replacing its curved edges by a flat, sloping edge. Show how you can exploit the symmetry to save work. Sum the estimates of volumes of the disks and compare this sum with the actual volume of the sphere. Repeat this for $n = 4, 6, 8, 10, \ldots$ and plot your error as a function of n.

Experiment 11.4   For amusement, the fly from Experiment 9.6 passes through a narrow tunnel into a spherical room of radius 1 m, and waits for the spider in a diametrically opposite position. What are the first few moves after the spider comes through the tunnel? You may use the fact that the shortest distance between two points on the surface of a sphere is along the great circle joining them.

Chapter 12

Loci

Commands used in this chapter

- allvalues
- arctan
- convert
- cos
- evalf
- geometry
- geometry[altitude]
- geometry[are_tangent]
- geometry[bisector]
- geometry[center]
- geometry[centroid]
- geometry[circumcircle]
- geometry[coordinates]
- geometry[excircle]
- geometry[incircle]
- geometry[inter]
- geometry[line]
- geometry[median]
- geometry[orthocenter]
- geometry[parallel]

- geometry[point]
- geometry[triangle]
- op
- plot
- plot[options]
- plots
- plots[display]
- plots[implicitplot]
- plots[polarplot]
- plots[polygonplot]
- plots[textplot]
- polar
- rand
- readlib
- sin
- solve
- student
- student[isolate]
- subs
- with

12.1   Locus

A *locus* is the set of all points satisfying a certain condition. For example, a circle centred at the origin with radius r can be described by the set of all points P in the x, y plane located r units from the origin, the defining equation of which is $x^2 + y^2 = r^2$. A sphere is the set of all points located at a fixed distance from a given point in space.

Exercise 12.1   Let a circle of diameter b have centre at $x = \frac{b}{2}$. Take a straight line $\overline{PQ}$ of any length $2a$ and move it such that its midpoint M moves round the given circle and the line $\overline{PQ}$ passes through the origin O. Then the points P and Q have as loci the curve $r = \overline{MP} + \overline{OP} = a + b\cos(\theta)$. This curve is called a *limaçon*; in particular it is called a *cardioid* if $a = b$. Plot a family of limaçons for a range of values of a and b.

Exercise 12.2   Let $\overline{AB}$ be the diameter of a circle and P a point on the upper semicircle. The tangent to the circle at B is a vertical line and suppose that the line $\overline{AP}$ extends to meet this vertical line at T. Now choose Q on $\overline{AT}$ such that $\overline{AQ} = \overline{PT}$ and the locus of Q is given by $y^2(2a - x) = x^3$ when A is the origin and $\overline{AB}$ lies along the x-axis. This is called a *cissoid* and has polar equation $r = \frac{2a\sin^2\theta}{\cos\theta}$. Plot a family of cissoids for a range of values of a.

Experiment 12.1   Consider the *ellipse*

$$\frac{x^2}{a^2} + \frac{y^2}{b^2} = 1 \text{ for } a > b$$

The parametric equations $x = a\cos(u)$ and $y = b\sin(u)$ give this ellipse for $0 \le u \le 2\pi$. Plot the graph of an ellipse with $a = 2$, $b = 1$; then its *eccentricity* is $\sqrt{a^2 - b^2}/a = \sqrt{3}/2$, and its area is $\pi ab = 2\pi$. Imagine slicing the graph into parts using vertical cuts at $x = \pm\frac{a}{2}$ and $x = 0$. Solve for the coordinates of the cuts and use straight lines to approximate the ellipse between cuts. Calculate the length of the line segments to give approximations to the perimeter and area of the ellipse. Proceed by cutting the ellipse with $5, 7, \ldots$ vertical lines and so obtain a sequence of approximations to the perimeter and area. You might try first the case of a circle which has $a = b = 1$.

12.2   Equations and Inequations of a Locus

Find the values of x for which y is greater than zero if $y = x + 3$.

```
> y:=x+3>0;
```
$$y := 0 < x + 3$$

```
> solve(y,{x});
```
$$\{-3 < x\}$$

For all values of x greater than $x = -3$, y is greater than zero. On the plot of $y = x + 3$ (Figure 12.1(a)), you can see that the line crosses the y-axis at that point.

```
> plot(x+3,x=-6..6);
```

Now let's examine the inequation of a parabola $y = x^2 + 3x < 4$.

```
> y:=x^2+3*x<4;
```
$$y := x^2 + 3x < 4$$

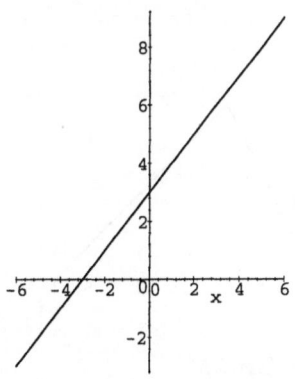 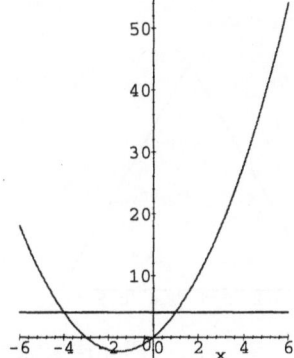

Figure 12.1. (a) The graph of $y = x + 3$, on $[-6, 6]$. (b) The graph of $y = x^2 + 3x$, on $[-6, 6]$

```
> solve(y,{x});
```
$$\{\, x < 1, -4 < x \,\}$$

Over the interval $x = -4$ to $x = 1$, the graph of the function (Figure 12.1(b)) is below the line $y = 4$.

```
> plot({x^2+3*x,4},x=-6..6);
```

12.3 Circles Associated with a Triangle

In the subsequent sections dealing with the circles associated with a triangle, two triangles will be studied. One is the equilateral triangle $\triangle ABC$ and the other is the generic triangle $\triangle KLM$. Both triangles are shown in Figure 12.2.

$\triangle ABC$ is an equilateral triangle with sides 1 unit in length.

```
> with(geometry):
> triangle(ABC,[point(A,0,0),point(B,1,0),
> point(C,0.5,1/2*3^(1/2))]);
```
$$ABC$$

```
> with(plots):
> polygonplot([coordinates(A),coordinates(B),coordinates(C)],
> scaling=CONSTRAINED);
```

Now we will define $\triangle KLM$ with vertices at $K(0,0)$, $L(3,0)$, and $M(1,2)$.

```
> triangle(KLM,[point(K,0,0),point(L,3,0),point(M,1,2)]);
```
$$KLM$$

```
> polygonplot([coordinates(K),coordinates(L),coordinates(M)],
> scaling=CONSTRAINED);
```

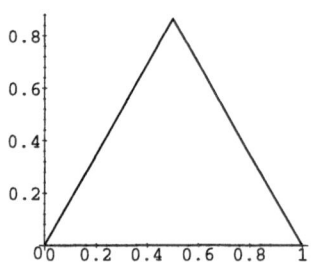 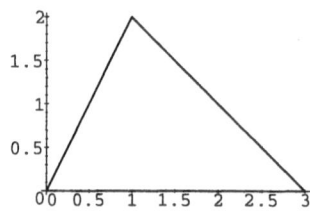

Figure 12.2. (a) $\triangle ABC$. (b) $\triangle KLM$

12.3.1   Circumcircle and Circumcentre

The *circumcircle* of a triangle is a circle through the vertices of the triangle with its centre at the *circumcentre*. To determine the equation of the circumcircle, we simply use the `circumcircle` command.

```
> circumcircle(ABC,ACC);
```
$$ACC$$

```
> ACC[equation];
```
$$x^2 + y^2 + .1\,10^{-9} - x - .3333333334\,\sqrt{3}\,y = 0$$

The circumcentre is the centre of the circumcircle, and it is found using the `center` command.

```
> center(ACC);
```
$$center\_ACC$$

```
> coordinates(center(ACC));
```
$$\left[\frac{1}{2}, .1666666667\,\sqrt{3}\right]$$

```
> at:=plot([coordinates(A),coordinates(B),coordinates(C),
> coordinates(A)]):
> acc:=implicitplot(ACC[equation],x=-1..2,y=-1..2,numpoints=1000):
> ac:=textplot([1/2,1/6*sqrt(3),'CC']):
> ap1:=textplot([0,0,'A'],align=ABOVE):
> ap2:=textplot([1,0,'B'],align=LEFT):
> ap3:=textplot([0.5,1/2*3^(1/2),'C'],align=RIGHT):
> display({at,acc,ac,ap1,ap2,ap3},scaling=CONSTRAINED);
```

The circumcircle and circumcentre of $\triangle ABC$ can be seen in Figure 12.3(a).

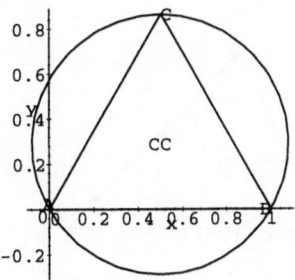 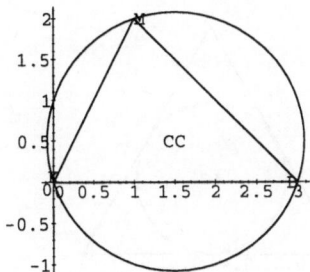

Figure 12.3. Circumcircle and circumcentre of (a) $\triangle ABC$, (b) $\triangle KLM$

We use the same procedure to find the circumcircle and circumcentre of $\triangle KLM$.

```
> circumcircle(KLM,KCC);
```
$$KCC$$

```
> KCC[equation];
```
$$x^2 + y^2 - 3x - y = 0$$

```
> center(KCC);
```
$$center\_KCC$$

```
> coordinates(center(KCC));
```
$$\left[\frac{3}{2}, \frac{1}{2}\right]$$

```
> kt:=plot([coordinates(K),coordinates(L),coordinates(M),
> coordinates(K)]):

> kcc:=implicitplot(KCC[equation],x=-1..4,y=-1.5..2.5,
> numpoints=1000):

> kc:=textplot([3/2,1/2,'CC']):

> kp1:=textplot([0,0,'K'],align=ABOVE):

> kp2:=textplot([3,0,'L'],align=LEFT):

> kp3:=textplot([1,2,'M'],align=RIGHT):

> display({kt,kcc,kc,kp1,kp2,kp3},scaling=CONSTRAINED);
```

Figure 12.3(b) shows the circumcircle and circumcentre of $\triangle KLM$.

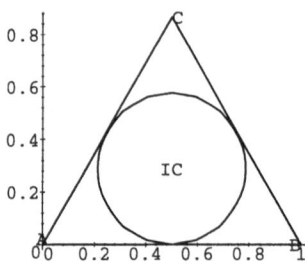 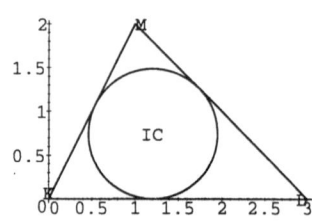

Figure 12.4. Inscribed circle and incentre of (a) $\triangle ABC$, (b) $\triangle KLM$

12.3.2   Inscribed Circle and Incentre

The *inscribed circle* of a triangle is the circle tangent to the sides of the triangle.

```
> incircle(ABC,AIC);
```
$$AIC$$

```
> AIC[equation];
```
$$x^2 + y^2 + .2500000000 - 1.000000000\,x - .5773502694\,y = 0$$

The centre of the inscribed circle is the *incentre*. It is also the point at which the bisectors of the interior angles of a triangle are *concurrent* (or the point at which the bisectors intersect). Section 12.3.3 deals with bisectors.

```
> coordinates(center(AIC));
```
$$[.5000000000, .2886751347]$$

```
> aic:=implicitplot(AIC[equation],x=-1..2,y=0..1,numpoints=1000):
> ac:=textplot([0.5,0.166667*sqrt(3),'IC']):
> display({at,aic,ac,ap1,ap2,ap3},scaling=CONSTRAINED);
```

The inscribed circle and incentre of $\triangle ABC$ can be seen in Figure 12.4(a).

And now for $\triangle KLM$,

```
> incircle(KLM,KIC);
```
$$KIC$$

```
> KIC[equation];
```

$$x^2 + y^2 + 9\,\frac{\left(\sqrt{5}+1\right)^2}{\left(\sqrt{5}+3+2\sqrt{2}\right)^2} - 6\,\frac{\left(\sqrt{5}+1\right)\,x}{\sqrt{5}+3+2\sqrt{2}} - 12\,\frac{y}{\sqrt{5}+3+2\sqrt{2}} =$$

0

```
> coordinates(center(KIC));
```
$$\left[3\,\frac{\sqrt{5}+1}{\sqrt{5}+3+2\sqrt{2}},6\,\frac{1}{\sqrt{5}+3+2\sqrt{2}}\right]$$

```
> kic:=implicitplot(KIC[equation],x=0..2,y=0..2,numpoints=1000):
> kc:=textplot([3*(sqrt(5)+1)/(sqrt(5)+3+2*sqrt(2)),6/
> (sqrt(5)+3+2*sqrt(2)),'IC']):
> display({kt,kic,kc,kp1,kp2,kp3},scaling=CONSTRAINED);
```

See Figure 12.4(b) for the inscribed circle and incentre of $\triangle KLM$.

12.3.3   Bisectors of Interior Angles

The line through an interior angle of a triangle which is equidistant from the sides
of the triangle which enclose the angle is called a bisector. There is a bisector
for each of the three interior angles of a triangle.

```
> bisector(ABC,A,bisector_A);
```
$$bisector\_A$$

```
> bisector_A[form];
```
$$line$$

```
> bisector_A[equation];
```
$$-.2500000000\,\sqrt{3}\,x + .7500000000\,y = 0$$

```
> bisector(ABC,B,bisector_B);
```
$$bisector\_B$$

```
> bisector_B[equation];
```
$$-.2500000000\,\sqrt{3}\,x - .7500000000\,y + .2500000000\,\sqrt{3} = 0$$

```
> bisector(ABC,C,bisector_C);
```
$$bisector\_C$$

```
> bisector_C[equation];
```
$$.5000000000\,\sqrt{3}\,x - .2500000000\,\sqrt{3} = 0$$

All three bisectors of a triangle intersect at the incentre.
```
> i1:=evalf(coordinates(inter(bisector_A,bisector_B)),5);
> i2:=evalf(coordinates(inter(bisector_B,bisector_C)),5);
```

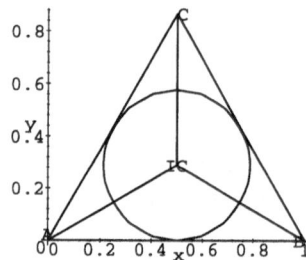 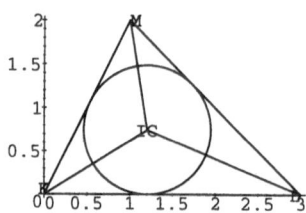

Figure 12.5. Bisectors with incircle and incentre of (a) $\triangle ABC$, (b) $\triangle KLM$

```
> i3:=evalf(coordinates(inter(bisector_C,bisector_A)),5);
> evalf(coordinates(center(AIC)),5);
                    [.50000,.28868]
```

```
> ab1:=implicitplot(bisector_A[equation],x=0..0.5,y=0..0.28869):
> ab2:=implicitplot(bisector_B[equation],x=0.5..1,y=0..0.28869):
> ab3:=implicitplot(bisector_C[equation],x=0..1,
> y=0.28869..1/2*sqrt(3)):
> display({at,aic,ac,ab1,ab2,ab3,ap1,ap2,ap3},
> scaling=CONSTRAINED);
```

This plot can be seen in Figure 12.5(a).

And now for $\triangle KLM$,

```
> bisector(KLM,K,bisector_K);
                    bisector_K
```

```
> bisector(KLM,L,bisector_L);
                    bisector_L
```

```
> bisector(KLM,M,bisector_M);
                    bisector_M
```

```
> kb1:=implicitplot(bisector_K[equation],x=0..3*(sqrt(5)+1)/
> (sqrt(5)+3+2*sqrt(2)),y=0..6/(sqrt(5)+3+2*sqrt(2))):
> kb2:=implicitplot(bisector_L[equation],x=3*(sqrt(5)+1)/
> (sqrt(5)+3+2*sqrt(2))..3,y=0..6/(sqrt(5)+3+2*sqrt(2))):
> kb3:=implicitplot(bisector_M[equation],x=0..3,y=6/
> (sqrt(5)+3+2*sqrt(2))..2):
> display({kt,kic,kc,kb1,kb2,kb3,kp1,kp2,kp3},scaling=CONSTRAINED);
```

This plot can be seen in Figure 12.5(b).

12.3.4   Exscribed Circles and E-Centres

If you extend the sides of a triangle beyond its vertices and draw a circle which
is tangent to side $\overline{AB}$ of the triangle and to the extensions of the other two sides
$\overline{AC}$ and $\overline{BC}$, the circle which you have drawn is called the *exscribed circle* to $\overline{AB}$.
The centre of this circle is called an *E-centre*. Each triangle has three exscribed
circles and three E-centres.

All three exscribed circles of a triangle are calculated when the `excircle`
command is used. In *Maple's* notation, the $excircle\_of\_ABC\_A$ is the exscribed
circle to the line $\overline{BC}$, opposite $\angle BAC$.

```
> excircle(ABC);
```
$$excircle\_of\_ABC\_A,\ excircle\_of\_ABC\_B,\ excircle\_of\_ABC\_C$$

```
> excircle_of_ABC_A[equation];
```
$$x^2 + y^2 + 2.250000000 - 3.000000000\,x - 1.000000000\,\sqrt{3}\,y = 0$$

```
> center(excircle_of_ABC_A);
```
$$center\_1\_ABC$$

```
> coordinates(center(excircle_of_ABC_A));
```
$$\left[1.500000000, .5000000000\,\sqrt{3}\right]$$

```
> excircle_of_ABC_B[equation];
```
$$x^2 + y^2 + .2500000000 + 1.000000000\,x - 1.000000000\,\sqrt{3}\,y = 0$$

```
> center(excircle_of_ABC_B);
```
$$center\_2\_ABC$$

```
> coordinates(center(excircle_of_ABC_B));
```
$$\left[-.5000000000, .5000000000\,\sqrt{3}\right]$$

```
> excircle_of_ABC_C[equation];
```
$$x^2 + y^2 + .2500000000 - 1.000000000\,x + 1.000000000\,\sqrt{3}\,y = 0$$

```
> center(excircle_of_ABC_C);
```
$$center\_3\_ABC$$

```
> coordinates(center(excircle_of_ABC_C));
```
$$\left[.5000000000, -.5000000000\,\sqrt{3}\right]$$

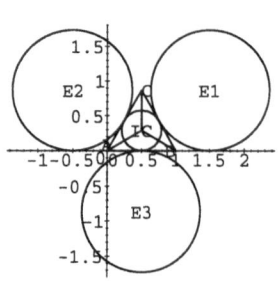 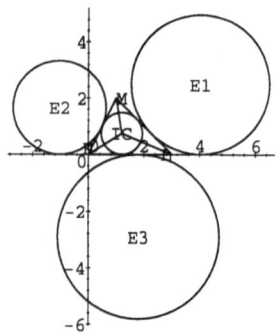

Figure 12.6. Exscribed circles and E-centres of (a) $\triangle ABC$, (b) $\triangle KLM$

```
> aec1:=implicitplot(excircle_of_ABC_A[equation],x=0.5..2.5,
> y=0..2):

> aec2:=implicitplot(excircle_of_ABC_B[equation],x=-1.5..0.5,
> y=0..2):

> aec3:=implicitplot(excircle_of_ABC_C[equation],x=-0.5..1.5,
> y=-2..0):

> atxt:=textplot({[1.5,0.86605,'E1'],[-0.5,0.86605,'E2'],
> [0.5,-.86605,'E3']}):

> display({at,aic,ac,ap1,ap2,ap3,ab1,ab2,ab3,aec1,aec2,
> aec3,atxt},scaling=CONSTRAINED);
```

Figure 12.6(a) is a plot of the exscribed circles and E-centres of $\triangle ABC$.

And now for $\triangle KLM$,

```
> excircle(KLM);
```
$excircle\_of\_KLM\_K, \; excircle\_of\_KLM\_L, \; excircle\_of\_KLM\_M$

```
> coordinates(center(excircle_of_KLM_K));
```
$$\left[\frac{3\sqrt{5}+\sqrt{9}}{\sqrt{5}+\sqrt{9}-\sqrt{8}}, 2\frac{\sqrt{9}}{\sqrt{5}+\sqrt{9}-\sqrt{8}}\right]$$

```
> coordinates(center(excircle_of_KLM_L));
```
$$\left[\frac{-3\sqrt{5}+\sqrt{9}}{-\sqrt{5}+\sqrt{9}+\sqrt{8}}, 2\frac{\sqrt{9}}{-\sqrt{5}+\sqrt{9}+\sqrt{8}}\right]$$

```
> coordinates(center(excircle_of_KLM_M));
```
$$\left[\frac{-3\sqrt{5}+\sqrt{9}}{-\sqrt{5}+\sqrt{9}-\sqrt{8}}, 2\frac{\sqrt{9}}{-\sqrt{5}+\sqrt{9}-\sqrt{8}}\right]$$

```
> kec1:=implicitplot(excircle_of_KLM_K[equation],x=1..7,
> y=-1..7):

> kec2:=implicitplot(excircle_of_KLM_L[equation],x=-7..1,
> y=0..5):
```

```
> kec3:=implicitplot(excircle_of_KLM_M[equation],x=-3..5,
> y=-6..0):
> ktxt:=textplot({[(3*sqrt(5)+sqrt(9))/(sqrt(5)+sqrt(9)
> -sqrt(8)),(2*sqrt(9))/(sqrt(5)+sqrt(9)-sqrt(8)),'E1'],
> [(-sqrt(9)+3*sqrt(5))/(sqrt(5)-sqrt(9)-sqrt(8)),-2*sqrt(9)/
> (sqrt(5)-sqrt(9)-sqrt(8)),'E2'],[(-sqrt(9)+3*sqrt(5))/
> (sqrt(5)-sqrt(9)+sqrt(8)),-2*sqrt(9)/(sqrt(5)-sqrt(9)+sqrt(8)),
> 'E3']}):
> display({kt,kic,kc,kp1,kp2,kp3,kb1,kb2,kb3,kec1,kec2,kec3,
> ktxt},scaling=CONSTRAINED);
```

Figure 12.6(b) is a plot of the exscribed circles and E-centres of $\triangle KLM$.

12.3.5   Centroid of a Triangle

The point at which *medians* of a triangle are concurrent is called the *centroid* of a triangle.

```
> centroid(ABC,centroid_ABC);
```
$$centroid\_ABC$$

```
> coordinates(centroid_ABC);
```
$$\left[.5000000000, \frac{1}{6}\sqrt{3}\right]$$

```
> median(ABC,A,med_A);
```
$$med\_A$$

```
> median(ABC,B,med_B);
```
$$med\_B$$

```
> median(ABC,C,med_C);
```
$$med\_C$$

For an equilateral triangle such as $\triangle ABC$, the median lines are the same as the bisectors, and the centroid is the same as the incentre and the circumcentre.

```
> am1:=implicitplot(med_A[equation],x=0..0.5,y=0..1/6*sqrt(3)):
> am2:=implicitplot(med_B[equation],x=0.5..1,y=0..1/6*sqrt(3)):
> am3:=implicitplot(med_C[equation],x=0..1,y=1/6*sqrt(3)..
> 1/2*sqrt(3)):
> acent:=textplot([0.5,1/6*sqrt(3),'C']):
> display({at,ap1,ap2,ap3,am1,am2,am3,acent},scaling=CONSTRAINED);
```

Figure 12.7(a) shows the medians and centroid of $\triangle ABC$.

And now for $\triangle KLM$,

```
> centroid(KLM,centroid_KLM);
```
$$centroid\_KLM$$

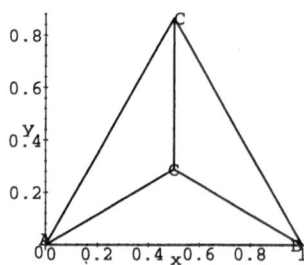 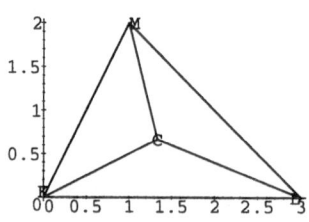

Figure 12.7. Medians and centroid of (a) $\triangle ABC$, (b) $\triangle KLM$

```
> coordinates(centroid_KLM);
```
$$\left[\frac{4}{3}, \frac{2}{3}\right]$$

```
> median(KLM,K,med_K);
```
$$med\_K$$

```
> median(KLM,L,med_L);
```
$$med\_L$$

```
> median(KLM,M,med_M);
```
$$med\_M$$

```
> km1:=implicitplot(med_K[equation],x=0..4/3,y=0..2/3):
> km2:=implicitplot(med_L[equation],x=4/3..3,y=0..2/3):
> km3:=implicitplot(med_M[equation],x=0..3,y=2/3..2):
> kcent:=textplot([4/3,2/3,'C']):
> display({kt,kp1,kp2,kp3,km1,km2,km3,kcent},scaling=CONSTRAINED);
```

Figure 12.7(b) shows the medians and centroid $\triangle KLM$.

12.3.6   Orthocentre of a Triangle

The point at which *altitudes* of a triangle are concurrent is called the *orthocentre* of a triangle.

```
> orthocenter(ABC,AO);
```
$$AO$$

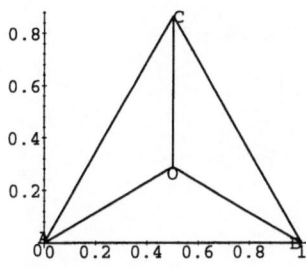 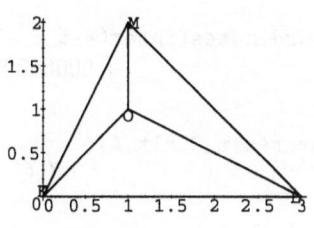

Figure 12.8. Altitudes and orthocentre of (a) $\triangle ABC$, (b) $\triangle KLM$

```
> coordinates(AO);
```
$$\left[.5000000001, .1666666667\sqrt{3}\right]$$

```
> altitude(ABC,A,alt_A);
```
$$alt\_A$$

```
> alt_A[equation];
```
$$.5\,x - \frac{1}{2}\sqrt{3}\,y = 0$$

```
> altitude(ABC,B,alt_B);
```
$$alt\_B$$

```
> alt_B[equation];
```
$$.5\,x + \frac{1}{2}\sqrt{3}\,y - .5 = 0$$

```
> altitude(ABC,C,alt_C);
```
$$alt\_C$$

```
> alt_C[equation];
```
$$x - .5 = 0$$

All altitudes of a triangle intersect at the orthocentre.

```
> inter(alt_A,alt_B);
```
$$alt\_A\_intersect\_alt\_B$$

```
> coordinates(inter(alt_A,alt_B));
```
$$[\,.5000000001, .2886751347\,]$$

```
> inter(alt_B,alt_C);
```
$$alt\_B\_intersect\_alt\_C$$

```
> coordinates(inter(alt_B,alt_C));
```
$$[.5000000001, .2886751347]$$

```
> inter(alt_C,alt_A);
```
$$alt\_C\_intersect\_alt\_A$$

```
> coordinates(inter(alt_C,alt_A));
```
$$[.5000000001, .2886751347]$$

As you may have guessed, for the equilateral triangle $\triangle ABC$, the altitudes have the same equation as the medians and bisectors, and the coordinates of the orthocentre are the same as the circumcentre, incentre, and centroid.

```
> aa1:=implicitplot(alt_A[equation],x=0..0.5,y=0..1/6*sqrt(3)):
> aa2:=implicitplot(alt_B[equation],x=0.5..1,y=0..1/6*sqrt(3)):
> aa3:=implicitplot(alt_C[equation],x=0..1,
> y=1/6*sqrt(3)..1/2*sqrt(3)):
> ao:=textplot([0.5,1/6*sqrt(3),'O'],align=BELOW):
> display({at,ap1,ap2,ap3,aa1,aa2,aa3,ao},scaling=CONSTRAINED);
```

The altitudes and orthocentre $\triangle ABC$ can be seen in Figure 12.8(a).

And now for $\triangle KLM$,

```
> coordinates(orthocenter(KLM,KO));
```
$$[1,1]$$

```
> altitude(KLM,K,alt_K);
```
$$alt\_K$$

```
> altitude(KLM,L,alt_L);
```
$$alt\_L$$

```
> altitude(KLM,M,alt_M);
```
$$alt\_M$$

```
> ka1:=implicitplot(alt_K[equation],x=0..1,y=0..1):
> ka2:=implicitplot(alt_L[equation],x=1..3,y=0..1):
> ka3:=implicitplot(alt_M[equation],x=0..3,y=1..2):
> ko:=textplot([1,1,'O'],align=BELOW):
> display({kt,kp1,kp2,kp3,ka1,ka2,ka3,ko},
> scaling=CONSTRAINED);
```

The altitudes and orthocentre $\triangle KLM$ can be seen in Figure 12.8(b).

Experiment 12.2   A regular tetrahedron in a 4-sided solid with each face an equilateral triangle, that is, a triangular pyramid. What is the side length of triangles for the inscribed and exscribed tetrahedra for a unit sphere? Try first the inscribed and exscribed *cubes* for a unit sphere.

12.4   Equations of Loci After a Transformation

A *transformation* is a mapping of coordinates, which may result in the translation, stretch, reflection, or dilatation of the original locus. In this section we shall be studying the effects of transformations in the plane, on *eq1*, defined below.

```
> eq1:=y=x^2+2;
```
$$eq1 := y = x^2 + 2$$

12.4.1   Translation

To perform the transformation $(x, y) \mapsto (x + 2, y)$ on *eq1*, we let $u = x + 2$, and $v = y$. Rearranging for x and y we get $x = u - 2$ and $y = v$, which we can substitute into *eq1* to get a new equation *eq2*.

```
> eq2:=subs(x=u-2,y=v,eq1);
```
$$eq2 := v = (u - 2)^2 + 2$$

Since it is more convenient to have all equations in terms of x and y, we make a change of variables $u = x$ and $v = y$, and substitute these expressions into *eq2* to create *eq3*.

```
> eq3:=subs(u=x,v=y,eq2);
```
$$eq3 := y = (x - 2)^2 + 2$$

```
> display({implicitplot({eq1,eq3},x=-2..4,y=0..10,numpoints=500),
> textplot([[2.8,10,'y=x^2+2'],[4,6,'y=(x-2)^2+2']])});
```

As you can see from the graphs of *eq1* and *eq3*, shown in Figure 12.9(a), *eq3* has been translated 2 units in the positive $x-$ direction.

We can translate functions in both directions. Now let's perform the transformation $(x, y) \mapsto (x+2, y+2)$ on *eq1*. As before, we let $u = x+2$ and $v = y+2$, and make the substitution $x = u - 2$ and $y = v - 2$ into *eq1* to make *eq4*, and then substitute $u = x$ and $v = y$ to make *eq5*.

```
> eq4:=subs(x=u-2,y=v-2,eq1);
```
$$eq4 := v - 2 = (u - 2)^2 + 2$$

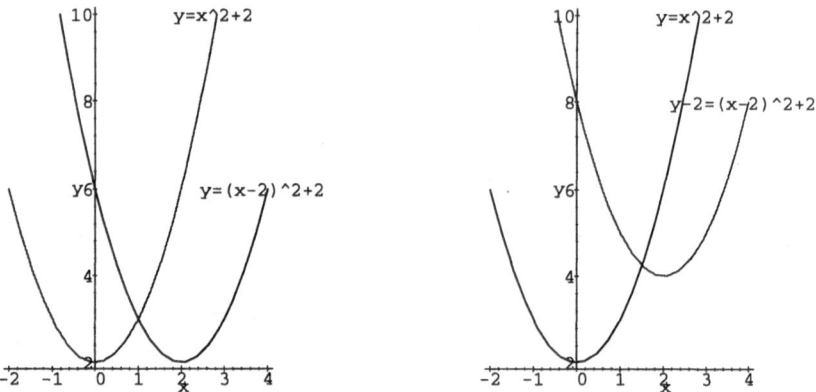

Figure 12.9. (a) Translation $(x, y) \mapsto (x + 2, y)$ of $y = x^2 + 2$. (b) Translation $(x, y) \mapsto (x + 2, y + 2)$ of $y = x^2 + 2$

```
> eq5:=subs(u=x,v=y,eq4);
```
$$eq5 := y - 2 = (x - 2)^2 + 2$$

```
> display({implicitplot({eq1,eq5},x=-2..4,y=0..10,numpoints=500),
> textplot([[2.8,10,'y=x^2+2'],[4,8,'y-2=(x-2)^2+2']])});
```

The graph of this translation is shown in Figure 12.9(b).

12.4.2   Stretch

The transformation $(x, y) \mapsto (2x, y)$ produces a one-way stretch in the x-axis. In this case we substitute $u = 2x$, or $x = \frac{u}{2}$ into the equation $eq1$ to form $eq6$, and then replace u with x to form $eq7$. Since y is unchanged, we needn't replace it.

```
> eq6:=subs(x=u/2,eq1);
```
$$eq6 := y = \frac{1}{4} u^2 + 2$$

```
> eq7:=subs(u=x,eq6);
```
$$eq7 := y = \frac{1}{4} x^2 + 2$$

```
> display({implicitplot({eq1,eq7},x=-2..2,y=0..10,numpoints=500),
> textplot([[2,6,'y=x^2+2'],[2,3,'y=1/4*x^2+2']])});
```

You can see the results of this stretch in Figure 12.10(a).

The transformation $(x, y) \mapsto (x, 2y)$ produces a one-way stretch in the y-axis. Now substitute $v = 2y$, or $y = \frac{v}{2}$, into the equation $eq1$ as before.

```
> eq8:=subs(y=v/2,eq1);
```
$$eq8 := \frac{1}{2} v = x^2 + 2$$

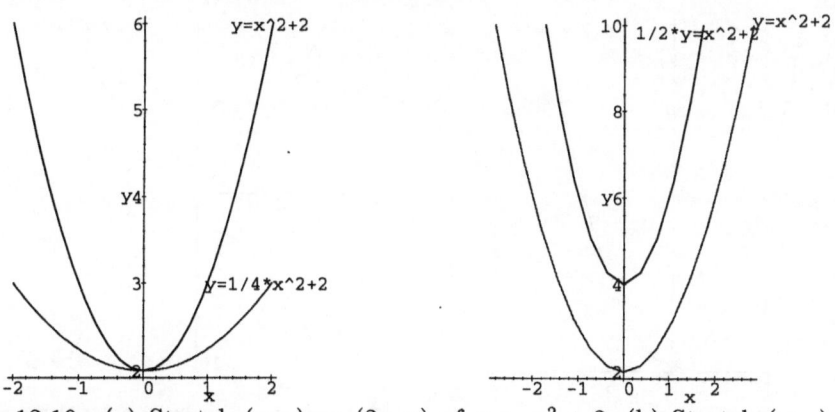

Figure 12.10. (a) Stretch $(x,y) \mapsto (2x, y)$ of $y = x^2 + 2$. (b) Stretch $(x, y) \mapsto$ $(x, 2y)$ of $y = x^2 + 2$

```
> eq9:=subs(v=y,eq8);
```

$$eq9 := \frac{1}{2} y = x^2 + 2$$

```
> display({implicitplot({eq1,eq9},x=-4..4,y=0..10,numpoints=500),
> textplot([[2.8,10,'y=x^2+2'],[1.8,10,'1/2*y=x^2+2']])});
```

See Figure 12.10(b) for the results of this transformation.

12.4.3   Reflection

The transformation $(x,y) \mapsto (-x, y)$ does not result in any change in the graph of $y = x^2 + 2$. Why? (See Figure 12.11(a).)

```
> eq10:=subs(x=-u,eq1);
```

$$eq10 := y = u^2 + 2$$

```
> eq11:=subs(u=x,eq10);
```

$$eq11 := y = x^2 + 2$$

```
> display({implicitplot({eq1,eq11},x=-4..4,y=-10..10,
> numpoints=500),textplot([2.8,10,'y=x^2+2'])});
```

Try the transformation $(x, y) \mapsto (-x, y)$ on a function of odd degree, say $f(x) = x^3$. Does the transformation result in a change in the graph of the function?

The transformation $(x, y) \mapsto (x, -y)$ produces a reflection of the graph in the x-axis. In this case, x is unchanged, so we make the substitution $v = -y$, or $y = -v$, in $eq1$.

```
> eq12:=subs(y=-v,eq1);
```

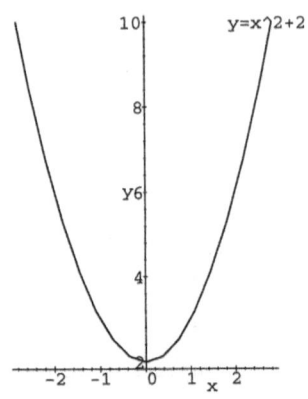
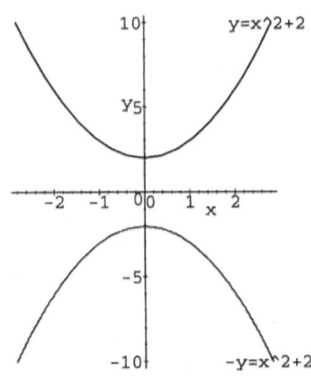

Figure 12.11.   (a) Reflection $(x,y) \mapsto (-x,y)$ of $y = x^2 + 2$. (b) Reflection $(x,y) \mapsto (x,-y)$ of $y = x^2 + 2$

$$eq12 := -v = x^2 + 2$$

```
> eq13:=subs(v=y,eq12);
```
$$eq13 := -y = x^2 + 2$$

```
> display({implicitplot({eq1,eq13},x=-4..4,y=-10..10,
> numpoints=500),textplot([[2.8,10,'y=x^2+2'],
> [2.8,-10,'-y=x^2+2']])});
```

This plot is shown in Figure 12.11(b).

12.4.4   Dilatation

To examine dilatations, we define a new relation *eqn1*.
```
> eqn1:=x^2+y^2=1;
```
$$eqn1 := x^2 + y^2 = 1$$

The transformation $(x,y) \mapsto (2x, 2y)$ will cause the graph of the loci *eqn1* to be dilated by a factor of two. This transformation is similar to a stretch. Rather than substitute $x = u/2$ and then change u to x and so on, we can substitute $x = \frac{x}{2}$ and $y = \frac{y}{2}$ directly into *eqn1* to get *eqn2*.
```
> eqn2:=subs(x=x/2,y=y/2,eqn1);
```
$$eqn2 := \frac{1}{4} x^2 + \frac{1}{4} y^2 = 1$$

```
> display({implicitplot({eqn1,eqn2},x=-2..2,y=-2..2,
> scaling=CONSTRAINED),textplot([[0.9,0.9,'x^2+y^2=1'],
> [1.7,1.7,'1/4*x^2+1/4*x^2=1']])});
```

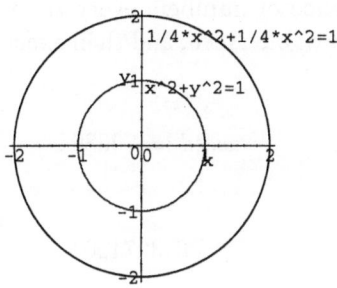

Figure 12.12. Dilatation $(x, y) \mapsto (2x, 2y)$ of unit circle

This plot is shown in Figure 12.12.

Exercise 12.3 Try plotting the *cycloid* curves with the following parameterizations:

(a) $x(t) = t - \sin(t), y(t) = 1 - \cos(t)$ (cycloid)

(b) $x(t) = t - 3\sin(t), y(t) = 1 - 3\cos(t)$ (prolate cycloid)

(c) $x(t) = 2t - \sin(t), y(t) = 2 - \cos(t)$ (curate cycloid)

The cycloid represents the trajectory of a point on the circumference of a wheel which is rolling along the x-axis; t represents time.

Experiment 12.3 Consider a solid triangle with vertices $v_0 = (-1, 0)$, $v_1 = (1, 0)$, and $v_2 = (x, y)$ where x and y are real numbers. The formula for the coordinates of the centre of gravity of a triangle with vertices v_0, v_1, and v_2 is $\frac{1}{3}(v_0 + v_1 + v_2)$. Plot its distace from the origin as a function of (x, y).

12.5 Simulations

Computer *simulations* can be helpful in modeling, visualizing and approximately solving problems which are difficult to describe analytically. One of the early ways in which computers were used to estimate solutions was the so called *Monte Carlo Method*, named after the famous casino. We illustrate it as a means of estimating the area enclosed by the closed loop of the Folium of Descartes $x^3 + y^3 = 3xy$, in the following experiment.

For such calculations, we often need to generate random numbers as sample points in a given region. *Maple* has a random number generator called **rand**.

Before we can use it, though, we must 'seed' the generator so that is does not produce the same sequence of numbers every time we use it. Simply assign a non-zero integer to the variable _seed, and then execute rand(); as shown below.

```
> _seed:=700806;
```
$$\_seed := 700806$$

```
> rand();
```
$$268613274204$$

To generate random numbers between 0 and 2, we type the following commands.

```
> random_number:=rand(0..2):
> random_number();
```
$$1$$

```
> random_number();
```
$$0$$

Notice that the numbers generated are positive integers. In order to generate random numbers in the range $[0, 2]$ with three significant figures, we must use rand to produce random numbers in the range $[0, 200]$ and then divide them by one hundred.

```
> random_number:=rand(0..200):
> evalf(random_number()/100);
```
$$1.710000000$$

```
> evalf(random_number()/100);
```
$$.3000000000$$

Experiment 12.4 From the *Maple* plot with commands

```
> p1:=implicitplot(x^3+y^3=3*x*y,x=-2..2,y=-2..2,
> scaling=CONSTRAINED,numpoints=1000):
> p2:=implicitplot(x=2,x=-1..2,y=0..2):
> p3:=implicitplot(y=2,x=0..2,y=-1..2):
> display({p1,p2,p3});
```

we see that the loop of the Folium of Descartes lies in the upper right quadrant, wholly contained in the square of side 2. Use *Maple's* random number generator to choose random coordinates (x_i, y_i) where x_i and y_i lie in the range $[0, 2]$. This will give points in the 2×2 square. If you repeat this for a large number of points, then the fraction of this number that lie in the loop gives an estimate of its area as a fraction of the area of the square.

Chapter 13

Sequences and Series

Commands used in this chapter
- `seq`
- `sum`
- `Sum`
- `value`

13.1 Sequences

A *sequence* is an ordered set, usually ordered by the natural numbers ($\mathsf{N} = \{1, 2, 3, \ldots\}$). Each element in the range of the sequence is a *term* of the sequence.

13.1.1 Arithmetic Sequences

An *arithmetic sequence* is a sequence defined by a linear function. For example:

```
> seq(2+n*3,n=1..6);
```
$$5, 8, 11, 14, 17, 20$$

What are the first five terms of the arithmetic sequence defined by $S_n = a + nd$, where a and d are constants?

```
> S:=[seq(a+n*d,n=1..5)];
```
$$S := [a + d, a + 2d, a + 3d, a + 4d, a + 5d]$$

```
> 'S[2]-S[1]':=S[2]-S[1];
```
$$S[2] - S[1] := d$$

```
> 'S[3]-S[2]':=S[3]-S[2];
```
$$S[3] - S[2] := d$$

```
> 'S[4]-S[3]':=S[4]-S[3];
```
$$S[4] - S[3] := d$$

```
> 'S[5]-S[4]':=S[5]-S[4];
```
$$S[5] - S[4] := d$$

Successive terms have a *common difference* of d.

```
> 'S[5]'=S[5];
```
$$S[5] = a + 5d$$

In general, $S_n = a + (n-1)d$ for an arithmetic sequence.

```
> A:=[seq(2*n+5,n=1..5)];
```
$$A := [7, 9, 11, 13, 15]$$

```
> A[2]-A[1];
```
$$2$$

```
> A[3]-A[2];
```
$$2$$

```
> A[4]-A[3];
```
$$2$$

```
> A[5]-A[4];
```
$$2$$

Exercise 13.1   Calculate the first 4 terms of the following sequences.

  (a) $f(n) = n - 1$
  (b) $f(n) = \frac{1}{2}n + 7$
  (c) $f(n) = -3n$

Exercise 13.2   Determine the nth term of the following sequences.

  (a) $5, 8, 11, 14, \ldots$
  (b) $\frac{5}{3}, \frac{4}{3}, 1, \frac{2}{3}, \frac{1}{3}, \ldots$
  (c) $5, 9, 13, 17, \ldots$

Exercise 13.3   The third term of an arithmetic sequence is 14 and the seventh is 26. Find the ninth term.

13.1.2   Geometric Sequences

A *geometric sequence* is a sequence defined by an exponential function. For example:

```
> seq(2*3^n,n=1..6);
```
$$6, 18, 54, 162, 486, 1458$$

What are the first five terms of the geometric sequence defined by $S_n = ar^n$, where a and r are constants?

```
> S:=[seq(a*r^n,n=1..5)];
```
$$S := [\, a\,r, a\,r^2, a\,r^3, a\,r^4, a\,r^5 \,]$$

```
> 'S[2]/S[1]':=S[2]/S[1];
```
$$S[2]/S[1] := r$$

```
> 'S[3]/S[2]':=S[3]/S[2];
```
$$S[3]/S[2] := r$$

```
> 'S[4]/S[3]':=S[4]/S[3];
```
$$S[4]/S[3] := r$$

```
> 'S[5]/S[4]':=S[5]/S[4];
```
$$S[5]/S[4] := r$$

Successive terms have a *common ratio* of r.

```
> 'S[5]':=S[5];
```
$$S[5] := a\,r^5$$

In general, $S_n = ar^{n-1}$ for a geometric sequence.

```
> G:=[seq(2*3^n,n=1..5)];
```
$$G := [\, 6, 18, 54, 162, 486 \,]$$

```
> G[2]/G[1];
```
$$3$$

```
> G[3]/G[2];
```
$$3$$

```
> G[4]/G[3];
```
$$3$$

```
> G[5]/G[4];
```
$$3$$

Exercise 13.4 Calculate the first four terms of the following sequences.

   (a) $f(n) = 7^n$
   (b) $f(n) = \frac{1}{2}(3)^n$
   (c) $f(n) = -5(2^n)$

Exercise 13.5 Calculate the nth term of the following sequences.

   (a) $6, 18, 54, 162, \ldots$
   (b) $1, 4, 16, 64, \ldots$
   (c) $-6, 12, -24, 96. \ldots$

Exercise 13.6 For the geometric sequence $-8, 16, -32, \ldots, t_n = 256$, find n.

13.2   Series

A *series* is the result of summing the terms of a sequence. Let $t_1, t_2, t_3, \ldots, t_k, \ldots$ be the terms of a sequence. Then the expression $s_k = t_1 + t_2 + t_3 + \ldots + t_k$ defines a new sequence, consisting of the *partial sums* of the original sequence. The kth term of the sequence is sometimes called the kth term of the series so care needs to be taken with labels!

13.2.1   Arithmetic Series

Like an arithmetic sequence, an *arithmetic series* is a series defined by a linear function.

```
> n:='n':
> Sum(2*n+5,n=1..3);
```
$$\sum_{n=1}^{3} (2n + 5)$$

```
> value(");
```
$$27$$

```
> s[1]:=sum(2*n+5,n=1..1);
```
$$s_1 := 7$$

```
> s[2]:=sum(2*n+5,n=1..2);
```
$$s_2 := 16$$

```
> s[3]:=sum(2*n+5,n=1..3);
```
$$s_3 := 27$$

```
> t[1]:=s[1];
```
$$t_1 := 7$$

```
> t[2]:=s[2]-s[1];
```
$$t_2 := 9$$

```
> t[3]:=s[3]-s[2];
```
$$t_3 := 11$$

```
> t[1]+t[2]+t[3];
```
$$27$$

In general, $s_n = \frac{n}{2}(t_1 + t_n)$,

```
> s[3];
```
$$27$$

```
> n:=3;
```
$$n := 3$$

```
> n/2*(t[1]+t[n]);
```
$$27$$

or $s_n = \frac{n}{2}(2a + (n-1)d)$.

```
> a:=7;
```
$$a := 7$$

```
> d:=2;
```
$$d := 2$$

```
> s[n]:=n/2*(2*a+(n-1)*d);
```
$$s_3 := 27$$

```
> n:='n':
```

Exercise 13.7 Calculate the sum of the following arithmetic series.

(a) $5 + 11 + 17 + \ldots + 59$

(b) $-9 + (-11) + (-13) + \ldots + (-25)$

(c) $-2 + 1 + 4 + \ldots + 295$

Exercise 13.8 Calculate the sum of the first 20 terms of the following arithmetic series.

(a) $4 + 5 + 6 + 7 + \ldots$

(b) $5 + 9 + 13 + 17 + \ldots$

(c) $-\frac{15}{2} - 7 - \frac{13}{2} - 6 - \ldots$

13.2.2 Geometric Series

A *geometric series* is a series defined by an exponential function.

```
> Sum(2*(3^n),n=1..3);
```

$$\sum_{n=1}^{3} (2\,3^n)$$

```
> value(");
```

$$78$$

```
> s[1]:=sum(2*(3^n),n=1..1);
```

$$s_1 := 6$$

```
> s[2]:=sum(2*(3^n),n=1..2);
```

$$s_2 := 24$$

```
> s[3]:=sum(2*(3^n),n=1..3);
```

$$s_3 := 78$$

```
> t[1]:=s[1];
```

$$t_1 := 6$$

```
> t[2]:=s[2]-s[1];
```

$$t_2 := 18$$

```
> t[3]:=s[3]-s[2];
```

$$t_3 := 54$$

```
> t[1]+t[2]+t[3];
```
$$78$$

In general, $s_n = a\frac{r^n-1}{r-1}$, where a is the first term of the geometric series and $r \neq 1$.

```
> s[3];
```
$$78$$

```
> n:=3;
```
$$n := 3$$

```
> a:=6;
```
$$a := 6$$

```
> r:=3;
```
$$r := 3$$

```
> s[n]:=a*(r^n-1)/(r-1);
```
$$s_3 := 78$$

Exercise 13.9   Find the sum of the following geometric series.

(a) $6 + 12 + 24 + \ldots + 384$

(b) $\frac{1}{4} + \frac{1}{16} + \frac{1}{64} + \ldots + \frac{1}{65536}$

(c) $-5 + 25 - 125 + \ldots + 15625$

Experiment 13.1   Plot the points (n, r^n) for $n = 0, 1, 2, \ldots, 100$ with

(a) $r = 0.9$

(b) $r = 1$

(c) $r = 1.1$

Experiment 13.2   It is possible to expand a *periodic function* as the sum of a trigonometric series, called a Fourier series. In each of the following cases, plot the function $f(x)$ for $n = 1, 2, \ldots, 20$ and animate the result.

(a) $f(x) = \pi - 2(\sin x + \dfrac{\sin 2x}{2} + \dfrac{\sin 3x}{3} + \ldots + \dfrac{\sin nx}{n})$

(b) $f(x) = \dfrac{4}{\pi}(\sin x + \dfrac{\sin 3x}{3} + \dfrac{\sin 5x}{5} + \ldots + \dfrac{\sin nx}{n})$

(c) $f(x) = 2(\sin x - \dfrac{\sin 2x}{2} + \dfrac{\sin 3x}{3} - \ldots \pm \dfrac{\sin nx}{n})$

(d) $f(x) = \dfrac{\pi^2}{3} - 4(\cos x - \dfrac{\cos 2x}{2^2} + \dfrac{\cos 3x}{3^2} - \ldots \pm \dfrac{\cos nx}{n^2})$

Can you guess the periodic functions being approximated? For example, (a) is the nth order approximation to the saw-tooth function $g(x) = x, 0 < x < 2\pi$ which is repeated on successive intervals of length 2π. How large must n be in (a) for $f(x)$ to approximate $g(x)$ with an average error of less than 0.001 over each interval of length 2π?

Chapter 14

Statistics and Probability

Commands used in this chapter

- binomial
- describe[mean]
- describe[median]
- describe[mode]
- describe[standarddeviation]
- evalf
- plots
- plots[display]
- solve

- sqrt
- stats
- stats[statplot]
- statplots[boxplot]
- statplots[histogram]
- statplots[scatter1d]
- statplots[scatter2d]
- sum
- transform[tallyinto]

14.1 Organizing and Presenting Data

Before we can draw any conclusions from data that we collect, we must be able
to present it in a useful way. The data we shall be examining is a set of numbers
assigned to *raw_data*. Actually, we chose this set by hitting keys at random.

```
> raw_data:=[5,7,2,5,4,5,0,9,10,10,2,5,4,3,4,7,8,0,8,5,3,5,8,9,7,
> 1,4,4,5,8,2,1,9,3,3,5,6,6,7,0,9,2,5,7,6,6,4,6]:
```

14.1.1 Plots

Without manipulating the data, we can produce a crude *histogram*, in which all
of the points of a certain value are stacked on top of one another. This plot is
shown in Figure 14.1; no choice of fontsize seems to be allowed!

```
> with(stats):
> with(plots):
> display(statplots[scatter1d[stacked]](raw_data),axes=FRAMED);
```

A histogram is similar to this, except that the stacks are replaced by bars, or
cells. The width of each cell is equal to the interval over which we group the data
(called the *class width*), and the height of each cell represents the frequency with
which the values in that interval or *class* occur (see Figure 14.2(a)). To produce
a histogram, we must first tally the number of data points within each interval.

```
> tallied_data1:=transform[tallyinto](raw_data,[0..1,1..2,2..3,
```

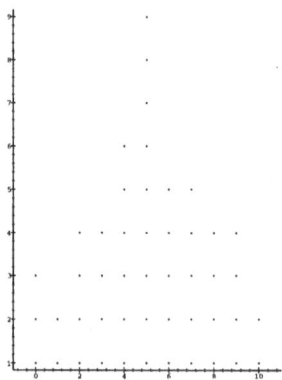

Figure 14.1. Crude histogram of raw data

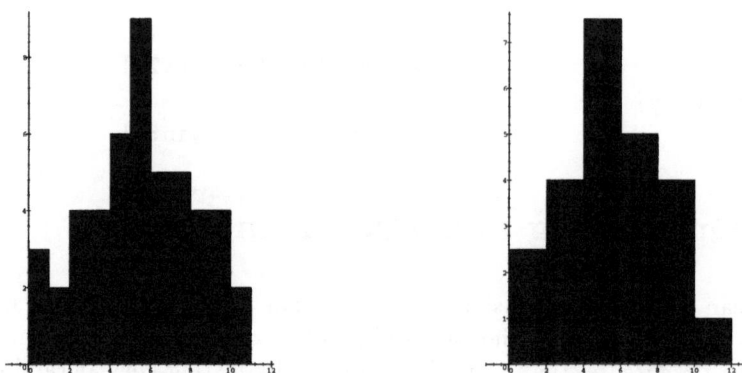

Figure 14.2. (a) Histogram with class width of 1. (b) Histogram with class width of 2

```
> 3..4,4..5,5..6,6..7,7..8,8..9,9..10,10..11]);
```

$$tallied\_data1 := [\text{Weight}(10..11, 2), \text{Weight}(0..1, 3),$$
$$\text{Weight}(1..2, 2), \text{Weight}(2..3, 4), \text{Weight}(3..4, 4),$$
$$\text{Weight}(4..5, 6), \text{Weight}(5..6, 9), \text{Weight}(6..7, 5),$$
$$\text{Weight}(7..8, 5), \text{Weight}(8..9, 4), \text{Weight}(9..10, 4)]$$

```
> statplots[histogram](tallied_data1);
```

Let's increase the width of the cells from 1 to 2, and observe the effect that has on the histogram. Compare the plots in Figure 14.2.

```
> tallied_data2:=transform[tallyinto](raw_data,[0..2,2..4,4..6,
> 6..8,8..10,10..12]);
```

$$tallied\_data2 := [\text{Weight}(6..8, 10), \text{Weight}(8..10, 8),$$
$$\text{Weight}(10..12, 2), \text{Weight}(0..2, 5), \text{Weight}(2..4, 8),$$
$$\text{Weight}(4..6, 15)]$$

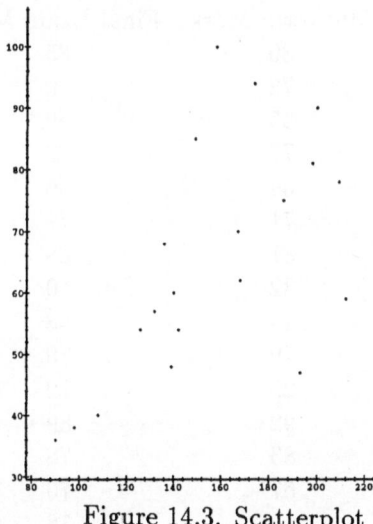

Figure 14.3. Scatterplot

```
> statplots[histogram](tallied_data2);
```

A useful method for examining the relationship between two data sets is the *scatterplot*. The data sets *height* and *mass* contain the heights (in centimetres) and masses (in kilograms) of 20 randomly selected test subjects. The scatterplot in Figure 14.3 shows that there is a *positive association* between the two data sets, in other words, the above-average values of one variable correspond to above-average values in the other variable.

```
> height:=[168,140,132,200,139,212,90,167,198,126,142,98,108,
> 193,158,149,209,174,186,136]:
```

```
> mass:=[62,60,57,90,48,59,36,70,81,54,54,38,40,47,100,85,78,
> 94,75,68]:
```

```
> statplots[scatter2d](height,mass);
```

We interpret a scatterplot as follows. Plot a point with coordinates the average, or *mean*, of the two data sets and imagine lines drawn through that *grand mean point* . These lines separate the points into four quadrants and we are interested in whether there are more in the south-west and north-east quadrants than in the south-east and north-west, or vice versa.

Exercise 14.1 The mid-term and final exam marks for a given class are as follows.

Student	Mid-term Mark	Final Exam Mark
1	80	85
2	72	70
3	55	50
4	77	82
5	64	66
6	71	77
7	81	88
8	32	40
9	48	58
10	79	70
11	29	20
12	92	89
13	83	78
14	67	70
15	45	78

Create a scatterplot of the data and determine if the two data sets are associated. Using a pencil and paper, try to find the equation of the line which best fits this data. Experiment 14.1 below will show you how to do this analytically.

14.1.2 Mean, Median, Mode, and Standard Deviation

Along with graphical methods, we can describe the properties of sets of data by calculating statistics such as the mean, *median, mode,* and *standard deviation.* These statistics tell us something about the centre and spread, or dispersion, of our data.

The mean is the average of all values in a set of data. That is, the total sum divided by the total number.

```
> mean:=describe[mean](raw_data);
```

$$mean := \frac{61}{12}$$

```
> evalf(");
```

$$5.083333333$$

The median is the middle value of a set of data, in order of increasing size; half are below this and half are above.

```
> describe[median](raw_data);
```

$$5$$

The mode is the value of the item that occurs most frequently in a set of data, easily remembered from à la mode for fashion.

```
> describe[mode](raw_data);
```
$$5$$

The standard deviation is a measure of the dispersion of values in a set of data. Precisely, the standard deviation is the *root mean square deviation*, that is the square root of the average square of the differences from the mean. So it measures the spread of values about the mean. The formula is easily written in *Maple*,

```
> sqrt(sum((raw_data[i]-mean)^2,i=1..48)/48);
```
$$\frac{1}{12}\sqrt{1019}$$

In fact, *Maple* has a built-in function standarddeviation to compute the standard deviation of a data set.

```
> describe[standarddeviation](raw_data);
```
$$\frac{1}{12}\sqrt{1019}$$

```
> evalf(");
```
$$2.660148283$$

Exercise 14.2 The circumferences of 40 trees living on a plot of land, given in centimetres, are listed below.

43.2	80.3	71.8	35.1	95.0	60.0	97.9	30.4	102.3	91.1
62.6	81.3	73.5	66.1	56.4	42.2	42.7	69.4	46.5	70.0
43.9	76.7	18.8	75.6	24.5	41.4	69.6	56.5	41.1	87.9
15.0	41.4	54.3	70.0	75.3	94.5	41.4	63.5	63.2	94.7

(a) Create a histogram of this data using a class width of 10. What is the modal frequency?

(b) What is the mean circumference of the trees?

(c) What is the standard deviation?

(d) What are the mean and standard deviation of the diameters of the trees?

Experiment 14.1 Construct a table of data (x_i, y_i) with $i = 1, 2, \ldots, N$ for which the scatterplot suggests that there may be a linear relation of the form $y = mx + c$. We wish to find the 'best' values for m and c. Obviously, we want the line to go through the grand mean and so we have $\bar{y} = m\bar{x} + c$, where

$$\bar{x} = \sum_{i=1}^{N} x_i \text{ and } \bar{y} = \sum_{i=1}^{N} y_i.$$

To find m it is convenient to subtract the means from the data, giving a new set of data (X_i, Y_i) where

$$X_i = x_i - \bar{x} \text{ and } Y_i = y_i - \bar{y}.$$

Now our equation is, by subtraction,

$$(y - \bar{y}) = (mx + c) - (m\bar{x} + c) = m(x - \bar{x}) \text{ or } Y = mX$$

To find m we shall arrange to minimize the mean vertical distance between the line $Y = mX$ and the data points. In fact we minimize the mean square of the difference for convenience. The expression to minimize with respect to m is

$$D = \frac{1}{N} \sum_{i=1}^{N} (Y_i - mX_i)^2$$

$$D = \frac{1}{N} \sum_{i=1}^{N} (Y_i^2 - 2mY_iX_i + m^2X_i^2)$$

$$D = \frac{1}{N} \sum_{i=1}^{N} Y_i^2 - \frac{2m}{N} \sum_{i=1}^{N} Y_iX_i + \frac{m^2}{N} \sum_{i=1}^{N} X_i^2$$

which is of the form

$$D = A - Bm + m^2K.$$

Now, we know that this is a parabola in m, with A, B, and K constants calculable from your data. The minimum of D is given when $m = \frac{B}{2K}$. So m is the ratio of the mean product of the X_i, Y_i values to the mean square of the X_i values. The process of finding the best fitting line is called *linear regression* and *Maple* has a function that does it, called fit[leastsquare] in the stats package.

14.2   Probability of Events

The *probability* of an event is the measure of the likelihood that the event will take place. The set of all possible *events* that may occur is called the *sample space*. Each event is a subset of the sample space.

The *probability distribution* on a given sample space S is a real-valued function P defined on subsets of S; it satisfies the following conditions.

1. P takes values in [0,1].

2. $P(S) = 1$, the sum of the probabilites of all possible outcomes in S must be one.

3. $P(A \text{ or } B) \leq P(A) + P(B)$, the probability of one of a number of events occurring is less than or equal to the sum of the individual probabilities. Equality occurs if the events are *disjoint*; that is if they have no outcomes in common.

Example 14.1   Out of a well-shuffled deck of playing cards, what is the probability of drawing

(a) a heart?

> p=13/52;

$$p = \frac{1}{4}$$

(b) a king or a queen?

> p=4/52+4/25;

$$p = \frac{2}{13}$$

(c) a red card?

> p=26/52;

$$p = \frac{1}{2}$$

(d) the ace of spades?

> p=1/52;

$$p = \frac{1}{52}$$

Two other properties of probability distributions are the multiplication rule and the complement rule. The first states that the probability of a number of events occurring at once is equal to the product of the individual probabilites, if the events are independent. Two events are independent if the occurrence of one does not affect the probability of the other occurring. The complement rule states that the probability of an event not occurring is equal to one minus the probability of the event occurring. When first trying to solve a probability problem, to find $P(A)$ say, it is worth checking to see if it is easier to find the probability of A not occurring, which is $1 - P(A)$.

Example 14.2 According to an insurance company, the probability of being hit by a car in a particular year is 0.05, while the probability of being robbed is 0.1. What is the probability of being hit by a car and being robbed in the same year?

> p=0.05*0.1;

$$p = .005$$

14.2.1   Binomial Theorem

The *binomial coefficient* is the number of ways of choosing k successes from n trials. It is equal to $\frac{n!}{k!(n-k)!}$. (Remember that $n!$ means the product $1 \times 2 \times 3 \times 4 \times \ldots \times n$, so $3! = 6$, $4! = 24$ etc.) The *Maple* function `binomial` will calculate the value of the binomial coefficient for you, given n and k.

Example 14.3 How many 13-card bridge hands are possible from a normal deck of 52 cards?

> n:=52;k:=13;

$$n := 52$$

$$k := 13$$

> binomial(n,k);

$$635013559600$$

Exercise 14.3 Calculate the probability that a 13-card bridge hand contains

(a) 4 hearts, 3 clubs, 3 diamonds, 3 spades

(b) 5 hearts, 4 clubs, 2 diamonds, 2 spades

(c) 6 hearts, 3 clubs, 3 diamonds, 1 spade

Exercise 14.4 A computer operator has to invent a seven character password from the letters of the alphabet and the numbers $0, 1, 2, \ldots, 9$. At least one letter must be a capital letter and exactly one character must be a number. In how many ways can the password by chosen? Another computer operator uses a random character selector to create a password meeting the same conditions; what is the probability that this password will contain at least two capital letters?

Exercise 14.5 In how many ways can 3 flags be flown on 2 flagpoles? What about r flags on n flagpoles?

To calculate the probability of choosing k successes from n trials, we must incorporate the multiplication rule of probabilities.

Example 14.4 If a coin is tossed 4 times, what is the probability that two of the tosses will produce heads?

The probability of obtaining three heads is $P(heads) \times P(heads) \times P(tails) \times P(tails)$, or $P(heads)^2 \times P(odd)^{4-2}$. We include the binomial coefficient to take into account the number of ways in which two heads may be obtained.

```
> n:=4;k:=2;p:=0.5;
```

$$n := 4$$

$$k := 2$$

$$p := .5$$

```
> P:=binomial(n,k)*(1-p)^(n-k)*(p)^(k);
```

$$P := .3750$$

Example 14.5 If a die is tossed five times, what is the probability that at least three of the tosses will produce a one?

In order to solve this problem, we must sum the probabilities that three, four, or five of the five tosses produce a one.

```
> n:=5;p:=1/6;
```

$$n := 5$$

$$p := \frac{1}{6}$$

```
> P:=binomial(n,3)*(1-p)^(n-3)*(p)^(3)+
> binomial(n,4)*(1-p)^(n-4)*(p)^(4)+
> binomial(n,5)*(1-p)^(n-5)*(p)^(5);
```

$$P := \frac{23}{648}$$

```
> evalf(");
```

$$.03549382716$$

Because of the complement rule, this is equivalent to one minus the sum of the probabilities that zero, one, or two of the five tosses produce a one.

```
> P:=1-(binomial(n,2)*(1-p)^(n-2)*(p)^(2)+
> binomial(n,1)*(1-p)^(n-1)*(p)^(1)+
```

```
> binomial(n,0)*(1-p)^(n-0)*(p)^(0));
```
$$P := \frac{23}{648}$$

```
> evalf(");
```
$$.03549382716$$

Exercise 14.6 In a manufacturing process N computer chips are made per day. Of these, on average d are defective and $(N - d)$ are good. A sample of r chips is chosen at random from the production on a given day. What is the probability that this group will contain exactly k defectives where k varies from 0 to the smaller of d and r? For the case $N = 1000, r = 100, d = 10$, plot this probability for $k = 1, 2, \ldots, 10$.

Experiment 14.2 Consider the two functions

$$P(r) = e^{-nq}\frac{(nq)^r}{r!} \text{ and } B(r) = \frac{n!}{(n-r)!r!}q^r(1-q)^{n-r}$$

for $r = 0, 1, \ldots, n$ and $0 < q < 1$. It is claimed that $P(r)$, the Poisson probability distribution, is a good approximation to $B(r)$, the Binomial probability distribution, when q is small and n is large. Investigate this claim for $n = 10, 100, 1000, 10000$ and $q = \frac{1}{2}, \frac{1}{3}, \frac{1}{4}, \ldots$ Both distribution functions can represent the number of occurrences of 'random' events when the average number of occurrences is nq. Construct histograms of the two distributions.

Experiment 14.3 A queue for service at a post office counter has an average waiting time of T seconds. The probability of having to wait at least t seconds is $p(t) = e^{-t/T}$. Plot this for a range of typical T values and animate the graphs. Find the probability of having to wait at least t_1 seconds but no more than t_2 seconds. This probability is a function of t_1 and t_2; use plot3d to illustrate this function and animate it over a range of T values. You might investigate whether other disordered phenomena, such as a slowly dripping tap for example, have negative exponentially distributed intervals between events.

Experiment 14.4 In a meeting of 4 people taken at random, estimate the probability that at least 2 of them will have a birthday on the same day of the year. Assume that the probability of each day in the year is $\frac{1}{365}$ and suppose that the four birthdays are a, b, c, d. Let A be the event that $b \neq a$, B be the event that $c \neq b$ and $c \neq a$, and let C be the event that $d \neq a$, $d \neq b$, and $d \neq c$. Assuming that A, B, and C are independent, then the probability that A and B

and C occur is the product of their probabilities

$$p(A)p(B)p(C) = \frac{364}{365} \times \frac{363}{365} \times \frac{362}{365}$$

The probability we require is $1 - p(A)p(B)p(C)$, the probability that *not* all of a, b, c, d are different. Repeat this for a variable number N of people in the meeting and plot the probability of at least two among N having the same birthday as a function of N. How large must N be in order for the probability of at least two people having the same birthday to be greater than 0.5?

Part II

Beginning Calculus

Chapter 15

Secants and Tangents

Commands used in this chapter

- `assign`
- `factor`
- `plot`
- `plot[options]`
- `simplify`
- `student`
- `student[isolate]`
- `student[showtangent]`
- `student[slope]`
- `subs`
- `with`

15.1 Slope of a Line

The general equation of a line is $y = mx + b$, where m is the slope of the line and b is the y-intercept. For example, find the slope of the line $f(x) = 2x - 1$, shown in Figure 15.1(a).

```
> f:=x->2*x-1;
```
$$f := x \rightarrow 2x - 1$$

```
> plot(f(x),x);
```

Compare $f(x) = 2x - 1$ to the general equation of a line. By observation, the slope of the line is 2; note that the axes are different scaling.

```
> with(student):
```

```
> slope(f(x));
```
$$2$$

Since $f(x)$ is a straight line, the slope between any two points on the line is always 2.

```
> slope([1,f(1)],[2,f(2)]);
```
$$2$$

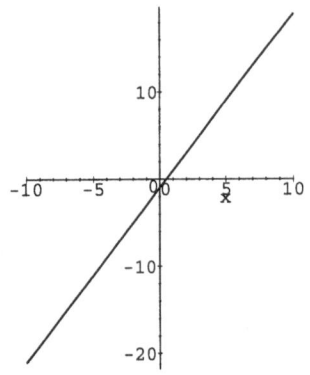 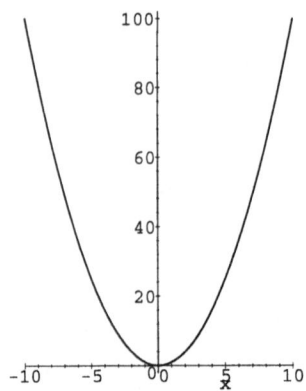

Figure 15.1. (a) The graph of $f(x) = 2x - 1$, on $[-10, 10]$. (b) The graph of $f(x) = x^2$, on $[-10, 10]$

There are two pathological cases for lines: horizontal lines have zero slope; vertical lines have infinite slope.

15.2 Slope of a Secant

A secant is a line which intersects a curve at two points. The slope of a secant is calculated from the points of intersection. Clearly, it gives the average slope of the curve between the chosen two points:

$$\text{average slope} = \frac{f(x_2) - f(x_1)}{x_2 - x_1}$$

Consider the parabola shown in Figure 15.1(b).

```
> f:=x->x^2;
```

$$f := x \to x^2$$

```
> plot(f(x),x);
```

Let's calculate the slope of the secant intersecting the points $(0.5, f(0.5))$ and $(2.5, f(2.5))$. See Figure 15.2(a).

```
> with(student):
> slope([0.5,f(0.5)],[2.5,f(2.5)]);
                    3.000000000
```

```
> plot({[0.5,f(0.5),2.5,f(2.5)],f(x)},x=0..3);
```

Now let's calculate the slope of the secant intersecting the points $(4, f(4))$ and $(5, f(5))$. See Figure 15.2(b).

```
> slope([4,f(4)],[5,f(5)]);
```

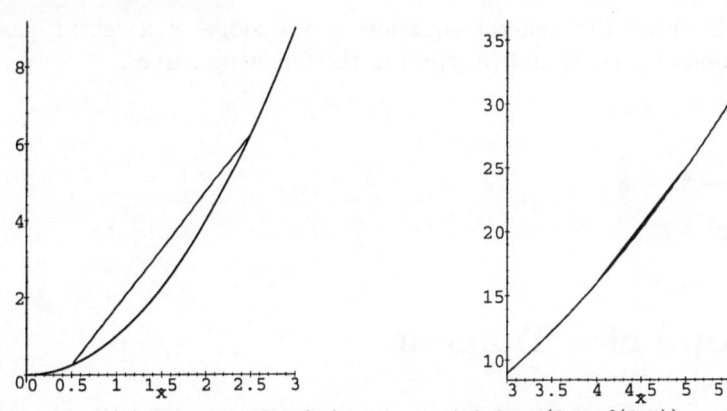

Figure 15.2. (a) The graph of the secant joining $(0.5, f(0.5))$ and $(2.5, f(2.5))$ on $f(x) = x^2$. (b) The graph of the secant joining $(4, f(4))$ and $(5, f(5))$ on $f(x) = x^2$

```
> plot({[4,f(4),5,f(5)],f(x)},x=3..6);
```

If we calculate the slope of a secant to the curve $f(x) = x^2$ at $(a, f(a))$ and $(x, f(x))$ we obtain the general equation for the slope of a secant anywhere on the curve.

```
> slope([x,f(x)],[a,f(a)]);
```

$$\frac{x^2 - a^2}{x - a}$$

```
> factor(");
```

$$x + a$$

In general, the slope, m, of the secant through the points $(x, f(x))$ and $(a, f(a))$ on the parabola $y = x^2$ is $m = x + a$. The same method can be used to determine the general equation for the slope of a secant to any second order function.

Exercise 15.1   Find the slope of a secant to $y = 2x^2 - 3x$ which passes through the following points:

    (a) $x = 1$ and $x = 4$

    (b) $x = 0$ and $x = -1$

    (c) $x = -2$ and $x = 2$

Exercise 15.2 Find the general equation of the slope of a secant passing through the points $(x, f(x))$ and $(a, f(a))$ to the following curves:

  (a) x^2

  (b) $3x^2 - 5x + 8$

  (c) $-7x^2 + 1$

15.3 Slope of a Tangent

A tangent to a point $(x, f(x))$ on a curve is the limiting line for all the secants between the points on both sides of the given point. The slope of a tangent is determined from the limit of the slope of a secant as the point $(a, f(a))$ approaches $(x, f(x))$.

```
> f:=x->x^2;
```
$$f := x \rightarrow x^2$$

```
> with(student):
> slope([x,f(x)],[a,f(a)]);
```
$$\frac{x^2 - a^2}{x - a}$$

```
> factor(");
```
$$x + a$$

```
> subs(a=x,");
```
$$2x$$

The slope of a tangent at the point (x, x^2) on the parabola $y = x^2$ is $2x$. The same method can be used to determine the general equation for the slope of a tangent to any second order function.

```
> f:=x->2*x^2+x;
```
$$f := x \rightarrow 2x^2 + x$$

```
> slope([x,f(x)],[a,f(a)]);
```
$$\frac{2x^2 + x - 2a^2 - a}{x - a}$$

```
> simplify(");
```
$$2x + 1 + 2a$$

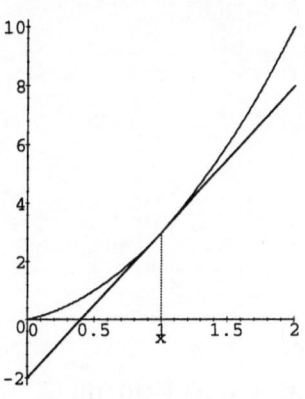 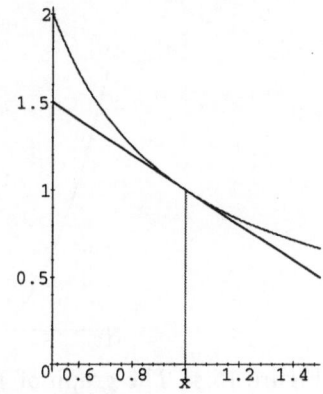

Figure 15.3. (a) The graph of the tangent to $f(x) = 2x^2 + x$ at $x = 1$. (b) The graph of the tangent to $f(x) = \dfrac{1}{x}$ at $x = 1$

```
> subs(a=x,");
```
$$4x + 1$$

The output from the following **showtangent** command is shown in Figure 15.3(a).

```
> showtangent(f(x),x=1,x=0..2);
```

We shall see later that the *derivative* (or derived function) of the function $y(x) = cx^n$ at x is $y'(x) = cnx^{n-1}$. For example, the slope at any point on the function $y(x) = x^2$ is $y'(x) = 2x$. Note also that the slope of the normal line, perpendicular to the tangent at a point, is $-\dfrac{1}{m}$ where m is the slope of the tangent.

Consider the function $f(x) = \dfrac{1}{x}$ and the tangent to the curve at $x = 1$, shown in Figure 15.3(b).

```
> f:=x->1/x;
```
$$f := x \to \frac{1}{x}$$

```
> slope([x,f(x)],[a,f(a)]);
```
$$\frac{\dfrac{1}{x} - \dfrac{1}{a}}{x - a}$$

```
> simplify(");
```
$$-\frac{1}{x\,a}$$

```
> subs(a=x,");
```
$$-\frac{1}{x^2}$$

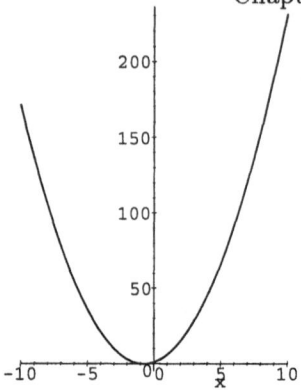

Figure 15.4. The graph of $f(x) = 2x^2 + 3x + 1$, on $[-10, 10]$

```
> showtangent(f(x),x=1,x=0.5..1.5);
```

Exercise 15.3 Find the equation of the tangent at $x = 2$ to the following curves. Find also the equations of the normal lines to these curves.

(a) x^2

(b) $3x^2 - 5x + 8$

(c) $-7x^2 + 1$

Exercise 15.4 Find the point on the graph $y = 3x^2 + 1$ at which the slope is

(a) 2

(b) $\frac{1}{5}$

(c) -4

15.4 Equation of the Tangent to a Curve

Knowing the slope of a tangent, it is a simple matter to determine its equation. For example, find the slope of the tangent at $x = 2$ to the function $f(x) = 2x^2 + 3x + 1$, as shown in Figure 15.4.

```
> f:=x->2*x^2+3*x+1;
```
$$f := x \rightarrow 2x^2 + 3x + 1$$

```
> plot(f(x),x);
> with(student):
> slope([x,f(x)],[a,f(a)]);
```
$$\frac{2x^2 + 3x - 2a^2 - 3a}{x - a}$$

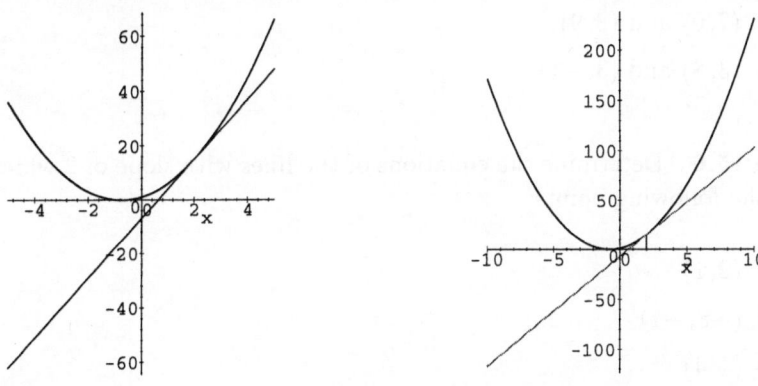

Figure 15.5. The graph of $f(x) = 2x^2 + 3x + 1$ and tangent at $x = 2$, on $[-5, 5]$

```
> subs(a=x,");
```
$$4x + 3$$

```
> m:=subs(x=2,");
```
$$m := 11$$

By definition, the slope of a line is the ratio of pairs of coordinates (x_1, y_1) and (x_2, y_2). Next we substitute the point $(2, f(2))$ into the expression for the slope, and solve for y.

```
> m=(y-f(2))/(x-2);
```
$$11 = \frac{y - 15}{x - 2}$$

```
> isolate(",y);
```
$$y = 11x - 7$$

```
> assign(");
```

Now that we know the equation of the tangent line, we can plot it with the function. This plot and the output from the **showtangent** command are shown in Figure 15.5.

```
> plot({f(x),y},x=-5..5);
> showtangent(f(x),x=2,x=-5..5);
```

Exercise 15.5 Determine the equations of the lines through the following points:

(a) $(0, 1)$ and $(4, 1)$

(b) $(7, 0)$ and $(8, 9)$

(c) $(3, 8)$ and $(3, -1)$

Exercise 15.6 Determine the equations of the lines with slope of 2 which pass through the following points:

(a) $(2, 1)$

(b) $(-8, -1)$

(c) $(4, 4)$

Exercise 15.7 Determine the equations of the lines which pass through the point $(2, 4)$ and have the following slopes. Find also the equations of the lines perpendicular to these.

(a) 3

(b) $\frac{1}{2}$

(c) $-\frac{3}{8}$

Exercise 15.8 The owner of a grocery store finds that he can sell 980 litres of milk at \$1.69/l and 1220 litres of milk each week at \$1.49/l. Assume a linear relationship between selling price and demand. How many litres could he sell weekly at \$1.56/l? (p22, Edwards and Penney [6])

Exercise 15.9 In hilly areas, reception for both television and radio is frequently poor. We consider a situation where an FM transmitter is located behind a hill, and a radio receiver is on the opposite side of the hill. How far from the base of the hill should the radio be located so that its reception is not obstructed? Obviously, the height of the transmitter, compared with the height of the hill is crucial, as is the specific positioning of the base of the transmitter. The radio can receive a clear signal if located far from the hill, provided the signal is strong enough. What is the closest that the radio can be to the hill so that reception is not obstructed? (p18, Goldenberg and Greenwald [7])

We consider an idealized situation: the contour of the hill is a semicircle of radius 1 unit. The transmitter is at the base of the hill and has a height of 2 units. The radio is on the side of the hill opposite the transmitter.

Figure 15.6 illustrates the situation. The position of the tangent line $\overline{TR}$ is the key element, where T is the top of the transmitter and R is the radio receiver. P is the point where the tangent line intersects the circle.

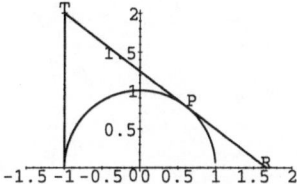

Figure 15.6. Transmitter problem

(a) We wish to find R. In other words, how far from the base of the hill should the radio be placed in order for it to receive an unobstructed signal from the transmitter?

(b) Solve the problem as given in the example if the transmitting tower (of height one unit) is placed on the top of the hill.

(c) Consider the problem as given in the example, but suppose the height of the transmitter is unknown. If it is known that the radio must be placed at least one unit from the base of the hill in order to receive an unobstructed signal, determine the height of the transmitter.

(d) Solve the problem as given in the example if the hill contour is described by the equation $y = x - x^2$.

(e) Solve the problem as given in the example, but suppose the transmitter is located at the point (-1,4).

Experiment 15.1 Draw the graph of $y = x^4$. Choose any two points on the graph and find the equation of the secant joining them. Now find a point on the graph at which the tangent has the same slope as the secant. Such a property also holds for any other polynomial graph; try a more complicated example.

Experiment 15.2 Draw the graph of $y = \mid x \mid$ for $-1 \le x \le 1$. Show that for every pair of points $x_1 < x_2$ there is a well-defined mean slope of the graph $s(x_1, x_2)$. Use `plot3d` to show the functions $s(x_1, x_2)$ and explain what goes wrong at $x_1 = x_2 = 0$.

Chapter 16

Sequences and Limits

Commands used in this chapter
- combinat[fibonacci]
- evalf
- expand
- iscont
- limit
- limit[dirs]
- Limit
- piecewise
- plot
- plot[options]
- plots
- plots[polygonplot]
- read
- readlib
- seq
- simplify
- sin
- sqrt
- student
- student[Limit]
- subs
- sum
- value
- with

16.1   Sequences

As we learned in Chapter 13, a sequence is an ordered set, usually ordered by the set of natural numbers.

Example 16.1   Calculate the first five members of the sequence $f(n) = 2n - 1$ where n is a member of the set of whole numbers.

```
> f:=n->2*n-1;
```
$$f := n \rightarrow 2n - 1$$

```
> seq(f(n),n=0..4);
```
$$-1, 1, 3, 5, 7$$

Example 16.2   Calculate the first five members of the sequence $f(m) = \frac{1}{m}$ where m is a member of the set of natural numbers.

```
> f:=m->1/m;
```
$$f := m \rightarrow \frac{1}{m}$$

```
> seq(f(m),m=1..5);
```

$$1, \frac{1}{2}, \frac{1}{3}, \frac{1}{4}, \frac{1}{5}$$

Exercise 16.1   Find the first six terms of the sequence with the following nth terms:

(a) $f(n) = 2n - 1$, where $n \in \mathsf{N}$

(b) $f(n) = 3^{n+1} + n$, where $n \in \mathsf{N}$

(c) $f(n) = 1 + \frac{7}{n}$, where $n \in \mathsf{N}$

Exercise 16.2   Find the stated terms of sequences defined by the following functions:

(a) 7th term of $f(n) = -5^n$, where $n \in \mathsf{N}$

(b) 11th term of $f(n) = 3n^2$, where $n \in \mathsf{N}$

(c) 100th term of $f(n) = \frac{3}{n^2}$, where $n \in \mathsf{N}$

16.2   Limit of an Infinite Sequence

The *limit*, if it exists, for an infinite sequence is the limit of the nth term, t_n, as n approaches infinity. To test a value for a limit analytically, you have to show that, for any small number $\epsilon > 0$, you can find a number N_ϵ such that t_n is within ϵ of l whenever $n > N_\epsilon$.

Example 16.3   Consider the sequence $f(n) = 1 + \frac{1}{n}$, shown in Figure 16.1.

```
> f:=n->1+1/n;
```

$$f := n \rightarrow 1 + \frac{1}{n}$$

```
> seq(f(n),n=1..10);
```

$$2, \frac{3}{2}, \frac{4}{3}, \frac{5}{4}, \frac{6}{5}, \frac{7}{6}, \frac{8}{7}, \frac{9}{8}, \frac{10}{9}, \frac{11}{10}$$

```
> plot(f(n),n=-infinity..infinity);
```

As n increases without bound, what happens to the value of $f(n)$?

```
> evalf(f(10));
```

$$1.100000000$$

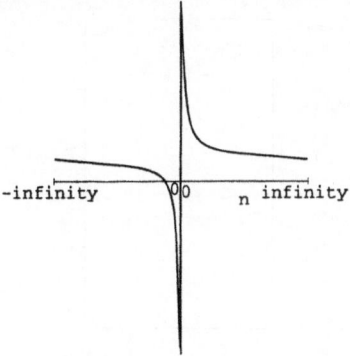

Figure 16.1. The graph of $f(n) = 1 + \dfrac{1}{n}$, on $(-\infty, \infty)$

```
> evalf(f(100));
```
$$1.010000000$$

```
> evalf(f(1000));
```
$$1.001000000$$

As you can see, the limit of this sequence as $n \to \infty$ is 1. What happens to the value of $f(n)$ as n becomes increasingly negative? (These terms are in right order of decreasing value of n).

```
> seq(f(-n),n=1..10);
```
$$0, \frac{1}{2}, \frac{2}{3}, \frac{3}{4}, \frac{4}{5}, \frac{5}{6}, \frac{6}{7}, \frac{7}{8}, \frac{8}{9}, \frac{9}{10}$$

```
> evalf(f(-10));
```
$$.9000000000$$

```
> evalf(f(-100));
```
$$.9900000000$$

```
> evalf(f(-1000));
```
$$.9990000000$$

The limit of this sequence as $n \to -\infty$ is also 1.

Golden Ratio

The Fibonacci Sequence is the sequence of numbers in which each member is equal to the sum of the two previous members, except for the first two members

Figure 16.2. The Golden Rectangle

0 and 1. The first 15 Fibonacci numbers are as follows:

```
> with(combinat, fibonacci):
```
```
> seq(fibonacci(i),i=0..15);
```
$$0, 1, 1, 2, 3, 5, 8, 13, 21, 34, 55, 89, 144, 233, 377, 610$$

The limit of the nth plus one Fibonacci number divided by the nth Fibonacci number results in a quadratic equation with an interesting solution.

```
> Limit(fibonacci(n+1)/fibonacci(n),n=infinity);
```
$$\lim_{n \to \infty} \frac{\text{fibonacci}(n+1)}{\text{fibonacci}(n)}$$

After some algebraic manipulation, the following quadratic results.

```
> alpha^2-alpha-1=0;
```
$$\alpha^2 - \alpha - 1 = 0$$

The positive root of this quadratic is known as the Golden Ratio.

```
> alpha=[solve(alpha^2-alpha-1=0,alpha)];
```
$$\alpha = \left[\frac{1}{2} + \frac{1}{2}\sqrt{5}, \frac{1}{2} - \frac{1}{2}\sqrt{5} \right]$$

```
> assign(");
```
```
> evalf(op(1,alpha));
```
$$1.618033989$$

```
> evalf(1/op(1,alpha));
```
$$.6180339887$$

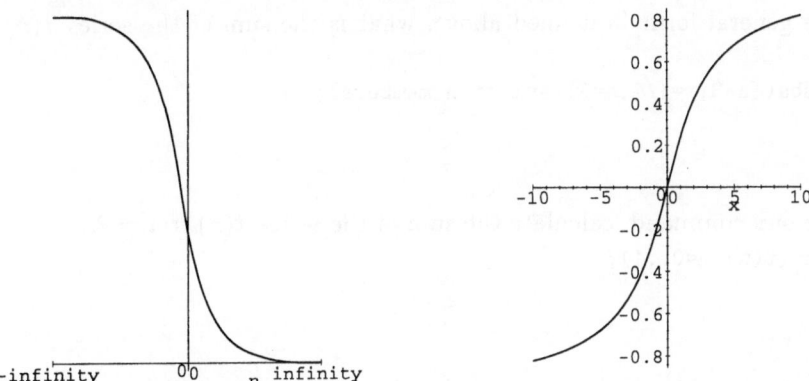

-infinity   0 0   n   infinity

Figure 16.3. (a) The graph of $f(n) = 3(\frac{1}{2})^n$, on $(-\infty, \infty)$. (b) The graph of $f(x) = \dfrac{(4+x^2)^{1/2} - 2}{x}$, on $[-10, 10]$

The numbers after the decimal are the same for α as for $\frac{1}{\alpha}$. The ancient Greeks used this ratio in architecture; it is believed to give pleasing proportions. The rectangle with aspect ratio $1 : \alpha$ is called the Golden Rectangle. See Figure 16.2.

```
> with(plots):

> polygonplot([[0,0],[0,op(1,alpha)],[1/op(1,alpha),op(1,alpha)],
> [1/op(1,alpha),0],[0,0]],scaling=CONSTRAINED,axes=NONE);
```

16.3   Sum of an Infinite Geometric Series

The general formula for the sum of an infinite geometric series of the form $f(n) = ar^{n-1}$ is $s(n) = \frac{a(r^n-1)}{r-1}$.

Example 16.4   Consider the function $f(n) = 3(\frac{1}{2})^n$, shown in Figure 16.3(a).

```
> f:=n->3*(1/2)^n;
```
$$f := n \rightarrow 3\left(\frac{1}{2}\right)^n$$

```
> plot(f(n),n=-infinity..infinity);

> sum_of_n_members:=a*(r^n-1)/(r-1);
```
$$sum\_of\_n\_members := \frac{a\left(r^n - 1\right)}{r - 1}$$

Using the general formula defined above, what is the sum of the series $f(n)$ if $n = 2$?

```
> subs({a=3,r=1/2,n=2},sum_of_n_members);
```
$$\frac{9}{2}$$

Using the **sum** command, calculate the sum of the series $f(n)$ if $n = 2$.

```
> sum(f(n),n=0..1);
```
$$\frac{9}{2}$$

Naturally, the results are the same. Let's calculate the sum of the first 10 members in the series.

```
> sum(f(n),n=0..9);
```
$$\frac{3069}{512}$$

```
> evalf(");
```
$$5.994140625$$

Now calculate the sum of the series from $n = 0$ to $n = \infty$.

```
> sum(f(n),n=0..infinity);
```
$$6$$

Experiment 16.1 Investigate the series
$$s(n) = 1 + \frac{1}{2^k} + \frac{1}{3^k} + \frac{1}{4^k} + \ldots + \frac{1}{n^k} \text{ for } k = 1, 1.1, 1.2, \ldots, 2$$
by plotting $s(n)$ against $n = 1, 2, \ldots, 1000$. It is actually known that, for $k = 2$, $s(n)$ tends to a limit as n tends to infinity; can you find this limit? It is known also that, for $k = 1$, $s(n)$ does *not* tend to a limit as n tends to infinity. What can you discover about the cases $1 < k < 2$?

16.4   Limit of a Function

We shall begin our investigation of the concept of limits with the function $f(x) = \frac{\sqrt{4-x^2}-2}{x}$, shown in Figure 16.3(b). What is the value of $f(x)$ at $x = 0$? By direct substitution we get,

```
> f:=x->((4+x^2)^(1/2)-2)/x;
> plot(f(x),x);
```
$$f := x \rightarrow \frac{\sqrt{4 + x^2} - 2}{x}$$

```
> f(0);
Error, (in f) division by zero
```

Let's look at values of f(x) very close to $x = 0$.
```
> f(1.0);
```
$$.2360679780$$

```
> f(0.1);
```
$$.02498439000$$

```
> f(0.01);
```
$$.002500000000$$

```
> f(0.001);
```
$$.0002500000000$$

It appears that $f(x)$ approaches 0 as x approaches 0. What about negative values of x very close to 0?
```
> f(-1.0);
```
$$-.2360679780$$

```
> f(-0.1);
```
$$-.02498439000$$

```
> f(-0.01);
```
$$-.002500000000$$

```
> f(-0.001);
```
$$-.0002500000000$$

So, $f(x)$ does indeed approach 0 as x approaches 0. We say that the limit $\lim_{x \to 0} f(x)$ is 0. Limits can be calculated directly using `limit`.
```
> limit(f(x),x=0);
```
$$0$$

Exercise 16.3 Try to prove analytically that $\lim_{x \to 0} \dfrac{\sqrt{4 - x^2} - 2}{x} = 0$. Later in this chapter we give a number of rules for limits but your intuition will probably guide you here since all of the rules are statements of what you would expect to be true! Here's a hint: on any exam about limits you can expect to need

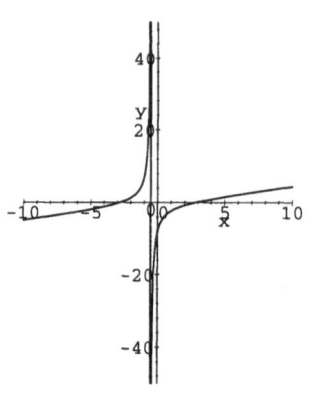
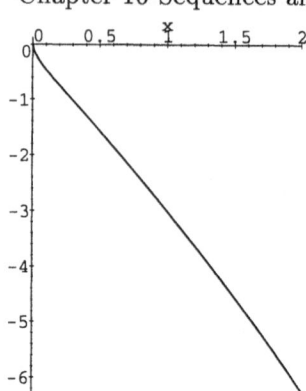

Figure 16.4. (a) The graph of $f(x) = \dfrac{x^2 - 8}{2x + 1}$, on $[-10, 10]$. (b) The graph of $f(x) = \dfrac{x^2 - x^{1/2}}{1 - x^{1/2}}$, on $[0, 2]$

the formula for the difference of two squares. Try multiplying top and bottom of $\frac{\sqrt{4-x^2}-2}{x}$ by $\sqrt{4 - x^2} + 2$ and simplify. Then find the limit of the result as x tends to zero. The multiplication is quite legitimate because as $x \to 0$ so $(\sqrt{4 - x^2} + 2) \to 4$.

Example 16.5 Consider the function $f(x) = \frac{x^2-8}{2x+1}$, shown in Figure 16.4(a). What is the limit $\lim_{x \to -0.5} f(x)$?

```
> f:=x->(x^2-8)/(2*x+1);
```

$$f := x \to \frac{x^2 - 8}{2\,x + 1}$$

```
> plot(f(x),x,y=-50..50);
> f(-0.501);
```
$$3874.499500$$

```
> f(-0.499);
```
$$-3875.499500$$

```
> limit(f(x),x=-1/2);
```
$$\textit{undefined}$$

In order for a limit to exist at $-\frac{1}{2}$, the limit of the function as $x \to -\frac{1}{2}$ from the left must be equal to the limit of the function as $x \to -\frac{1}{2}$ from the right.

```
> limit(f(x),x=-1/2,right);
```
$$-\infty$$

```
> limit(f(x),x=-1/2,left);
```
$$\infty$$

Example 16.6 What is the limit $\lim\limits_{x\to 1} f(x)$ of $f(x) = \frac{x^2-\sqrt{x}}{1-\sqrt{x}}$, shown in Figure 16.4(b)? (Looks like the difference of two squares would help here!)

```
> f:=x->(x^2-sqrt(x))/(1-sqrt(x));
```
$$f := x \to \frac{x^2 - \mathrm{sqrt}(x)}{1 - \mathrm{sqrt}(x)}$$

```
> plot(f(x),x=0..2);
> f(1.001);
```
$$-3.003000750$$

```
> f(0.999);
```
$$-2.996999951$$

```
> limit(f(x),x=1);
```
$$-3$$

To get more insight into this function, try to put it in a simpler form. The argument sqrt(x) in the factor command tells *Maple* that $\sqrt{x}$ is an acceptable factor of x.

```
> factor(f(x),sqrt(x));
```
$$-x^{3/2} - x - \sqrt{x}$$

```
> subs(x=1,");
```
$$-3$$

Example 16.7 Consider the function $f(x) = x\sin(\frac{1}{x})$. What is the limit $\lim\limits_{x\to 0} f(x)$?

```
> f:=x->x*sin(1/x);
```
$$f := x \to x\sin\left(\frac{1}{x}\right)$$

This function has a very interesting graph. Let's first examine the graph over the interval $-10 \le x \le 10$ and then zero in on the origin. See Figures 16.5 and 16.6.

```
> plot(f(x),x);
> plot(f(x),x=-1..1);
```

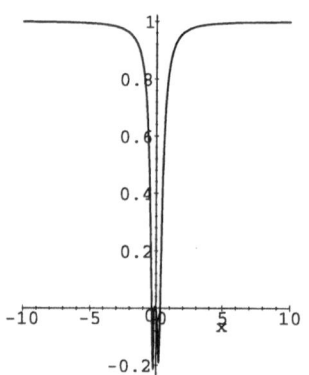 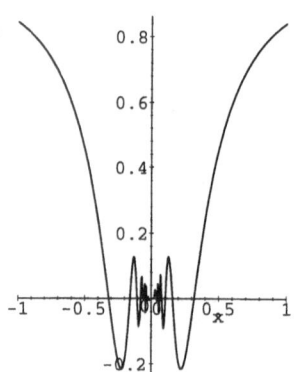

Figure 16.5. (a) The graph of $f(x) = x\sin(\frac{1}{x})$, on $[-10, 10]$. (b) The graph of $f(x) = x\sin(\frac{1}{x})$, on $[-1, 1]$

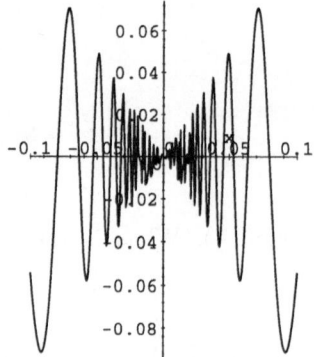 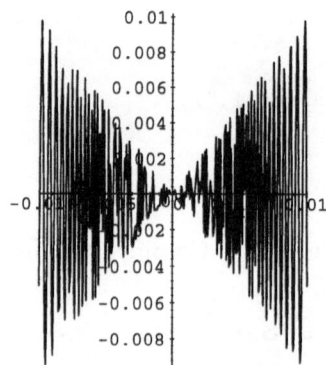

Figure 16.6. (a) The graph of $f(x) = x\sin(\frac{1}{x})$, on $[-0.1, 0.1]$. (b) The graph of $f(x) = x\sin(\frac{1}{x})$, on $[-0.01, 0.01]$

```
> plot(f(x),x=-0.1..0.1);
> plot(f(x),x=-0.01..0.01);
> f(-0.0001);
                        -.00003056143889

> f(0.0001);
                        -.00003056143889

> limit(f(x),x=0);
                        0
```

Exercise 16.4 Find the limiting value, as x increases without bound, of the following functions, or explain why no real limit exists.

(a) $f(x) = \frac{x^2-1}{x^2+1}$

(b) $f(x) = \frac{3+x}{x}$

(c) $f(x) = e^x$

Exercise 16.5 Tabulate the sequence $(1 + \frac{1}{n})^n$ for $n = 1, 2, \ldots$ and estimate its limiting value correct to five decimal places.

Exercise 16.6 Let $x = 1 + z$ and use the Binomial Theorem

$$(1 + z)^n = 1 + nz + \frac{n(n-1)z^2}{2!} + \frac{n(n-1)(n-2)z^3}{3!} + \ldots \text{ for } \mid x \mid < 1$$

to prove that $\lim_{x \to 1} \frac{x^n - 1}{x - 1} = n$.

Exercise 16.7 Use the Binomial Theorem to expand $(1 + \frac{1}{n})^n$ and deduce that

$$\lim_{n \to \infty} (1 + \frac{1}{n})^n = 1 + 1 + \frac{1}{2!} + \frac{1}{3!} + \ldots$$

Experiment 16.2 Let $x_1 = \sqrt{2}$, $x_2 = \sqrt{x_1 + \sqrt{2}}$, $x_3 = \sqrt{x_2 + \sqrt{2}}, \ldots$ Plot this sequence. Now show that for all n we have $2 - x_n = 4 - x_{n+1}^2$ and $0 < x_n < x_{n+1} < 2$. Next consider $(2 - x_n)(2 + x_n) = 2 - x_n$ to show that $x_{n+1} > x_n$ for all n. So what do you expect the limit of x_n to be as n increases?

16.5   Rules for Limits

The most important thing to remember, in trying to simplify functions to find their limits, is that performing *continuous operations* preserves limits: $g($limit of $f(x)) = $ limit of $g(f(x))$ for continuous functions g. We give a precise definition in Section 16.6.

The following set of rules is used to calculate the limit of a function. Try to come up with an additional example for each.

16.5.1   Constants Rule

The limit of a constant times a function equals the constant times the limit of the function.

```
> with(student):
> Limit(c*f(x),x=a);
```

$$\lim_{x \to a} c\, f(\, x\,)$$

```
> expand(");
```

$$c \left(\lim_{x \to a} f(\, x\,) \right)$$

```
> limit(2*x^2,x=1)=2*limit(x^2,x=1);
```

$$2 = 2$$

Also,

```
> Limit(c,x=a);
```

$$\lim_{x \to a} c$$

```
> expand(");
```

$$c$$

16.5.2   Sum Rule

The limit of a sum of functions is equal to the sum of the limits of the functions.

```
> Limit(f(x)+g(x),x=a);
```

$$\lim_{x \to a} f(\, x\,) + g(\, x\,)$$

```
> expand(");
```

$$\left(\lim_{x \to a} f(x)\right) + \left(\lim_{x \to a} g(x)\right)$$

```
> limit(x^2+x,x=1)=limit(x^2,x=1)+limit(x,x=1);
```
$$2 = 2$$

16.5.3   Product Rule

The limit of a product of functions is equal to the product of the limits of the functions.

```
> Limit(f(x)*g(x),x=a);
```

$$\lim_{x \to a} f(x) \, g(x)$$

```
> expand(");
```

$$\left(\lim_{x \to a} f(x)\right) \left(\lim_{x \to a} g(x)\right)$$

```
> limit((x-2)*x^2,x=1)=limit(x-2,x=1)*limit(x^2,x=1);
```
$$-1 = -1$$

16.5.4   Quotient Rule

The limit of a quotient of functions is equal to the quotient of the limits of the functions as long as the limit of the denominator is not zero.

```
> Limit(f(x)/g(x),x=a);
```

$$\lim_{x \to a} \frac{f(x)}{g(x)}$$

```
> expand(");
```

$$\frac{\lim_{x \to a} f(x)}{\lim_{x \to a} g(x)}$$

```
> limit(x/(x^2+1),x=3)=limit(x,x=3)/limit(x^2+1,x=3);
```
$$\frac{3}{10} = \frac{3}{10}$$

16.5.5   Exponential Rule

The limit of an exponential equals the exponential of the limits.

```
> Limit(f(x)^g(x),x=a);
```

$$\lim_{x \to a} f(x)^{g(x)}$$

```
> expand(");
```

$$\left(\lim_{x \to a} f(x) \right)^{(\lim_{x \to a} g(x))}$$

```
> limit((x+2)^x,x=2)=limit(x+2,x=2)^(limit(x,x=2));
```

$$16 = 16$$

16.5.6   Inequality Rule

If $f(x) \le g(x)$, then $\lim_{x \to a} f(x) \le \lim_{x \to a} g(x)$ if the limit exists. See Figure 16.7(a).

```
> f:=x->x^2+3*x-1;
```

$$f := x \to x^2 + 3x - 1$$

```
> g:=x->2*x^2+4*x+1;
```

$$g := x \to 2x^2 + 4x + 1$$

```
> plot({f(x),g(x)},x=-5..5);
> f(a)<g(a);
```

$$a^2 + 3a - 1 < 2a^2 + 4a + 1$$

```
> Limit(f(x),x=a)<Limit(g(x),x=a);
```

$$\lim_{x \to a} x^2 + 3x - 1 < \lim_{x \to a} 2x^2 + 4x + 1$$

```
> limit(f(x),x=1)<limit(g(x),x=1);
```

$$3 < 7$$

```
> limit(f(x),x=10)<limit(g(x),x=10);
```

$$129 < 241$$

16.5.7   Sandwich Rule

If $f(x) \le g(x) \le h(x)$, then $\lim_{x \to a} f(x) \le \lim_{x \to a} g(x) \le \lim_{x \to a} h(x)$ if the limit exists. See Figure 16.7(b).

```
> f:=x->x^2+3*x-1;
```

$$f := x \to x^2 + 3x - 1$$

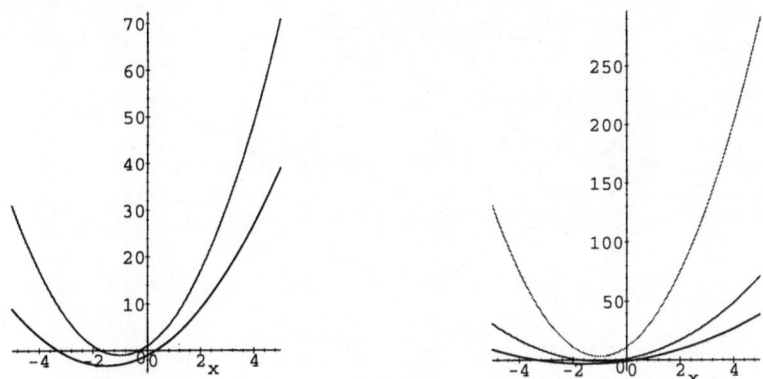

Figure 16.7. (a) Inequality Rule. (b) Sandwich Rule

```
> g:=x->2*x^2+4*x+1;
```
$$g := x \rightarrow 2\,x^2 + 4\,x + 1$$

```
> h:=x->2*(2*x+2)^2+3;
```
$$h := x \rightarrow 2\,(2\,x + 2)^2 + 3$$

```
> plot({f(x),g(x),h(x)},x=-5..5);
> Limit(f(x),x=a)<Limit(g(x),x=a);
```
$$\lim_{x \to a} x^2 + 3\,x - 1 < \lim_{x \to a} 2\,x^2 + 4\,x + 1$$

```
> Limit(g(x),x=a)<Limit(h(x),x=a);
```
$$\lim_{x \to a} 2\,x^2 + 4\,x + 1 < \lim_{x \to a} 2\,(2\,x + 2)^2 + 3$$

```
> limit(f(x),x=1)<limit(g(x),x=1);
```
$$3 < 7$$

```
> limit(g(x),x=1)<limit(h(x),x=1);
```
$$7 < 35$$

16.5.8 Squeeze Rule

If $f(x) \leq g(x) \leq h(x)$ and $\lim_{x \to a} f(x) = \lim_{x \to a} h(x) = L$ then $\lim_{x \to a} g(x) = L$. See Figure 16.8.

```
> f:=x->-3*(x-1)^2;
```
$$f := x \rightarrow -3\,(x - 1)^2$$

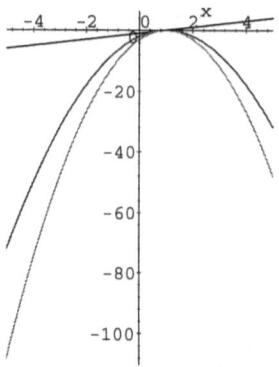

Figure 16.8. Squeeze Rule

```
> g:=x->-2*(x-1)^2;
```
$$g := x \to -2\,(\,x - 1\,)^2$$

```
> h:=x->x-1;
```
$$h := x \to x - 1$$

```
> plot({f(x),g(x),h(x)},x=-5..5);
> Limit(f(x),x=a)<Limit(g(x),x=a);
```
$$\lim_{x \to a} -3\,(\,x - 1\,)^2 < \lim_{x \to a} -2\,(\,x - 1\,)^2$$

```
> Limit(g(x),x=a)<Limit(h(x),x=a);
```
$$\lim_{x \to a} -2\,(\,x - 1\,)^2 < \lim_{x \to a} x - 1$$

```
> limit(f(x),x=1)=limit(h(x),x=1);
```
$$0 = 0$$

```
> limit(g(x),x=1);
```
$$0$$

Experiment 16.3 Let x be a positive real number. For each $n = 1, 2, 3, \ldots$
plot the graphs of $\dfrac{\frac{x}{n}}{1 + \frac{x}{n}}$, $\log(1 + \frac{x}{n})$, and $\frac{x}{n}$. Now, assuming that

$$\frac{\frac{x}{n}}{1 + \frac{x}{n}} \le \log(1 + \frac{x}{n}) \le \frac{x}{n}$$

for all n, multiply through by n to obtain

$$\frac{x}{1 + \frac{x}{n}} \le \log(1 + \frac{x}{n})^n \le x.$$

Plot these three components for $n = 1, 2, 3 \ldots$ and then deduce what will be their limiting value as n increases.

Experiment 16.4   Consider a unit circle, its inscribed square, and its exscribed square. Their respective areas are π, 2, and 4; thus $2 < \pi < 4$. Show that the inscribed regular n-sided polygon has area $\frac{n}{2} \sin(\frac{2\pi}{n})$ and the exscribed regular n-sided polygon has area $n \tan(\frac{\pi}{n})$. Hence

$$\frac{n}{2} \sin(\frac{2\pi}{n}) < \pi < n \tan(\frac{\pi}{n})$$

Archimedes used this method to determined the value π. Plot these components as a function of n for $n = 1, 2, 3, \ldots, 1000$.

16.6   Continuous Functions

Many functions are only defined on part of the real line. It is most important to know where a given function is defined and over what part of its domain it is continuous. To be *continuous* on an interval, a function must

1. be well-defined on the interval, and

2. have left and right limits, which agree, at every point.

We find some examples that violate these two conditions in turn.

Looking at the plot of the function $y = \frac{1}{x}$ in Figure 16.9(a), it is obvious that this function is undefined at $x = 0$, so it violates the first condition.

```
> f:=x->1/x;
```

$$f := x \rightarrow \frac{1}{x}$$

```
> plot(1/x,x=-2..2,y=-10..10);
```

The command `iscont` will tell us if the function is continuous over a given interval.

```
> readlib(iscont):
> iscont(f(x),x=1..2);
```
$$true$$

```
> iscont(f(x),x=-1..1);
```
$$false$$

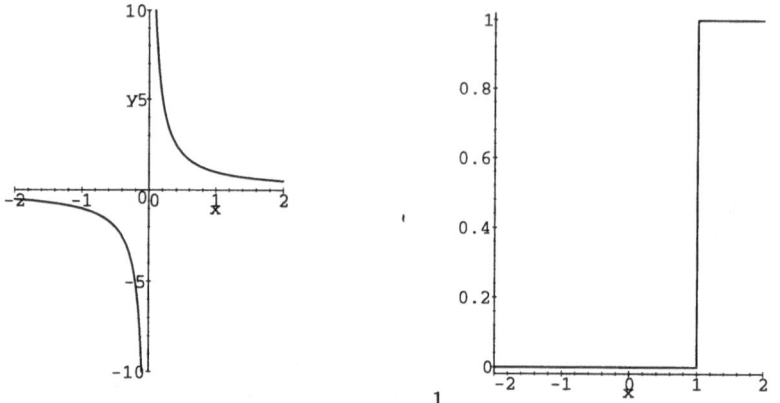

Figure 16.9. (a) The graph of $y = \dfrac{1}{x}$, on $[-2, 2]$. (b) Step function

> `limit(f(x),x=0);`

$$undefined$$

$f(x)$ is not continuous on $[-1, 1]$ since it doesn't exist on all of $[-1, 1]$. However, this function is continuous everywhere it is defined—that is everywhere except $x = 0$.

An example of a function which satisfies the first condition but breaks the second is the *step function* $s(x)$ defined by

$$s(x) = \begin{cases} 0, & \text{if } x \leq 1 \\ 1, & \text{if } x > 1 \end{cases}$$

This function is graphed in Figure 16.9(b); note that the value of $x = 1$ is $s(1) = 0$.

> `readlib(piecewise);`
> `s:=x->piecewise(0,x<=1,0,x>=1,1);`

$$s := x \rightarrow \text{piecewise}(\,0, x \leq 1, 0, 1 \leq x, 1\,)$$

> `plot(s(x),x=-2..2,axes=FRAME);`

Some discontinuous functions are not as easily detected. By studying the function $y = \frac{\sqrt{4+x^2}-2}{x}$, shown in Figure 16.10(a), we observe that at the point $x = 0$ the value of y is undefined. However, the plot of the function appears to be continuous, even though it violates the first condition, so we need to be careful interpreting graphs!

> `f:=x->((4+x^2)^(1/2)-2)/x;`

$$f := x \rightarrow \frac{\sqrt{4+x^2}-2}{x}$$

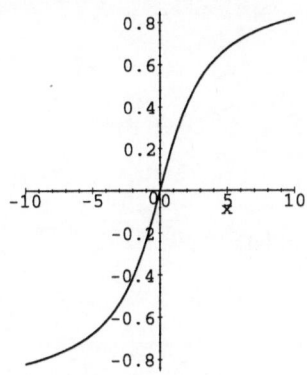 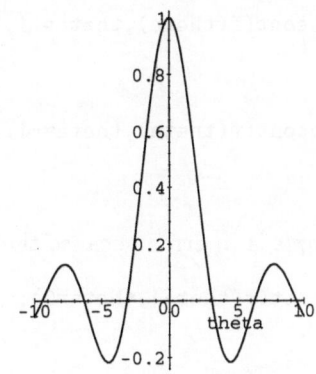

Figure 16.10. (a) The graph of $f(x) = \dfrac{(4 + x^2)^{1/2} - 2}{x}$, on $[-10, 10]$. (b) The graph of $f(\theta) = \dfrac{\sin(\theta)}{\theta}$, on $[-10, 10]$

```
> plot(f(x),x);
> iscont(1/x,x=0..1);
```
$$true$$

```
> iscont(1/x,x=-1..0);
```
$$true$$

```
> iscont(1/x,x=-1..1);
```
$$false$$

Maple is in error in this case because the limit $\lim_{x \to 0} f(x)$ exists.

```
> limit(f(x),x=0);
```
$$0$$

Later, in Chapter 21, we shall see the function $y = \frac{\sin(\theta)}{\theta}$ used in the development of the derivatives of trigonometric functions. This function appears to be discontinuous at $\theta = 0$. See Figure 16.10(b).

```
> f:=theta->sin(theta)/theta;
```
$$f := \theta \to \frac{\sin(\theta)}{\theta}$$

```
> plot(f(theta),theta);
```

In fact, $f(\theta)$ is a continuous function.

```
> iscont(f(theta),theta=0..1);
```
$$true$$

```
> iscont(f(theta),theta=-1..0);
```
$$true$$

```
> iscont(f(theta),theta=-1..1);
```
$$false$$

Again *Maple* is in error because the limit $\lim_{\theta \to 0} f(\theta)$ exists.

```
> limit(f(theta),theta=0);
```
$$1$$

Can you find another function for which `iscont` returns the wrong answer?

Exercise 16.8   Use *Maple's* `series` function to investigate the limits of $\frac{\sin x}{x}$ and $\frac{\tan x}{x}$ as $x \to 0$.

Exercise 16.9   Determine whether the following functions are continuous over the whole real line.

(a) $f(x) = \frac{x^3 + 3x^2 - 9x - 27}{x + 3}$

(b) $f(\theta) = \frac{\cos(\theta)}{\theta}$

(c) $f(x) = \frac{2}{x^2}$

Exercise 16.10   For which values of the real number x is the function $\log(\log x)$ defined?

Before leaving this section, remember the rule $g(\lim f(x)) = \lim g(f(x))$ if g is a continuous function.

16.7   Limits Involving $\frac{0}{0}$

The quotient $\frac{0}{0}$ is said to be *indeterminate*, whereas quotients of the form $\frac{a}{0}$ are called *undefined*.

Example 16.8   Find the limit $\lim_{x \to 0} \dfrac{x + x^2}{3x}$.

By direct substitution of $x = 0$ into the function, we get a limit of $\frac{0}{0}$. However, by reducing the function to a simpler form, we obtain a limit of $\frac{1}{3}$. The graph of

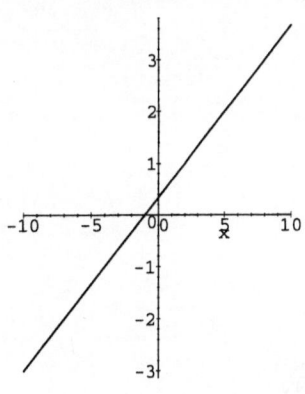 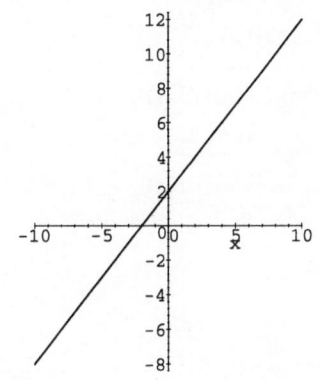

Figure 16.11. (a) The graph of $f(x) = \dfrac{x + x^2}{3x}$, on $[-10, 10]$. (b) The graph of $g(x) = \dfrac{x^2 - 4}{x - 2}$, on $[-10, 10]$

the function is shown in Figure 16.11(a).

```
> f:=x->(x+x^2)/(3*x);
```
$$f := x \rightarrow \frac{1}{3} \frac{x + x^2}{x}$$

```
> limit(f(x),x=0);
```
$$\frac{1}{3}$$

```
> simplify(f(x));
```
$$\frac{1}{3} x + \frac{1}{3}$$

```
> plot(f(x),x);
```

This rule is very simple: divide out polynomial ratios to investigate limits near zero.

Example 16.9 Find the limit $\lim\limits_{x \to 2} \dfrac{x^2 - 4}{x - 2}$. See Figure 16.11(b).

```
> g:=x->(x^2-4)/(x-2);
```
$$g := x \rightarrow \frac{x^2 - 4}{x - 2}$$

```
> limit(g(x),x=2);
```

4

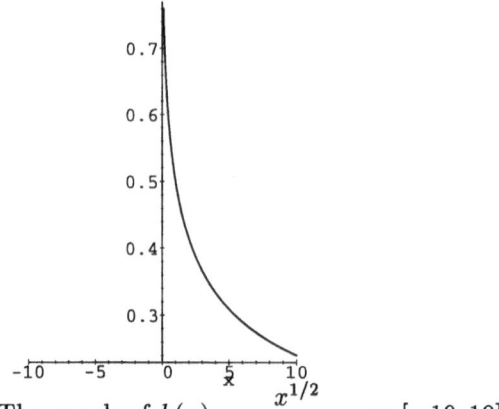

Figure 16.12. The graph of $h(x) = \dfrac{x^{1/2}}{x + x^{1/2}}$, on $[-10, 10]$

This function can be simplified by factoring the numerator and cancelling like terms in the demonator.

```
> simplify(g(x));
```
$$x + 2$$

```
> plot(g(x),x);
```

Example 16.10   The function $h(x) = \frac{\sqrt{x}}{x+\sqrt{x}}$, graphed in Figure 16.12, also has a common factor which can be cancelled.

```
> h:=x->x^(1/2)/(x+x^(1/2));
```
$$h := x \rightarrow \frac{\sqrt{x}}{x + \sqrt{x}}$$

```
> limit(h(x),x=0);
```
$$1$$

```
> factor(h(x),sqrt(x));
```
$$\frac{\sqrt{x} - 1}{-1 + x}$$

```
> plot(h(x),x);
```

Experiment 16.5   The function $f(\theta) = \frac{\sin(\theta)}{\theta}$, from Section 16.6, has an indeterminate quotient at $\theta = 0$. Try to find another function with an indeterminant quotient which cannot be reduced by factoring.

Chapter 17

Derivatives of Functions

Commands used in this chapter

- convert
- convert[D]
- convert[Diff]
- D
- diff
- Diff
- dsolve
- lhs
- limit
- Limit

- plot
- plot[options]
- rhs
- simplify
- student
- student[isolate]
- student[slope]
- subs
- value
- with

17.1 Derivative

The derivative of a function $y = f(x)$ is defined as follows:

```
> with(student):
> Limit(slope([x+delta,f(x+delta)],[x,f(x)]),delta=0);
```

$$\lim_{\delta \to 0} \frac{f(x + \delta) - f(x)}{\delta}$$

Notice that this expression is very similar to the one we used to find the tangent to a function. In fact, the derivative of a function is equal to the slope of its graph at each point along the curve.

17.2 Differentiating from First Principles

Using the formula presented above, we can find, in principle, the derivative of any function; however, as we have seen, some limits are quite hard to find! Later we assemble some rules which can help to avoid the explicit finding of limits for derivatives.

Example 17.1 Find the derivative of the function $f(x) = x^2 + 3x + 1$. The function and its derivative are graphed in Figure 17.1.

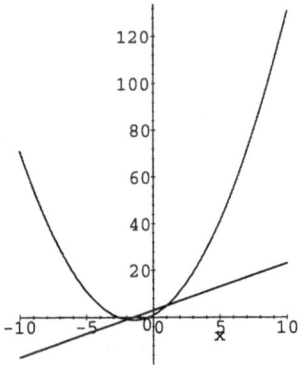

Figure 17.1. The graphs of $f(x) = x^2 + 3x + 1$ and first derivative $f'(x) = 2x + 3$, on $[-10, 10]$

```
> f:=x->x^2+3*x+1;
```
$$f := x \rightarrow x^2 + 3x + 1$$

```
> slope([x+delta,f(x+delta)],[x,f(x)]);
```
$$\frac{(x+\delta)^2 + 3\delta - x^2}{\delta}$$

```
> limit(",delta=0);
```
$$2x + 3$$

```
> plot({f(x),",x},x);
```

Functions are more easily differentiated using the **diff** command.

```
> diff(f(x),x);
```
$$2x + 3$$

Exercise 17.1   Differentiate the following functions from first principles using the concept of limits.

(a) $y = 3x^3 + 2x^2 + 1$

(b) $y = \frac{\sin(x)}{x^2}$

(c) $y = \frac{x+2}{x^2+7}$

Experiment 17.1   Find examples of functions $f(x)$ satisfying the following conditions on the derivative $f'(x)$.

(a) $f'(x) = 3$

(b) $f'(x) = 3x + 4$

(c) $f'(x) = 3x^2 + 4x + 5$

(d) $f'(x) = 1 + x + \frac{x^2}{2!} + \frac{x^3}{3!}$

(e) $f'(x) = 1 - \frac{x^2}{2!} + \frac{x^4}{4!} - + \frac{x^6}{6!}$

(f) $f'(x) = x - \frac{x^3}{3!} + \frac{x^5}{5!} - + \frac{x^7}{7!}$

Note that an equation in terms of the derivative will have many possible solutions for the function; an expression for a derivative defines a *family* of functions. Usually we describe the family by using constants whose values may be chosen arbitrarily. For example, the *general solution* to (a) is the family of functions $f(x) = 3x + a$, where a is an arbitrary constant. Find the other general solutions. What is the general solution of the equation $f'(x) = g(x)$ where $g'(x) = 3$? Try the cases when $g'(x)$ is given by other functions from the list above.

17.3   Rules of Differentiation

The difficulty of finding limits each time we want a derivative becomes very tedious. So we are keen to develop some labour-saving tools.

17.3.1   Derivative of a Constant

The graph of the function $y = c$ is a straight line parallel to the x-axis. Since the slope of $y = c$ is zero at every point, the derivative of a constant is zero.

```
> f:=x->c;
```
$$f := x \to c$$

```
> with(student):
> Limit(slope([x+delta,f(x+delta)],[x,f(x)]),delta=0);
```
$$\lim_{\delta \to 0} 0$$

```
> value(");
```
$$0$$

```
> Diff(c,x)=diff(c,x);
```
$$\frac{\partial}{\partial x} c = 0$$

Chapter 17 Derivatives of Functions

17.3.2   Power Rule

To find the general derivative of functions of the form $f(x) = ax^n$, we must use the concept of limits as before.

```
> f:=x->a*x^n;
```
$$f := x \rightarrow a\, x^n$$

```
> with(student):
> slope([x+delta,f(x+delta)],[x,f(x)]);
```
$$\frac{a\,(x+\delta)^n - a\,x^n}{\delta}$$

```
> Limit(",delta=0);
```
$$\lim_{\delta \to 0} \frac{a\,(x+\delta)^n - a\,x^n}{\delta}$$

```
> value(");
```
$$\frac{a\,x^n\, n}{x}$$

```
> simplify(");
```
$$a\,x^{(n-1)}\, n$$

Using the `diff` command, we arrive at the same result.

```
> Diff(f(x),x)=simplify(diff(f(x),x));
```
$$\frac{\partial}{\partial x}\, a\,x^n = a\,x^{(n-1)}\, n$$

For example, find the first derivative of the function $f(x) = 3x^7$.

```
> f:=x->3*x^7;
```
$$f := x \rightarrow 3\,x^7$$

```
> Diff(f(x),x)=diff(f(x),x);
```
$$\frac{\partial}{\partial x}(3\,x^7) = 21\,x^6$$

17.3.3   Derivative of a Sum of Functions

The derivative of a sum of functions is equal to the sum of the derivatives of the functions.

```
> f:=x->5*x^2;
```
$$f := x \rightarrow 5\,x^2$$

```
> g:=x->3*x+1;
```
$$g := x \rightarrow 3\,x + 1$$

```
> f(x)+g(x);
```
$$5\,x^2 + 3\,x + 1$$

```
> Diff(f(x)+g(x),x)=diff(f(x)+g(x),x);
```
$$\frac{\partial}{\partial x}\left(5\,x^2 + 3\,x + 1\right) = 10\,x + 3$$

```
> Diff(f(x),x)+Diff(g(x),x)=diff(f(x),x)+diff(g(x),x);
```
$$\left(\frac{\partial}{\partial x}\left(5\,x^2\right)\right) + \left(\frac{\partial}{\partial x}\left(3\,x + 1\right)\right) = 10\,x + 3$$

Experiment 17.2   From the formula for the sum of a geometric progression we know that for $\mid x \mid < 1$

$$1 + x + x^2 + x^3 + \ldots = \frac{1}{1 - x}$$

Differentiate both sides to find a new series. Find expressions for the series

(a) $x + 2x^2 + 3x^3 + \ldots$

(b) $1 + x^2 + x^4 + x^6 + \ldots$

(c) $2x + 4x^3 + 6x^5 + \ldots$

(d) $\frac{d}{dx}(1 + x + \frac{x^2}{2!} + \frac{x^3}{3!} + \ldots)$

17.3.4   Chain Rule

For functions of the form $f(x) = (1 + ax)^n$, the chain rule is used to find the derivative.

```
> f:=x->(1+a*x)^n;
```
$$f := x \rightarrow (1 + a\,x)^n$$

```
> Diff(f(x),x)=simplify(diff(f(x),x));
```
$$\frac{\partial}{\partial x}(1 + a\,x)^n = (1 + a\,x)^{(n-1)}\,n\,a$$

An example of such a function is $f(x) = (1 + 2x)^3$, shown in Figure 17.2.

```
> f:=x->(1+2*x)^3;
```
$$f := x \rightarrow (1 + 2\,x)^3$$

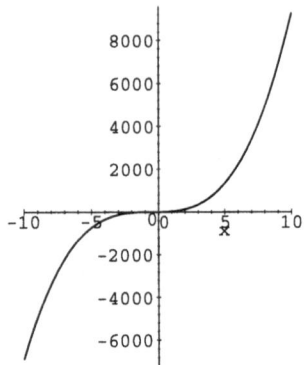

Figure 17.2. The graph of $f(x) = (1 + 2x)^3$, on $[-10, 10]$

```
> plot(f(x),x);
> diff(f(x),x);
```
$$6 (1 + 2x)^2$$

Exercise 17.2 Find the derivatives of the following functions.

(a) $y = x^2 + 2$

(b) $y = \frac{1}{2+3x}$

(c) $y = \frac{6}{x^3}$

17.3.5 Product Rule

The derivative of a product of functions is found using the product rule. The general formula of the product rule is presented below.

```
> 'Diff(f(x)*g(x),x)'=g(x)*Diff(f(x),x)+f(x)*Diff(g(x),x);
```
$$\frac{\partial}{\partial x} f(x)\, g(x) = g(x)\left(\frac{\partial}{\partial x} f(x)\right) + f(x)\left(\frac{\partial}{\partial x} g(x)\right)$$

To illustrate this rule, let's find the derivative of the product of the functions $f(x) = x^2 + 9$ and $g(x) = \sin(x)$, graphed in Figure 17.3.

```
> f:=x->x^2+9;
```
$$f := x \to x^2 + 9$$

```
> g:=x->sin(x);
```
$$g := \sin$$

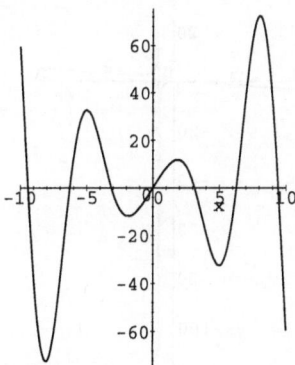

Figure 17.3. The graph of $f(x) = x^2 + 9$ times $g(x) = \sin(x)$, on $[-10, 10]$

```
> 'f(x)*g(x)'=f(x)*g(x);
```
$$f(x)\, g(x) = (x^2 + 9)\sin(x)$$

```
> plot(f(x)*g(x),x);
> 'Diff(f(x)*g(x),x)'=g(x)*Diff(f(x),x)+f(x)*Diff(g(x),x);
```
$$\frac{\partial}{\partial x} f(x)\, g(x) = \sin(x)\left(\frac{\partial}{\partial x}(x^2 + 9)\right) + (x^2 + 9)\left(\frac{\partial}{\partial x}\sin(x)\right)$$

```
> value(rhs("));
```
$$2\sin(x)\, x + (x^2 + 9)\cos(x)$$

Exercise 17.3 Differentiate the product of the functions $f(x)$ and $g(x)$ using the product rule.

(a) $f(x) = (3x - 1)$ and $g(x) = (x^2 + 2)$

(b) $f(x) = \ln(x)$ and $g(x) = 2x^2$

(c) $f(x) = \frac{1}{x}$ and $g(x) = (5x - 6)$

17.3.6   Quotient Rule

The derivative of a quotient of functions is found using the quotient rule. The general formula of the quotient rule is presented below.

```
> 'Diff(f(x)/g(x),x)'=(g(x)*Diff(f(x),x)-f(x)*Diff(g(x),x))/g(x)^2;
```
$$\frac{\partial}{\partial x}\frac{f(x)}{g(x)} = \frac{g(x)\left(\frac{\partial}{\partial x} f(x)\right) - f(x)\left(\frac{\partial}{\partial x} g(x)\right)}{g(x)^2}$$

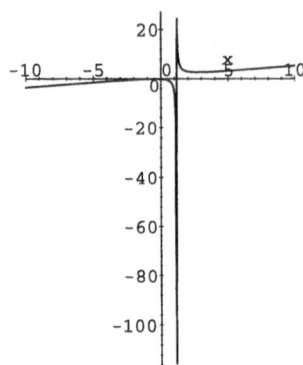

Figure 17.4. The graph of $f(x) = 3x^2 + 2x + 1$ divided by $g(x) = 7x - 8$, on $[-10, 10]$

For example, what is the derivative of the quotient of functions $f(x) = 3x^2 + 2x + 1$ and $g(x) = 7x - 8$, shown in Figure 17.4?

```
> f:=x->3*x^2+2*x+1;
```
$$f := x \rightarrow 3x^2 + 2x + 1$$

```
> g:=x->7*x-8;
```
$$g := x \rightarrow 7x - 8$$

```
> f(x)/g(x);
```
$$\frac{3x^2 + 2x + 1}{7x - 8}$$

```
> plot(f(x)/g(x),x);
> 'Diff(f(x)/g(x),x)'=(g(x)*Diff(f(x),x)-f(x)*Diff(g(x),x))/g(x)^2;
```
$$\frac{\partial}{\partial x}\frac{f(x)}{g(x)} = \left(\right.$$
$$(7x - 8)\left(\frac{\partial}{\partial x}(3x^2 + 2x + 1)\right) - (3x^2 + 2x + 1)\left(\frac{\partial}{\partial x}(7x - 8)\right)\Big)\Big/$$
$$(7x - 8)^2$$

```
> value(rhs("));
```
$$\frac{(7x - 8)(6x + 2) - 21x^2 - 14x - 7}{(7x - 8)^2}$$

```
> simplify(");
```
$$\frac{21x^2 - 48x - 23}{(7x - 8)^2}$$

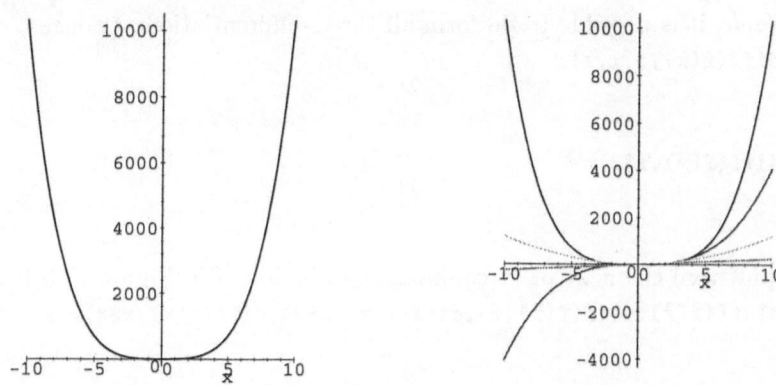

Figure 17.5. (a) The graph of $f(x) = x^4 + 3x^2$, on $[-10, 10]$. (b) The graphs of $f(x) = x^4 + 3x^2$ and its first, second, and third derivatives

Exercise 17.4 Differentiate the quotient of the functions $f(x)$ and $g(x)$ using the quotient rule.

   (a) $f(x) = x^2 - 1$ and $g(x) = 6x - 8$

   (b) $f(x) = \sin(x)$ and $g(x) = x^3 + x$

   (c) $f(x) = 11x + 9$ and $g(x) = e^x$

17.4 Higher Order Derivatives

Find the third derivative of the function $f(x) = x^4 + 3x^2$, graphed in Figure 17.5(a).

```
> f:=x->x^4+3*x^2;
```
$$f := x \rightarrow x^4 + 3x^2$$

```
> plot(f(x),x);
```

One way is to differentiate the function $f(x)$, then differentiate the resulting function $f'(x)$, and so on until we find the third derivative $f'''(x)$.

```
> diff(f(x),x);
```
$$4x^3 + 6x$$

```
> diff(",x);
```
$$12x^2 + 6$$

```
> diff(",x);
```
$$24x$$

Using *Maple*, it is possible to perform all three differentiations at once.
```
> diff(f(x),x,x,x);
```
$$24\,x$$

```
> diff(f(x),x$3);
```
$$24\,x$$

The output from the next `plot` command can be found in Figure 17.5(b).
```
> plot({f(x),diff(f(x),x),diff(f(x),x$2),diff(f(x),x$3)},x);
```

Exercise 17.5   Find the first and second derivatives of the following functions and plot them on the same graph.

 (a) $y = -2x$
 (b) $y = x^3 - 8$
 (c) $y = \frac{9}{x}$

17.5   Notation

The two types of notation for the derivative used in *Maple* are D operator notation and Leibniz notation. The D operator is used to represent differential expressions.
```
> D(f)(x);
```
$$D(\,f\,)(\,x\,)$$

Leibniz notation is used to represent functions.
```
> Diff(f(x),x);
```
$$\frac{\partial}{\partial x}\,\mathrm{f}(\,x\,)$$

Another common notation is $f'(x)$ ('f prime of x'). It is not used in *Maple* except in the Help browser.

The **convert** command allows you to do many things, including convert from one derivative notation to another.
```
> convert(Diff(f(x),x),D);
```
$$D(\,f\,)(\,x\,)$$

```
> convert(D(f)(x),diff);
```
$$\frac{\partial}{\partial x}\,\mathrm{f}(\,x\,)$$

Compare these differential equations, one written in D operator notation and the other in Leibniz notation.

```
> diff(y(x),x) - k*y(x) = 0;
```

$$\left(\frac{\partial}{\partial x} \mathrm{y}(x)\right) - k\,\mathrm{y}(x) = 0$$

```
> D(y)(x)-k*y=0;
```

$$D(y)(x) - k\,y = 0$$

17.6   Implicit Differentiation

The differentiation of expressions which are not explicit in y is called *implicit differentiation*. Typically, we differentiate both sides of an equation then rearrange to isolate the derivative we want.

```
> with(student):
> x*y=y+5;
```

$$x\,y = y + 5$$

```
> solve(",{y});
```

$$\left\{y = 5\,\frac{1}{x-1}\right\}$$

First, we shall solve this implicit expression using Leibniz notation. Remember that you must specify that y is a function of x.

```
> diff(x*y(x)=y(x)+5,x);
```

$$\mathrm{y}(x) + x\left(\frac{\partial}{\partial x}\mathrm{y}(x)\right) = \frac{\partial}{\partial x}\mathrm{y}(x)$$

Now we isolate $\frac{\partial}{\partial}\mathrm{y}(x)$ on the left hand side of the equation. (Within the `isolate` command, the symbol '$\frac{\partial}{\partial x}\mathrm{y}(x)$' is referred to as `diff(y(x),x)`.)

```
> isolate(",diff(y(x),x));
```

$$\frac{\partial}{\partial x}\mathrm{y}(x) = -\frac{\mathrm{y}(x)}{x-1}$$

To solve for $\frac{\partial}{\partial x}\mathrm{y}(x)$ explicitly, we must substitute the value of $y(x)$ into the right hand side of the equation.

```
> lhs(")=subs(y(x)=5/(x-1),rhs("));
```

$$\frac{\partial}{\partial x}\mathrm{y}(x) = -5\,\frac{1}{(x-1)^2}$$

If we compare this result with the result obtained by solving for $y(x)$ in the first place and differentiating that, we notice that they are exactly the same.

> diff(y(x)=5/(x-1),x);

$$\frac{\partial}{\partial x} y(x) = -5 \frac{1}{(x-1)^2}$$

If you can isolate the dependent variable easily, do so. Otherwise, use implicit differentiation.

Exercise 17.6 Differentiate the following equations implicitly and solve for $y(x)$.

(a) $xy(x) = 6$

(b) $xy(x)^2 = 12$

(c) $x^2 y(x) = 18$

Exercise 17.7 By now, pictures of astronauts and their equipment floating with their space vehicle are familiar. Here the concept of weight and weightlessness are quite visibly demonstrated. We know that an object at a distance from the earth has less weight than if measured on the earth's surface. But what of the rate of change of weight, as an object rises? Let's consider a rocket carrying a satellite of known weight on earth. To determine the rate of change of the weight of an object, we need to know the force, F, between the earth and the satellite as a function of the distance, r, between them. This is described by $F(r) = \frac{K}{r^2}$ where K is a constant. Using the Chain Rule $\frac{dF}{dt} = \frac{dF}{dr} \cdot \frac{dr}{dt}$ and calculating $\frac{dF}{dr}$ and $\frac{dr}{dt}$, we can determine $\frac{dF}{dt}$, the quantity we wish to find. (p22, Goldenberg and Greenwald [7])

(a) A satellite weighs 10 kg on the surface of the earth. A rocket carrying the satellite is leaving the earth at a rate of 2000 km/hr. Assume that the radius of the earth is approximately 6373 km and that the force describing the rocket's weight varies inversely as the square of the distance from the centre of the earth. Determine the rate of change of the weight of the satellite with respect ot time, when the satellite is 60 km above the surface of the earth.

(b) Rework the example with the position of the satellite given by $r = -16t^2 + 21120000$ m. Find $\frac{dF}{dt}$ when $t = 1$.

(c) The satellite in the example weighs about 1.66 kg on the surface of the moon. If the radius of the moon is approximately 1733 km, and if the speed of the rocket is 2000 km/hour, find the rate of change of the weight of the satellite when the satellite is 60 km above the surface of the moon.

(d) The rocket in the example is rising at a constant velocity. If the change in the weight of the satellite is given by $\frac{dF}{dt} = -10$ kg/hr determine the rate at which the rocket is rising when the rocket is 160 km above the surface of the earth.

(e) Consider the satellite in the example except assume that the weight of the satellite is unknown and the change in the weight of the satellite is given by $\frac{dF}{dt} = -20$ kg/hr when the satellite is 60 km above the surface of the earth, determine the weight of the satellite on the surface of the earth.

Experiment 17.3   Plot the ellipse $x^2 + 4y^2 = 1$ and on it show the equation of the tangent lines at points (x, y) where $\tan \frac{y}{x} = \frac{\pi}{6}, \frac{\pi}{4}, \frac{\pi}{3}, \frac{\pi}{2}$.

Chapter 18

Functions and Graphs

Commands used in this chapter
- denom
- diff
- divide
- evalf
- limit
- numer
- plot
- plot[options]
- plots
- plots[implicitplot]
- quo
- solve
- subs
- with

18.1 Plotting Functions

With the help of *Maple,* or another computer algebra system or plotting package, it is very easy to plot any function without knowing anything about its shape. But there is a lot you can learn about a function by performing a few mathematical manipulations. In this chapter, we shall be studying the function $f(x) = 4x^3 - 6x^2 + 1$. It will not be plotted until the end of the chapter.

```
> f:=x->4*x^3-6*x^2+1;
```
$$f := x \rightarrow 4\,x^3 - 6\,x^2 + 1$$

18.2 X- and Y-Intercepts

The *y*-intercept of a function is the point at which $x = 0$.

```
> f(0);
```

The x-intercept of a function is the set of all points at which $f(x) = 0$. Because $f(x)$ is a third order function it has three roots and can have a maximum of three x-intercepts. If all of the roots of $f(x)$ are real, then the graph of $f(x)$ will have three x-intercept points. But, for example, if two of the roots are complex, then the graph will have only one x-intercept point.

```
> solve(f(x),{x});
```

$$\left\{x = \frac{1}{2}\right\}, \left\{x = \frac{1}{2} + \frac{1}{2}\sqrt{3}\right\}, \left\{x = \frac{1}{2} - \frac{1}{2}\sqrt{3}\right\}$$

Exercise 18.1   Determine the $x-$ and $y-$intercepts for the following functions.

(a) $y = 2x^2 - 9$

(b) $y = -x^3$

(c) $y = \frac{x+11}{6x^2+5x+1}$

18.3   Asymptotes

18.3.1   Vertical Asymptotes

The vertical asymptote of a function is the line $x = a$ at which the limit of $f(x)$ approaches infinity as x approaches a, or $\lim_{x \to a} f(x) = \infty$.

```
> solve(limit(f(x),x=a)=infinity,{a});
```

$$\left\{a = \%1^{1/3} + \frac{1}{4}\frac{1}{\%1^{1/3}} + \frac{1}{2}\right\},$$

$$\left\{a = -\frac{1}{2}\%1^{1/3} - \frac{1}{8}\frac{1}{\%1^{1/3}} + \frac{1}{2} + \frac{1}{2}I\sqrt{3}\left(\%1^{1/3} - \frac{1}{4}\frac{1}{\%1^{1/3}}\right)\right\},$$

$$\left\{a = -\frac{1}{2}\%1^{1/3} - \frac{1}{8}\frac{1}{\%1^{1/3}} + \frac{1}{2} - \frac{1}{2}I\sqrt{3}\left(\%1^{1/3} - \frac{1}{4}\frac{1}{\%1^{1/3}}\right)\right\}$$

$$\%1 := \frac{1}{8}\infty + \frac{1}{8}\sqrt{-1+\infty^2}$$

```
> limit(f(x),x=infinity);
```

$$\infty$$

What the above output tells us is that the limit $\lim_{x \to \infty} f(a) = \infty$. In other words, it has no vertical asymptote. In contrast, the function $g(x) = \frac{1}{x^2}$, shown in Figure 18.1(a), does have a vertical asymptote at $y = 0$.

```
> g:=x->1/x^2;
```

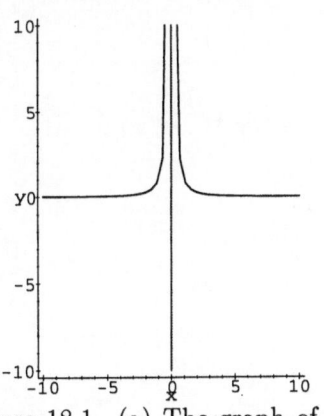
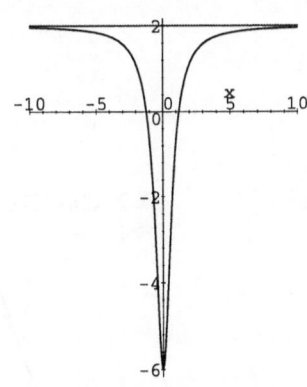

Figure 18.1. (a) The graph of $g(x) = \frac{1}{x^2}$ and its vertical asymptote $x = 0$, on $[-10, 10]$. (b) The graph of $g(x) = \dfrac{4x^2 - 6}{2x^2 + 1}$ and its horizontal asymptote $y = 2$, on $[-10, 10]$

$$g := x \rightarrow \frac{1}{x^2}$$

```
> limit(g(x),x=0);
```
$$\infty$$

```
> with(plots):
> implicitplot({x=0,y=1/x^2},x=-10..10,y=-10..10,numpoints=2000,
> axes=FRAME);
```

18.3.2   Horizontal Asymptotes

A horizontal asymptote of a function is a horizontal limit line approached as $x \rightarrow \infty$, or $\lim_{x \to \infty} f(x) = a$.

```
> limit(f(x),x=infinity);
```
$$\infty$$

Our function $f(x)$ has no horizontal asymptote. The function $g(x) = \frac{4x^2-6}{2x^2+1}$, shown in Figure 18.1(b), has a horizontal asymptote at $y = 2$.

```
> g:=x->(4*x^2-6)/(2*x^2+1);
```
$$g := x \rightarrow \frac{4\,x^2 - 6}{2\,x^2 + 1}$$

```
> limit(g(x),x=infinity);
```
$$2$$

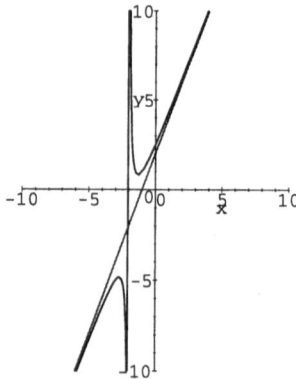

Figure 18.2. The graph of $g(x) = \dfrac{2x^2 + 6x + 5}{x + 2}$ and its oblique asymptote $y = 2x + 2$, on $[-1, 10]$

```
> plot({g(x),2},x,);
```

18.3.3 Oblique Asymptotes

An oblique asymptote of a function is a line which is neither vertical nor horizontal but to which the limit of the function approaches as $x \to \infty$, or $\lim\limits_{x \to \infty} f(x) = mx + b$.

```
> limit(f(x),x=infinity);
```
$$\infty$$

Clearly, as $x \to \infty$, $f(x) \to \infty$, therefore $f(x)$ has no oblique asymptote.

The function $g(x) = \frac{2x^2+6x+5}{x+2}$, shown in Figure 18.2, has an oblique aymptote at $y = 2x + 2$.

```
> g:=x->(2*x^2+6*x+5)/(x+2);
```
$$g := x \to \frac{2\,x^2 + 6\,x + 5}{x + 2}$$

First we divide through by $x + 2$.

```
> divide(numer(g(x)),denom(g(x)));
```
$$false$$

```
> quotient:=quo(numer(g(x)),denom(g(x)),x,'remain');
```
$$quotient := 2\,x + 2$$

```
> remain;
```

$$1$$

```
> g(x):=quotient+remain/denom(g(x));
```

$$g(x) := 2x + 2 + \frac{1}{x+2}$$

We can see that as x increases, the term $\frac{1}{x+2}$ will become negligible compared to $2x+2$. Therefore, the function $g(x)$ has as oblique asymptote the line $y = 2x+2$.

```
> oblique_asymptote:=2*x+2;
```

$$oblique\_asymptote := 2x + 2$$

```
> plot({g(x),oblique_asymptote},x=-10..10,y=-10..10);
```

Exercise 18.2 Find the asymptotes of the following functions.

(a) $f(x) = \frac{x}{\sqrt{4x-1}}$

(b) $f(x) = \frac{3x^2+5x-6}{x+2}$

(c) $f(x) = \frac{x^2}{7-x}$

18.4 Symmetry

A function is said to be an even function if it is symmetric about the y-axis, that is $f(-x) = f(x)$. The even powers of x are even functions of x.

```
> f(-x);
```

$$-4x^3 - 6x^2 + 1$$

Since $f(-x)$ is not equal to $f(x)$, the function is not even. The function $g(x) = -2x^2 + 3$, shown in Figure 18.3(a), is an even function.

```
> g:=x->-2*x^2+3;
```

$$g := x \rightarrow -2x^2 + 3$$

```
> g(-x);
```

$$-2x^2 + 3$$

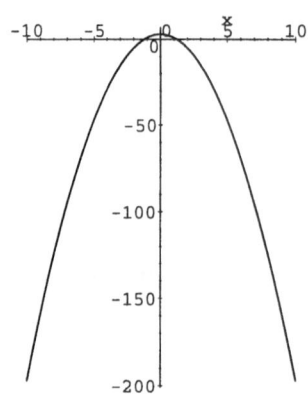 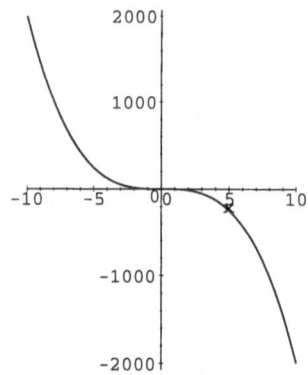

Figure 18.3. (a) The graph of even function $g(x) = -2x^2 + 3$, on $[-10, 10]$. (b) The graph of odd function $g(x) = -2x^3 + x$, on $[-10, 10]$

```
> plot(g(x),x);
```

A function is said to be an odd function if it is symmetric about the origin, that is $f(-x) = -f(x)$. So, the odd powers of x are odd functions of x.

```
> f(-x);
```

$$-4\,x^3 - 6\,x^2 + 1$$

Since $f(-x)$ is not equal to $-f(x)$, the function is not odd. The function $g(x) = -2x^3 + x$, shown in Figure 18.3(b), is an odd function.

```
> g:=x->-2*x^3+x;
```

$$g := x \rightarrow -2\,x^3 + x$$

```
> g(-x);
```

$$2\,x^3 - x$$

```
> plot(g(x),x);
```

Experiment 18.1   Show that every function is a sum of an odd part and an even part.

Exercise 18.3   Determine whether the following functions are even, odd, or neither.

(a) $y = 3x^4 - x^2 + 2$

(b) $y = \frac{x^2-4}{x+2}$

(c) $y = \frac{x^3-x}{3}$

(d) $y = \sin(x)$

18.5   Increasing and Decreasing Functions

A function is increasing on an interval if $x < y$ implies $f(x) < f(y)$ everywhere on the interval. This property can be detected by finding a positive derivative, if it exists.

```
> diff(f(x),x);
```
$$12\,x^2 - 12\,x$$

```
> solve(">0);
```
$$\{\,x < 0\,\},\{\,1 < x\,\}$$

A function is decreasing on an interval if $x < y$ implies $f(x) > f(y)$ everywhere on the interval. This property can be detected by finding a negative derivative, if it exists.

```
> solve("'<0);
```
$$\{\,0 < x, x < 1\,\}$$

18.6   Concavity

In the interval where $f''(x) > 0$, the function is concave (Figure 18.4(a)).

```
> diff(f(x),x$2);
```
$$24\,x - 12$$

```
> solve(">0);
```
$$\left\{\frac{1}{2} < x\right\}$$

In the interval where $f''(x) < 0$, the function is convex (Figure 18.4(b)).

```
> solve("'<0);
```
$$\left\{x < \frac{1}{2}\right\}$$

Exercise 18.4   Calculate the intervals over which the following functions are increasing, decreasing, concave, and convex.

(a) $y = x^3 - 2x$

(b) $y = x^4 + 8x$

(c) $y = \frac{1}{x} + x$

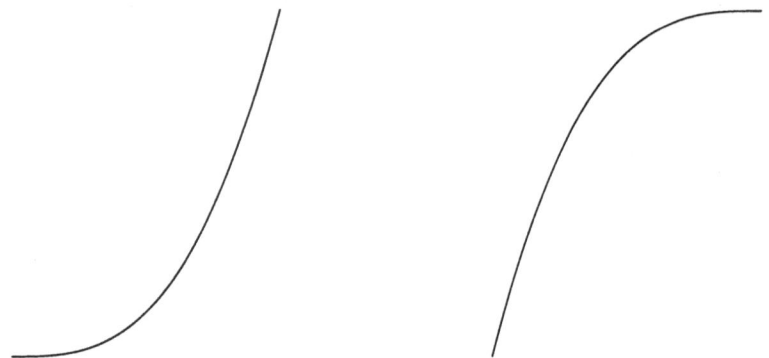

Figure 18.4. (a) Example of a concave function. (b) Example of a convex (or concave down) function

18.7   Relative Maxima and Minima

The points at which $f'(a) = 0$ are relative, or local, extrema.
> diff(f(x),x);
$$12\,x^2 - 12\,x$$

> solve(diff(f(x),x)=0,{x});
$$\{\,x = 0\,\},\{\,x = 1\,\}$$

If $f''(a) < 0$, the point $x = a$ is a relative maximum. If $f''(a) > 0$, the point $x = a$ is a relative minimum.
> diff(f(x),x$2);
$$24\,x - 12$$

The point $(0, f(0))$ is a local maximum.
> subs(x=0,diff(f(x),x$2));
$$-12$$

The point $(1, f(1))$ is a local minimum.
> subs(x=1,diff(f(x),x$2));
$$12$$

18.8   Inflection Point

The point at which $f''(a) = 0$ is an inflection point if $f'(a) \neq 0$.

```
> solve(diff(f(x),x$2)=0,{x});
```
$$\left\{ x = \frac{1}{2} \right\}$$

```
> diff(f(x),x);
```
$$12\,x^2 - 12\,x$$

```
> subs(x=1/2,");
```
$$-3$$

Therefore, the function $f(x)$ has an inflection point at $(\frac{1}{2}, f(\frac{1}{2}))$.

Exercise 18.5   Find the local extrema of the following functions and determine if each is a maximum, minimum, or inflection point.

(a) $y = -5x^2 + 8x - 1$

(b) $y = x^3 - 7x$

(c) $y = 2x^4 - x^3 + 6x^2 + 9$

18.9   Plot of $f(x)$

To summarize, we know that the function $f(x) = 4x^3 - 6x^2 + 1$

- has a y-intercept at $y = 1$

- has x-intercepts at $x = \frac{1}{2}$, $x = \frac{1}{2} + \frac{1}{2}\sqrt{3}$, and $x = \frac{1}{2} - \frac{1}{2}\sqrt{3}$

- has no asymptotes

- is neither even nor odd

- is increasing for $x < 0$ and $x > 1$

- is decreasing for $0 < x < 1$

- is concave for $x > \frac{1}{2}$

- is convex for $x < \frac{1}{2}$

- has a maximum at $(0, f(0))$

- has a minimum at $(1, f(1))$

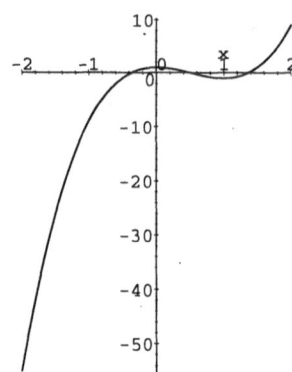

Figure 18.5. The graph of $f(x) = 4x^3 - 6x^2 + 1$, on $[-2, 2]$

- has an inflection point at $\left(\frac{1}{2}, f\left(\frac{1}{2}\right)\right)$

The plot of $f(x) = 4x^3 - 6x^2 + 1$ is shown in Figure 18.5.

```
> plot(f(x),x=-2..2);
```

Chapter 19

Rates

Commands used in this chapter
- abs
- assign
- diff
- Diff
- int
- Int
- lhs
- plot

- plot [options]
- rhs
- student
- student [isolate]
- student [slope]
- subs
- with

19.1   Position, Velocity, and Acceleration

The velocity of a particle is the rate of change of its position; the rate of change of its velocity is its acceleration.

Differentiating Position

Suppose the function $s(t) = t^4$, graphed in Figure 19.1(a), describes the position of an object with time.

```
> s:=t->t^4;
```
$$ s := t \rightarrow t^4 $$

```
> plot(s(t),t=0..5);
```

Then, the velocity of the object is the first derivative of position, $s(t)$, with respect to time. See Figure 19.1(b)

```
> Diff(s(t),t)=diff(s(t),t);
```
$$ \frac{\partial}{\partial t} t^4 = 4\,t^3 $$

```
> plot({s(t),diff(s(t),t)},t=0..5);
```

Acceleration is the derivative of velocity, $v(t)$, with repect to time, or the second derivative of position with respect to time. See Figure 19.2.

```
> Diff(s(t),t$2)=diff(s(t),t$2);
```

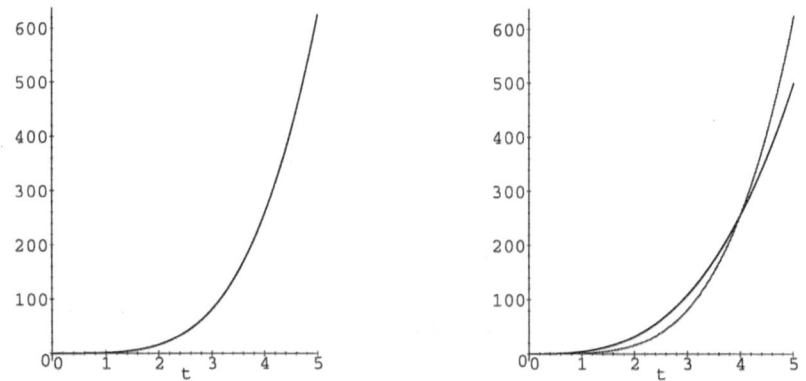

Figure 19.1. (a) The graph of $s(t) = t^4$, on $[0, 5]$. (b) The graphs of $s(t) = t^4$ and $v(t) = 4t^3$, on $[0, 5]$

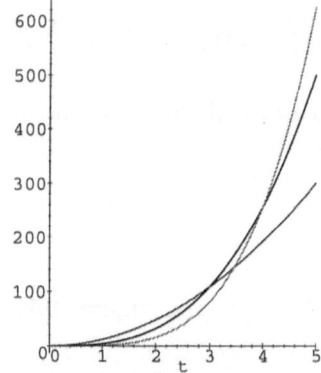

Figure 19.2. The graphs of $s(t) = t^4$, $v(t) = 4t^3$, and $a(t) = 12t^2$, on $[0, 5]$

$$\frac{\partial^2}{\partial t^2} t^4 = 12\, t^2$$

```
> plot({s(t),diff(s(t),t),diff(s(t),t$2)},t=0..5);
```

A useful hint for remembering how position, velocity and acceleration are related is to look at the units of each. In SI, position is measured in metres. The Leibniz notation for the derivative of position with respect to time is read 'ds/dt'. Substituting the units for each we get m/s, which are the units for velocity. Similarly, acceleration is the measure of change in the velocity with time. It is the derivative of velocity with respect to time, 'dv/dt', and has units of m/s/s or m/s^2 or ms^{-2}.

Example 19.1 A body falling under gravity, neglecting air resistance, has position given by

$$s(t) = \frac{1}{2}gt^2 + v_0 t + s_0$$

What is the position at $t = 0$?

```
> s:=t->1/2*g*t^2+v0*t+s0;
```

$$s := t \rightarrow \frac{1}{2}g\,t^2 + v0\,t + s0$$

```
> s(0);
```

$$s0$$

Show that the acceleration is constant and that its value is g.

```
> diff(s(t),t,t);
```

$$g$$

Given that $g = -9.8$ m/s^2, estimate the velocity with which an apple may have hit Newton's head if it fell from a branch 4 m above the ground. If he heard it break off the branch, how long did he have to move out of the way?

```
> g:=-9.8;v0:=0;s0:=0;
```

$$g := -9.8$$

$$v0 := 0$$

$$s0 := 0$$

```
> solve(s(t)=-4,t);
```

$$-.9035079028, .9035079028$$

```
> v:=diff(s(t),t);
```

$$v := -9.800000000\,t$$

```
> subs(t=0.904,v);
```

$$-8.859200000$$

Antidifferentiating Acceleration

Suppose the function $a(t) = t^3 + 3t + 2$, graphed in Figure 19.3(a), describes the acceleration of an object with time.

```
> a:=t->t^3+3*t+2;
```

$$a := t \rightarrow t^3 + 3\,t + 2$$

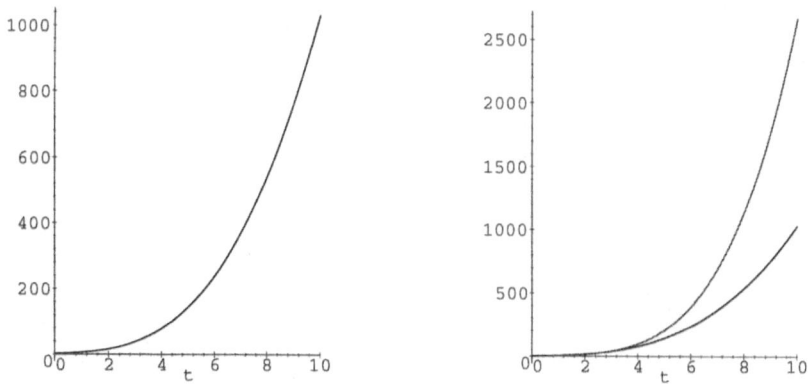

Figure 19.3. (a) The graph of $a(t) = t^3 + 3t + 2$, on $[0, 10]$. (b) The graphs of $a(t) = t^3 + 3t + 2$ and $v(t) = \frac{1}{4}t^4 + \frac{3}{2}t^2 + 2t$, on $[0, 10]$

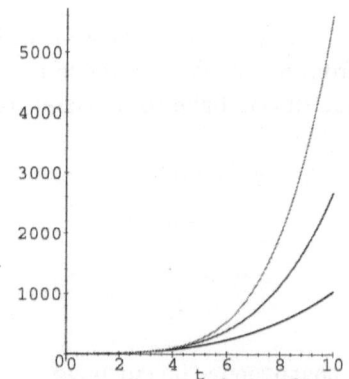

Figure 19.4. The graphs of $t^3 + 3t + 2$, $\frac{1}{4}t^4 + \frac{3}{2}t^2 + 2t$, and $\frac{1}{20}t^5 + \frac{1}{2}t^3 + t^2$ on $[0, 10]$

```
> plot(a(t),t=0..10);
```

Then the velocity of the object is the *integral* of $a(t)$ with respect to time. See Figure 19.3(b). We study integration in detail in Chapter 20.

```
> Int(a(t),t)=int(a(t),t);
```

$$\int t^3 + 3t + 2\, dt = \frac{1}{4}t^4 + \frac{3}{2}t^2 + 2t$$

```
> plot({a(t),int(a(t),t)},t=0..10);
```

Position is the integral of velocity with respect to time, or the double integral of acceleration, $a(t)$, with respect to time. See Figure 19.4.

```
> Int(rhs(""),t)=int(rhs(""),t);
```

$$\int \frac{1}{4}t^4 + \frac{3}{2}t^2 + 2t\, dt = \frac{1}{20}t^5 + \frac{1}{2}t^3 + t^2$$

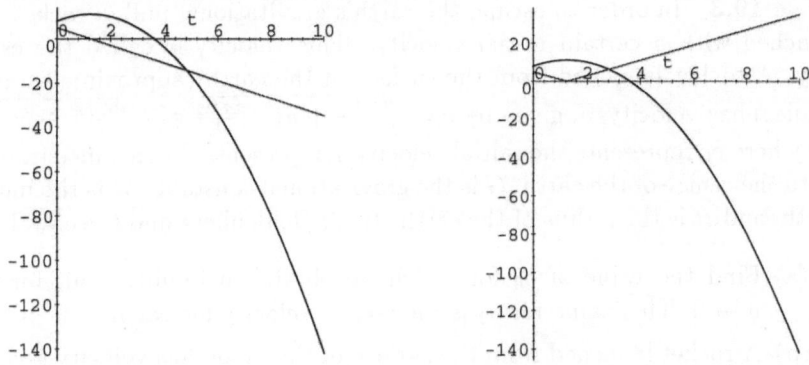

Figure 19.5. (a) Position and velocity. (b) Position and speed

```
> plot({a(t),int(a(t),t),int(int(a(t),t),t)},t=0..10);
```

Velocity and Speed

Speed is a measure of the distance travelled during a given time period, whereas velocity is a measure of the change in postition during a given time period. In other words, speed is a scalar and velocity is a vector. Negative velocity indicates a change of position in the negative direction; it is not possible to have negative speeds. Compare the plots in Figure 19.5 to get an idea of the difference between velocity and speed.

```
> s:=t->-2*t^2+5*t+8;
```

$$s := t \rightarrow -2t^2 + 5t + 8$$

```
> plot({s(t),diff(s(t),t)},t=0..10);
> plot({s(t),abs(diff(s(t),t))},t=0..10);
```

Exercise 19.1   A vehicle travelling at 100 km/h applies its breaks, giving a constant deceleration, and stops after a distance of 50 m is covered in a time of 4 s. What was the constant deceleration?

Exercise 19.2   A vehicle travelling at 150 km/h passes a stationary police car. It takes 10 s for the police car to start moving and then it continues at constant acceleration until it is adjacent to the vehicle, which happens 20 s after the initial passing. How far has the police car travelled and what is its speed when it catches up with the other vehicle? What was the constant acceleration of the police car?

Exercise 19.3 In order to escape the earth's gravitational pull, a rocket must be launched with a certain initial velocity. This velocity is called the escape velocity. A rocket launched from the surface of the earth (approximate radius: 4000 miles) has velocity, v, given by $v = \sqrt{\frac{2GM}{r} + v_0^2 - \frac{2GM}{R}} \approx \sqrt{\frac{192000}{r} + v_0^2 - 48}$ mi/sec where v_0 represents the initial velocity, r represents the distance from the rocket to the centre of the earth, G is the gravitational constant, M is the mass of the earth, and R is the radius of the earth. (p13, Goldenberg and Greenwald [7])

(a) Find the value of v_0 for which we obtain an infinite limit for r as $v \to 0$. This value for v_0 is the escape velocity for earth.

(b) A rocket launched from the surface of the moon has velocity given by $v = \sqrt{\frac{1920}{r} + v_0^2 - 2.17}$ mi/sec. Find the escape velocity for the moon.

(c) A rocket launched from the surface of Mars has velocity given by $v = \sqrt{\frac{20544}{r} + v_0^2 - 9.55}$ mi/sec. Find the escape velocity for Mars.

(d) A rocket launched from the surface of Planet X has velocity given by $v = \sqrt{\frac{10600}{r} + v_0^2 - 6.99}$ mi/sec. Find the escape velocity for Planet X.

(e) Is the planet larger or smaller than earth? (Assume that the mean density of Planet X is the same as that of the earth.)

19.2   Rate of Change

Consider the polynomial $f(x) = x^4$, shown in Figure 19.6(a).

```
> f:=x->x^4;
```

$$f := x \to x^4$$

```
> plot(f(x),x);
```

The average rate of change of a function is equal to the slope of the secant between two points. Calculate the average rate of change of $f(x)$ between $x = 1$ and $x = 3$. See Figure 19.6(b)

```
> with(student):
> sl1:=slope([1,f(1)],[3,f(3)]);
```

$$sl1 := 40$$

Now find the equation of the secant.

```
> secant-f(1)=sl1*(x-1);
```

$$secant - 1 = 40\,x - 40$$

```
> isolate(",secant);
```

$$secant = 40\,x - 39$$

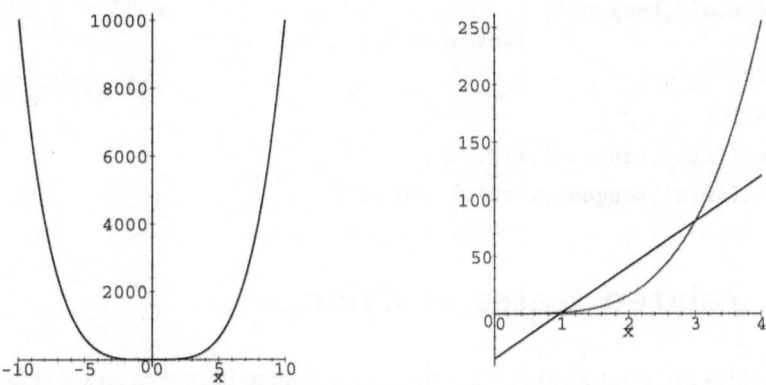

Figure 19.6. (a) The graph of $f(x) = x^4$, on $[-10, 10]$. (b) Average rate of change of $f(x) = x^4$ between $x = 1$ and 3

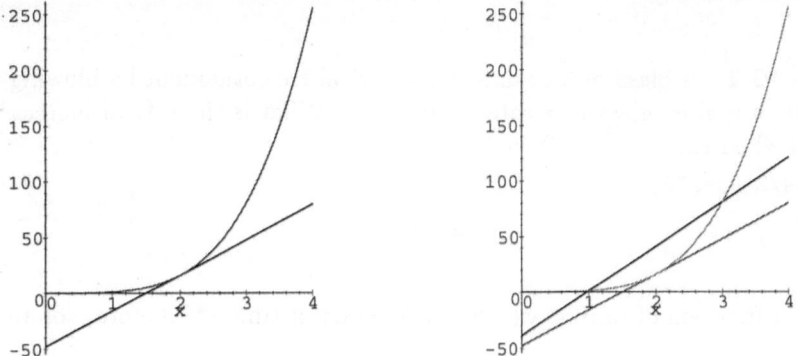

Figure 19.7. (a) Instantaneous rate of change of $f(x) = x^4$ at $x = 2$. (b) Average and instantaneous rate of change of $f(x) = x^4$

```
> assign(");
> plot({f(x),secant},x=0..4);
```

The instantaneous rate of change of a function is equal to the slope of the tangent (or derivative) at a point. Now consider the rate of change of $f(x)$ at $x = 2$. See Figure 19.7.

```
> diff(f(x),x);
```
$$4\,x^3$$

```
> sl2:=subs(x=2,");
```
$$\mathit{sl2} := 32$$

```
> tangent-f(2)=sl2*(x-2);
```
$$tangent - 16 = 32\,x - 64$$

```
> isolate(",tangent);
```
$$tangent = 32\,x - 48$$

```
> assign(");
> plot({f(x),tangent},x=0..4);
> plot({f(x),tangent,secant},x=0..4);
```

19.3   Related Rates of Change

Often variables are related to one another, so a change in one causes a change in the other. For example, if the radius of a circle increases, then the area, which is a function of the radius, will also increase. Here are some problems involving related rates of change.

Example 19.2   A glass blower makes a spherical tree ornament by blowing air into a bulb of molten glass at a rate of 10 cm^3/s. What is the rate of increase of the radius when the radius is 2 cm?

```
> V=4/3*Pi*r^3;
```
$$V = \frac{4}{3}\,\pi\,r^3$$

Volume is a function of radius, which is a function of time. Therefore, volume is a function of time.

```
> diff(V(t),t)=4/3*Pi*(diff(r(t)^3,t));
```
$$\frac{\partial}{\partial t}\,V(t) = 4\,\pi\,\mathrm{r}(t)^2\left(\frac{\partial}{\partial t}\mathrm{r}(t)\right)$$

We now substitute the rate of change of the volume, $\frac{\partial}{\partial t}V(t)$, and the length of the radius, $r(t)$, at which we wish to know the rate of change of the radius.

```
>  with(student):
> isolate("",diff(r(t),t));
```
$$\frac{\partial}{\partial t}\,\mathrm{r}(t) = \frac{1}{4}\,\frac{\frac{\partial}{\partial t}V(t)}{\pi\,\mathrm{r}(t)^2}$$

```
> lhs(")=subs({diff(V(t),t)=10,r(t)=2},rhs("));
```
$$\frac{\partial}{\partial t}\mathrm{r}(t) = \frac{5}{8}\,\frac{1}{\pi}$$

```
>  evalf(");
```
$$\frac{\partial}{\partial t}\,\mathrm{r}(t) = .1989436788$$

Example 19.3 Soda cans with a diameter of 6.5 cm are being filled in a packaging plant. If the volume of liquid in the can is increasing at a rate of 300 cm^2/s, at what rate is the liquid height increasing?

```
> V=Pi*(d/2)^2*l;
```

$$V = \frac{1}{4} \pi d^2 l$$

In this example, volume is a function of length which is a function of time. Again, volume is a function of time. Differentiating both sides of the equation with respect to time, we get,

```
> diff(V(t),t)=Pi*(d/2)^2*diff(l(t),t);
```

$$\frac{\partial}{\partial t} V(t) = \frac{1}{4} \pi d^2 \left(\frac{\partial}{\partial t} l(t) \right)$$

We now substitute the rate at which the volume is increasing, $\frac{\partial}{\partial t} V(t)$, and the radius, r.

```
> subs(diff(V(t),t)=300,lhs("))=subs(d=6.5,rhs("));
```

$$300 = 10.56250000 \pi \left(\frac{\partial}{\partial t} l(t) \right)$$

```
> with(student):
> isolate("",diff(l(t),t));
```

$$\frac{\partial}{\partial t} l(t) = 28.40236686 \frac{1}{\pi}$$

```
> evalf(");
```

$$\frac{\partial}{\partial t} l(t) = 9.040754160$$

Example 19.4 A bicycle courier picks up a package and travels 1 km north and then heads east at 20 km/h. At what rate is the courier moving away from the pick-up point when she has travelled 2 km east?

```
> d^2=x^2+y^2;
```

$$d^2 = x^2 + y^2$$

Let x represent the east direction and y represent the north direction. Because we are only interested in the time when the courier is going east, y is not a function of time.

```
> diff(d(t)^2,t)=diff(x(t)^2,t)+diff(y^2,t);
```

$$2 d(t) \left(\frac{\partial}{\partial t} d(t) \right) = 2 x(t) \left(\frac{\partial}{\partial t} x(t) \right)$$

We can now substitute the values we know, $x(t)$ and $\frac{\partial}{\partial t}x(t)$.

```
> lhs(")=subs({x(t)=2,diff(x(t),t)=20},rhs("));
```

$$2\,\mathrm{d}(t)\left(\frac{\partial}{\partial t}\,\mathrm{d}(t)\right) = 80$$

We determine the value of $d(t)$ using Pythagoras' Theorem.

```
> solve({d^2=x^2+y^2,x=2,y=1});
```

$$\{\,d = \mathrm{RootOf}(\_Z^2 - 5), x = 2, y = 1\,\}$$

```
> allvalues(");
```

$$\left\{x = 2, d = \sqrt{5}, y = 1\right\}, \left\{x = 2, d = -\sqrt{5}, y = 1\right\}$$

```
> isolate("'",diff(d(t),t));
```

$$\frac{\partial}{\partial t}\,\mathrm{d}(t) = 40\,\frac{1}{\mathrm{d}(t)}$$

```
> lhs(")=subs(d(t)=sqrt(5),rhs("));
```

$$\frac{\partial}{\partial t}\,\mathrm{d}(t) = 8\sqrt{5}$$

```
> evalf(");
```

$$\frac{\partial}{\partial t}\,\mathrm{d}(t) = 17.88854382$$

Exercise 19.4   The surface area of a sphere of radius r is $4\pi r^2$ and its volume is $\frac{4}{3}\pi r^3$. What is the rate of change of volume per unit change in area? What is the percentage change in area caused by a 5% increase in radius?

Exercise 19.5   An inverted cone has radius $r = \frac{z}{2}$ where z is the height above the x, y plane in metres. The top of the cone is at $z = 1$ m. Find the rate of increase of fluid height in the cone if fluid is pumped in at the apex at a rate of 0.1 m^3/s. (Recall that the volume of a cone is $\frac{1}{3}\pi r^2 h$.)

Exercise 19.6   Air is being pumped into a spherical balloon at the constant rate of 10 cm^3/s. At what rate is the surface area of the balloon increasing when its radius is 5 cm? (p170, Edwards and Penney [6])

Chapter 20

Integration

Commands used in this chapter

- assign
- convert
- convert[parfrac]
- diff
- Diff
- dsolve
- evalf
- expand
- fsolve
- int
- Int
- plot
- plot[options]
- read
- simplify

- sin
- solve
- student
- student[intparts]
- student[leftbox]
- student[leftsum]
- student[middlebox]
- student[middlesum]
- student[rightbox]
- student[rightsum]
- student[simpson]
- student[trapezoid]
- subs
- value
- with

20.1 Approximation of the Area Under a Curve

We can approximate the area under a curve by partitioning the region into rectangular sections and calculating the area of the rectangles (Figure 20.1(a)). The greater the number of rectangles, the more accurate the approximation of the area (Figure 20.1(b)). By inspection of the curve, we see that the area under it must be about $(4 \times 2) - 2 = 6$.

We know that the area of a rectangle is the product of the length and the

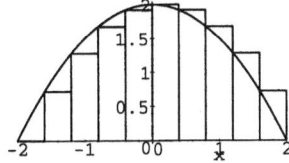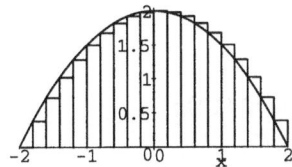

Figure 20.1. Estimation of area under $f(x) = -\frac{1}{2}x^2 + 2$ using (a) 10 rectangles, (b) 20 rectangles

width. In Figure 20.1(a), the width is equal to the total width of the interval (δ) divided by the number of rectangles (n). The length of the rectangle is equal to the value of the function at the left hand side of the rectangle. (Only 9 rectangles appear in Figure 20.1(a) because the first has a length of 0.)

```
> f:=x->-1/2*x^2+2;
```

$$f := x \rightarrow -\frac{1}{2}x^2 + 2$$

```
> width:=2-(-2);
```

$$width := 4$$

```
> n:=10;
```

$$n := 10$$

```
> delta:=width/n;
```

$$\delta := \frac{2}{5}$$

```
> area:=f(-2+1*delta)*delta+f(-2+2*delta)*delta+f(-2+3*delta)*
> delta+f(-2+4*delta)*delta+f(-2+5*delta)*delta+f(-2+6*delta)*
> delta+f(-2+7*delta)*delta+f(-2+8*delta)*delta+f(-2+9*delta)*
> delta+f(-2+10*delta)*delta;
```

$$area := \frac{132}{25}$$

Expressing the variable *area* as a series gives the same result.

```
> area:=delta*(sum(f(-2+delta*i),i=1..n));
```

$$area := \frac{132}{25}$$

```
> evalf(");
```

$$5.280000000$$

The integral of a function is the limit of the area as δ/n goes to zero. It is exactly equal to the area under the graph of the function.

```
> Int(f(x),x=-2..2);
```

$$\int_{-2}^{2} -\frac{1}{2}x^2 + 2\,dx$$

```
> evalf(");
```

$$5.333333333$$

Using built-in functions, we can examine the effect of increasing the number of rectangles thereby decreasing the width of the rectangles. We shall first approximate the area under the function $f(x) = -2x^2 + 6$ between the x-intercepts using 10 rectangles, as shown in Figure 20.2.

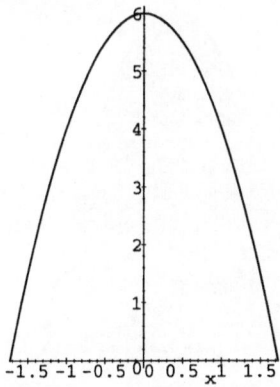 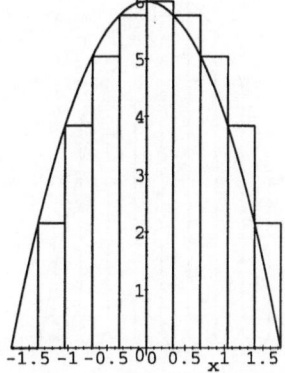

Figure 20.2. (a) The graph of $f(x) = -2x^2 + 6$, on $[-3^{1/2}, 3^{1/2}]$. (b) Estimation of area under $f(x) = -2x^2 + 6$ using 10 rectangles

```
> f:=x->-2*x^2+6;
```

$$f := x \rightarrow -2\,x^2 + 6$$

```
> root:=solve(f(x),x);
```

$$root := -\sqrt{3}, \sqrt{3}$$

```
> plot(f(x),x=root[1]..root[2]);
```

The function leftbox creates a plot of $f(x)$ with the rectangular subregions superimposed. It is shown in Figure 20.2(b). We then use leftsum to calculate the sum of the areas of the rectangles.

```
> with(student):
> leftbox(f(x),x=root[1]..root[2],10);
> leftsum(f(x),x=root[1]..root[2],10);
```

$$\frac{1}{5} \sqrt{3} \left(\sum_{i=0}^{9} \left(-2 \left(-\sqrt{3} + \frac{1}{5} i \sqrt{3} \right)^2 + 6 \right) \right)$$

```
> evalf(");
```

$$13.71784240$$

Now we shall double the number of rectangles (Figure 20.3(a)).

```
> leftbox(f(x),x=root[1]..root[2],20);
> leftsum(f(x),x=root[1]..root[2],20);
```

$$\frac{1}{10} \sqrt{3} \left(\sum_{i=0}^{19} \left(-2 \left(-\sqrt{3} + \frac{1}{10} i \sqrt{3} \right)^2 + 6 \right) \right)$$

```
> evalf(");
```

$$13.82176545$$

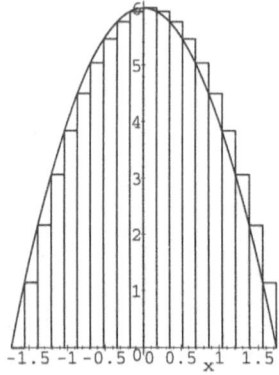 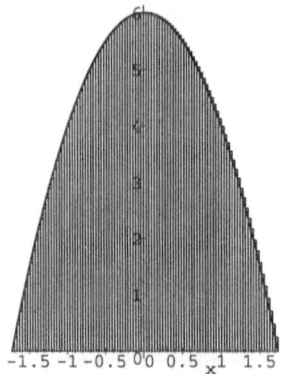

Figure 20.3. Estimation of area under $f(x) = -2x^2 + 6$ using (a) 20 rectangles, (b) 100 rectangles

Finally, we shall use 100 rectangles (Figure 20.3(b)).

```
> leftbox(f(x),x=root[1]..root[2],100);
> leftsum(f(x),x=root[1]..root[2],100);
```

$$\frac{1}{50}\sqrt{3}\left(\sum_{i=0}^{99}\left(-2\left(-\sqrt{3}+\frac{1}{50}i\sqrt{3}\right)^2+6\right)\right)$$

```
> evalf(");
```

$$13.85502082$$

To get an idea of the accuracy of these approximations, we must compare them to the integral of the function.

```
> Int(f(x),x=root[1]..root[2]);
```

$$\int_{-\sqrt{3}}^{\sqrt{3}} -2x^2 + 6 \, dx$$

```
> evalf(");
```

$$13.85640646$$

The approximation using 100 rectangles is quite a good approximation of the actual area under the curve.

There is no reason why we should define the length of the rectangles under the curve as the value of the function on the left hand side and not in the middle or on the right hand side. The accuracy of the approximation will depend on the curve being studied. Using the function $f(x) = -2x^2 + x + 6$, let's compare the three different approximations of the area under the curve over the domain $-1 \le x \le 1$. Figures 20.4 and 20.5 show the graphs the function and the rectangular subregions used to estimate the area.

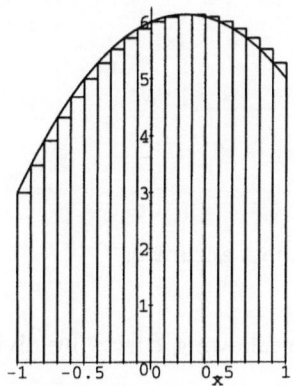 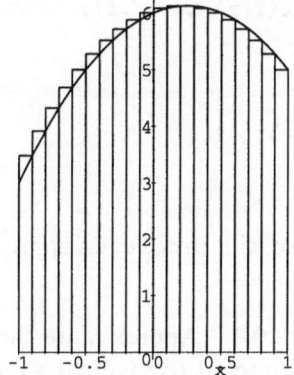

Figure 20.4. Estimation of area ander $f(x) = -2x^2 + x + 6$ asing (a) `leftbox`, (b) `rightbox`

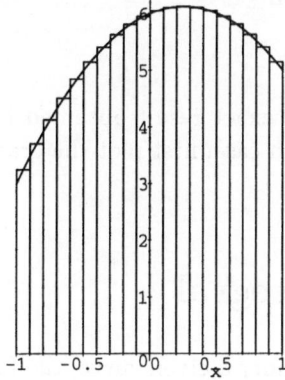

Figure 20.5. Estimation of area under $f(x) = -2x^2 + x + 6$ using `middlebox`

```
> f:=x->-2*x^2+x+6;
```
$$f := x \rightarrow -2\,x^2 + x + 6$$

```
> leftbox(f(x),x=-1..1,20);
> evalf(leftsum(f(x),x=-1..1,20));
                    10.56000000
```

```
> rightbox(f(x),x=-1..1,20);
> evalf(rightsum(f(x),x=-1..1,20));
                    10.76000000
```

```
> middlebox(f(x),x=-1..1,20);
> evalf(middlesum(f(x),x=-1..1,20));
                    10.67000000
```

```
> int(f(x),x=-1..1);
```

$$\frac{32}{3}$$

```
> evalf(");
```

$$10.66666667$$

In this case, the `middlesum` approximation is the most accurate.

Exercise 20.1   Approximate the area under the following curves by dividing the region into 10 rectangular subregions.

   (a) $y = -2x^2 + 6, -1 \le x \le 1$

   (b) $y = x^3 + x + 10, -2 \le x \le 2$

   (c) $y = \sqrt{1 - x^2}, -1 \le x \le 1$

Compare these results to the exact result obtained by integrating the function over the given bounds. In each case, first plot the graphs and estimate the areas by inspection.

20.2   Definite Integral

To calculate the exact area under a given curve, the function must be integrated over a desired interval. This is called a *definite integral*. The integral is defined as the limit of the sum of the rectangular subregions as the width of the rectangles approaches zero.

Example 20.1   Suppose we wish to know the area under the curve $f(x) = x^3 + 2x$ between $x = 1$ and $x = 3$. See Figure 20.6.

```
> f:=x->x^3+2*x;
```

$$f := x \rightarrow x^3 + 2x$$

```
> Int(f(x),x=1..3);
```

$$\int_{1}^{3} x^3 + 2x\, dx$$

```
> value(");
```

$$28$$

```
> plot(f(x),x=0..4);
```

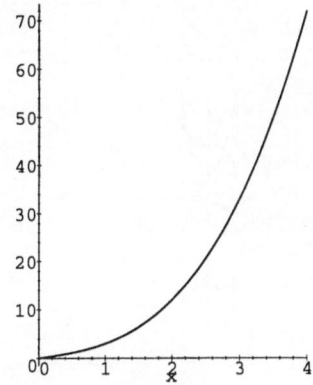

Figure 20.6. The graph of $f(x) = x^3 + 2x$, on $[0, 4]$

Exercise 20.2   Integrate the following functions.

(a) $y = x^2 + 3x + 2, 0 \leq x \leq 2$

(b) $y = x^5, -2 \leq x \leq 2$

(c) $y = 2^x, -\pi \leq x \leq \pi$

20.3   Indefinite Integral

An integral without specific upper or lower bounds is called an *indefinite integral.*

```
> int(f(x),x);
```

$$\int f(x)\, dx$$

```
> Int(3*x^2,x);
```

$$\int 3x^2\, dx$$

```
> value(");
```

$$x^3$$

Families of Curves

Because the derivative of a constant is zero, we must be careful when we antidifferentiate. Consider the function $f(x) = x^2 + 2$, the derivative of which is $f'(x) = 2x$. Now, if we integrate $2x$, we get x^2. For this reason, a constant of integration, C, is included with indefinite integrals, or *primitives*. Therefore, the

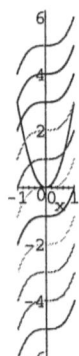

Figure 20.7. Family of curves $x^3 + C$, where $C = -5$ to 5; these are all antiderivatives of $3x^2$.

integral of $f'(x) = 2x$ is $f(x) = x^2 + C$. For example, the family of curves $x^3 + C$ is shown in Figure 20.7.

```
> read 'showAntiderivatives.m';
> showAntiderivatives(3*x^2,x=-1..1,scaling=CONSTRAINED);
```

Exercise 20.3 Integrate the following functions.

   (a) $y = x + 6$
   (b) $y = x^2 + \frac{x}{10}$
   (c) $y = 4e^{2x}$

20.4   Fundamental Theorem of Calculus

The Fundamental Theorem of Calculus states that, if $F(x)$ is the indefinite integral of $f(x)$, then the integral of $f(x)$ over the interval $a \le x \le b$ is equal to $F(b) - F(a)$. This is very important because it says that the area under a curve is given by an antiderivative of the function being plotted. So, while the derivative is a limit of differences, its inverse, the integral, is a limit of sums. Which is as it should be!

```
> Int(f(x),x=a..b);
```

$$\int_a^b f(x)\,dx$$

```
> Int(x^2,x)=int(x^2,x);
```

$$\int x^2\,dx = \frac{1}{3}x^3$$

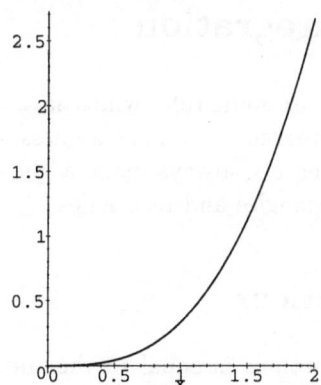

Figure 20.8. The graph of the integral of x^2, on $[0, 2]$

Find the integral of x^2, shown in Figure 20.8.

```
> plot(int(x^2,x),x=0..2);
> Int(x^2,x=0..2)-Int(x^2,x=0..1)=Int(x^2,x=1..2);
```

$$\int_0^2 x^2\,dx - \int_0^1 x^2\,dx = \int_1^2 x^2\,dx$$

```
> value(");
```

$$\frac{7}{3} = \frac{7}{3}$$

Experiment 20.1 Consider the function $f(x) = 3x^2$ and the area under its graph from $x = 0$ to $x = b$, where b is a parameter that we shall choose later. We know that the indefinite integral of $f(x)$ is $F(x) = x^3$ and so the area under the graph of the function is $A(b)$ where

$$A(b) = \int_0^b 3x^2\,dx = F(b) - F(0) = b^3$$

Show, from the definition of the derivative as a limit, that the derivative of A with respect to b is given by $A'(b) = f(b)$. This is in fact a very useful aspect of the Fundamental Theorem of Calculus. Repeat this exercise for some more functions $f(x)$ for which you do not know antiderivatives. In each case, plot the graph from $x = 0$ to $x = b + 0.01$ and show on it the vertical lines from $x = b$ and $x = b + 0.01$ for a range of b values. Then

$$\frac{\int_0^{b+0.01} f(x)dx - \int_0^b f(x)dx}{0.01}$$

is the average value near $x = b$ of the rate of change of the area with the right hand limit. Now you should be able to see intuitively why the Fundamental Theorem of Calculus is true.

20.5   Rules of Integration

As with differentiation, these are some rules which save time. However, the most popular method of finding integrals is to make a guess, check by differentiation, then adjust! For definite integrals, always make a guess at the area under the curve, using known area of triangles and rectangles.

20.5.1   Sum of Functions

The integral of a sum of functions is equal to the sum of the integrals of the functions.

```
> int(f(x)+g(x),x)=expand(int(f(x)+g(x),x));
```

$$\int f(\,x\,) + g(\,x\,)\, dx = \int f(\,x\,) + g(\,x\,)\, dx$$

20.5.2   Function Times a Constant

The integral of a function times a constant is equal to a constant times the integral of the function.

```
> int(a*f(x),x)=expand(int(a*f(x),x));
```

$$\int a\, f(\,x\,)\, dx = a \int f(\,x\,)\, dx$$

20.5.3   Power Rule

Recall that the derivative of $f(x) = x^n$ is $f'(x) = nx^{n-1}$. The integral of $f(x) = x^n$ is,

```
> Int(x^n,x)=int(x^n,x);
```

$$\int x^n\, dx = \frac{x^{(n+1)}}{n+1}$$

Experiment 20.2   Consider a unit square in which you choose two points, independently and at random. Investigate by simulation (see Chapter 12.5) the frequency of occurrence of separation r between the points. Construct a histogram of the measured frequency distribution of separations. Investigate the vailidity of the claim that for $r \ll 1$ the probability of occurrence of a gap with $r_1 < r < r_2$ is

$$\int_{r_1}^{r_2} 2\pi r - 8r^2 dr$$

Experiment 20.3   Mark in red on the x-axis the interval $[0, x]$. Choose at random two points A and B, distance r apart on the x-axis, and suppose that A lies on the red. What is the probability that B lies also on the red? If both A and B are chosen at random on the red, investigate by simulation the frequency distribution of separation r. You may be able to show analytically that the probability of occurrence of separation r with $r_1 < r < r_2$ is

$$\int_0^x \frac{2}{x}(1 - \frac{r}{x})dr$$

What do you expect to be the average separation?

Experiment 20.4   Consider a large square on which are drawn at random a large number of lines. Create such a system and investigate the following statistical results.

(a) The fraction of polygons that are triangles is $2 - \frac{\pi^2}{6}$.

(b) The average number of sides per polygon is 4.

(c) The probability of occurrence of a gap of length between x_1 and x_2 between successive crossings of a line is

$$\int_{x_1}^{x_2} \frac{1}{m} e^{-x/m} dx$$

where m is the mean gap length. Construct a histogram showing the measured frequency distribution of gaps.

20.6   Area Between Curves

To find the area between two curves, we must calculate the area under the higher curve and subtract the area under the lower curve. The difference between the two must be the area bounded by the upper and lower curves. This method can also be applied to find the area enclosed by a loop in a closed curve.

Example 20.2   Find the area between the curves $f(x) = x + 5$ and $g(x) = x^2$ between $x = 1$ and $x = 2$. See Figure 20.9.

```
> f:=x->x+5;
```

$$f := x \rightarrow x + 5$$

```
> g:=x->x^2;
```

$$g := x \rightarrow x^2$$

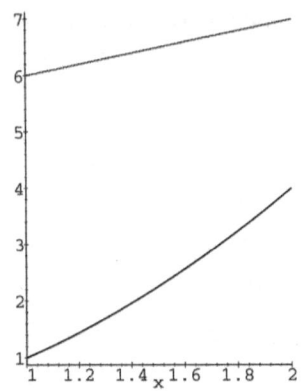

Figure 20.9. The graphs of $f(x) = x + 5$ and $g(x) = x^2$, on $[1, 2]$

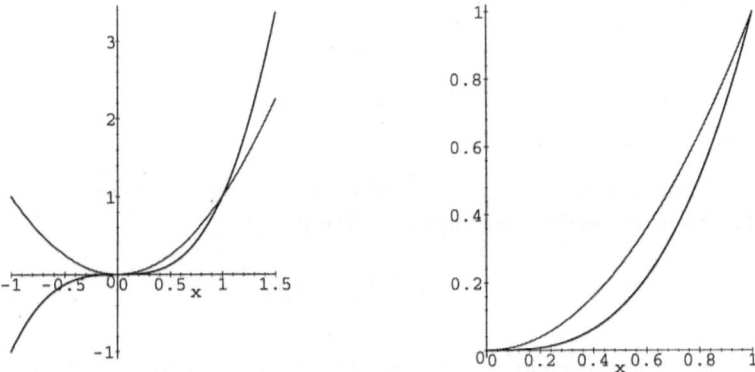

Figure 20.10. (a) The graphs of $f(x) = x^2$ and $g(x) = x^3$, on $[-1, 1.5]$. (b) The graphs of $f(x) = x^2$ and $g(x) = x^3$, on $[0, 1]$

```
> plot({f(x),g(x)},x=1..2);
> Int(f(x),x=1..2)-Int(g(x),x=1..2)=Int(f(x)-g(x),x=1..2);
```

$$\int_1^2 x + 5\, dx - \int_1^2 x^2\, dx = \int_1^2 x + 5 - x^2\, dx$$

```
> value(");
```

$$\frac{25}{6} = \frac{25}{6}$$

Example 20.3 Find the area between the curves $f(x) = x^2$ and $g(x) = x^3$, between their points of intersection. See Figure 20.10.

```
> f:=x->x^2;
```

$$f := x \to x^2$$

```
> g:=x->x^3;
```

$$g := x \rightarrow x^3$$

```
> plot({f(x),g(x)},x=-1..1.5);
> solve(f(x)=g(x),{x});
```

$$\{ x = 0 \}, \{ x = 0 \}, \{ x = 1 \}$$

```
> plot({f(x),g(x)},x=0..1);
> int(f(x)-g(x),x=0..1);
```

$$\frac{1}{12}$$

Exercise 20.4   Determine the area enclosed by the following curves.

(a) $f(x) = 2x^2 - 4, g(x) = -2x^2$

(b) $f(x) = x + 2, g(x) = x^2$

(c) $f(x) = \sqrt{x}, g(x) = x^2$

20.7   Differential Equations

An equation which contains a derivative is called a *differential equation*. The importance of differential equations lies in the fact that many real processes depend not only on the values of variables but also on their rates of change. Here's an example.

```
> diff(f(x),x)=2*x;
```

$$\frac{\partial}{\partial x} f(x) = 2x$$

In order to solve a differential equation using *Maple*, we must use `dsolve`.

```
> dsolve(",f(x));
```

$$f(x) = x^2 + \_C1$$

The `dsolve` command, unlike `int`, includes a constant of integration. If we know that at time $t = 0$, the value of the function is 2, then we can include this information in the `dsolve` command.

```
> dsolve({"",f(0)=2},f(x));
```

$$f(x) = x^2 + 2$$

Here's a more difficult example which contains a second derivative.

```
> diff(f(x),x$2)-2*sin(x)=0;
```

$$\left(\frac{\partial^2}{\partial x^2} f(x) \right) - 2\sin(x) = 0$$

Let's solve this differential equation with the initial conditions $f(0) = 0$ and $f(1) = 1$.

```
> dsolve({",f(0)=0,f(1)=1},f(x));
```

$$f(x) = -2\sin(x) + (2\sin(1) + 1)x$$

```
> evalf(");
```

$$f(x) = -2.\sin(x) + 2.682941970\,x$$

Experiment 20.5   We know that the differential equation $\frac{dy}{dx} = -y$, $y(0) = 1$ has solution $y(x) = e^{-x}$, but let us see how it arises from the derivative condition. Plot on a graph short line segments end to end beginning at $x = 0$, $y = 1$ and continuing over $x = 0.1, 0.2, \ldots$, each with slope given by the negative of the current value of y. Thus, the first segment has equation

$$y_0(x) = 1 - x \quad 0 \le x \le 0.1$$

The second segment begins where the first ended, at $y_0(0.1) = 0.9$, has slope -0.9 and extends over $0.1 \le x \le 0.2$. Gradually, you will develop an approximation to the solution. You can plot the true solution over the top in red, to see how good the approximation is. This method works for *any* differential equation and, though tedious to use over large ranges, it is commonly used to investigate the beginning part of a solution from different initial points. *Maple* has a function `fieldplot` in the `plots` package which plots for you the small line segments as arrows. Try the commands

```
> with(plots):
> fieldplot([x,-y],x=-1..1,y=-1..1,arrows=SLIM);
```

Observe that the input to `fieldplot` consists of the coordinate x and the function given for $\frac{dy}{dx}$, here x and $-y$. So at each point (x, y) it plots a segment of slope $\frac{dy}{dx}$ with an arrow in the forward direction. The range of x, y values covers the possible starting points for the solution. Try incorporating a particular solution in red on the `fieldplot`. Investigate other differential equations.

Exercise 20.5   Solve the following differential equations, first investigating them with the `fieldplot` function.

(a) $\frac{dy}{dx} = x, y(0) = 2$

(b) $\frac{dy}{dx} = x + y, y(0) = 0$

(c) $\frac{dy}{dx} = x^2 + y, y(0) = 1$

Newton's Law of Cooling

One practical example of differential equations is Newton's Law of Cooling. The rate at which an object at a high temperature cools to that of its surroundings can be described by the following equation.

```
> Diff(T(t),t)=k*(T(t)-Te);
```

$$\frac{\partial}{\partial t} T(t) = k(T(t) - Te)$$

Described in words, the rate of change of the temperature of an object with time is proportional to the difference in temperature between the object ($T(t)$) and the environment (T_e). This temperature difference is called the *driving force*. Let's solve the equation setting the temperature of the environment equal to $20°C$, and the initial temperature of the object, $T(0)$, equal to $95°C$.

```
> Diff(T(t),t)=k*(T(t)-20);
```

$$\frac{\partial}{\partial t} T(t) = k(T(t) - 20)$$

```
> dsolve({diff(T(t),t)=k*(T(t)-20),T(0)=95},T(t));
```

$$T(t) = 20 + 75 e^{(kt)}$$

```
> assign(");
```

To solve for the proportionality constant, k, assume that at $t = 30, T(t) = 50$.

```
> fsolve({t=30,T(t)=50});
```

$$\{ k = -.03054302440, t = 30. \}$$

Figure 20.11 shows the cooling curve we have just calculated.

```
> plot(20+75*exp(-0.03*t),t=0..200);
```

Experiment 20.6   Suppose a $200°C$ rod is left to cool in a room in which the air temperature is $25°C$. If after 5 minutes, the rod has cooled to $180°C$ what is the proportionality k? How long will it take for the rod to cool to $40°C$? If the air temperature was $30°C$, would the rod cool at a faster or slower rate, ie would k be larger or smaller? Use the `fieldplot` command to show a range of solutions depending on the rod's initial temperature in the range from $25°$ to $500°$.

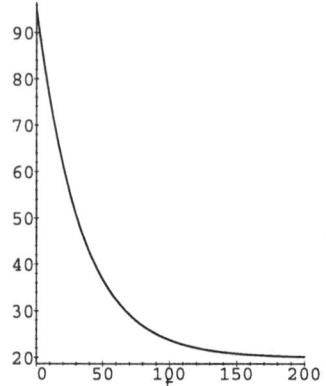

Figure 20.11. Newton's Law of Cooling example

20.8   Integration by Parts

The derivative of a product is easily remembered in the form $\frac{duv}{dx} = v\frac{du}{dx} + u\frac{dv}{dx}$.
Here we want to use it backwards to find the integral of a product. *Integration by parts* is a method of integrating a product of functions. Suppose the function to be integrated is the product of $f(x) = x$ and $g(x) = \sin(x)$.

```
> with(student):
> f:=x->x;
```
$$f := x \rightarrow x$$

```
> g:=x->sin(x);
```
$$g := \sin$$

The function $f(x)g(x)$ can be integrated directly using the **int** command.

```
> int(f(x)*g(x),x);
```
$$\sin(x) - x\cos(x)$$

To integrate this product by parts, we use the **intparts** command. Let's choose $f(x)$ to be the factor of the integrand to be differentiated.

```
> intparts(Int(f(x)*g(x),x),f(x));
```
$$-x\cos(x) - \int -\cos(x)\,dx$$

```
> value(");
```
$$\sin(x) - x\cos(x)$$

If we choose $g(x)$ to be the factor of the integrand to be differentiated, **intparts** returns a more complicated expression, but the final answer is the same.

```
> with(student):
> intparts(Int(f(x)*g(x),x),g(x));
```

$$\frac{1}{2}\sin(x)x^2 - \int \frac{1}{2}\cos(x)x^2\,dx$$

```
> value(");
```

$$\sin(x) - x\cos(x)$$

One way to remember the rule for integration of a product is to label the functions 1st and 2nd, then the integral is '1st times integral 2nd minus integral of derivative of 1st times integral 2nd.' You are free to choose which is 1st to reduce work. This is actually an integration of the rule $v\frac{du}{dx} = \frac{dvu}{dx} - u\frac{dv}{dx}$ in the form $\int vw = v\int w - \int(\frac{dv}{dx}\int w)$ with $w = \frac{du}{dx}$.

Exercise 20.6 Using the method of integration by parts, integrate the following functions.

(a) $y(x) = xe^x$

(b) $y(x) = \frac{1}{x}e^x$

(c) $y(x) = \ln(x)$

20.9 Partial Fractions

You can often simplify the integration of a division of polynomials by converting the function into *partial fractions*.

```
> f:=x->(11*x+7)/(x^2+4*x-5);
```

$$f := x \rightarrow \frac{11\,x + 7}{x^2 + 4\,x - 5}$$

Using *Maple* this function can be integrated directly.

```
> Int(f(x),x)=int(f(x),x);
```

$$\int \frac{11\,x + 7}{x^2 + 4\,x - 5}\,dx = 8\ln(x+5) + 3\ln(x-1)$$

To see how *Maple* arrived at this solution, convert the function into partial fractions.

```
> f(x)=convert(f(x),parfrac, x);
```

$$\frac{11\,x + 7}{x^2 + 4\,x - 5} = 8\,\frac{1}{x+5} + 3\,\frac{1}{x-1}$$

```
> int(convert(f(x),parfrac, x),x);
```
$$8\ln(x+5)+3\ln(x-1)$$

Exercise 20.7 Convert the following functions into partial fractions and then integrate them.

(a) $y(x) = \frac{x^2-4}{x^2+3x+2}$

(b) $y(x) = \frac{2x+8}{18x^2+52x-6}$

20.10 Numerical Integration

20.10.1 Trapezoidal Rule

The *Trapezoidal Rule* is a more sophisticated method of estimating the area under a curve than the methods we saw at the beginning of the chapter. With this method, the area is divided into trapezoidal segments. (Recall that a *trapezoid* is a quadrilateral having only two sides parallel.)

```
> with(student):
> trapezoid(f(x),x=a..b,n);
```

$$\frac{1}{2}(b-a)\left(\frac{11a+7}{a^2+4a-5} \right.$$

$$+2\left[\sum_{i=1}^{n-1} \frac{11a+11\frac{i(b-a)}{n}+7}{\left(a+\frac{i(b-a)}{n}\right)^2+4a+4\frac{i(b-a)}{n}-5} \right]$$

$$\left. +\frac{11b+7}{b^2+4b-5} \right)\Big/n$$

```
> f:=x->x^3+2*x^2;
```
$$f := x \rightarrow x^3 + 2x^2$$

For example, what is the approximation of the area under $f(x)$ between $x = 1$ and $x = 2$ using 6 sections?

```
> subs({a=1,b=2,n=6},trapezoid(f(x),x=a..b,n));
```

$$\frac{19}{12} + \frac{1}{6}\left(\sum_{i=1}^{5} \left(\left(1+\frac{1}{6}i\right)^3 + 2\left(1+\frac{1}{6}i\right)^2 \right) \right)$$

```
> evalf(");
```
$$8.446759261$$

```
> int(f(x),x=1..2);
```
$$\frac{101}{12}$$

```
> evalf(");
```
$$8.416666667$$

20.10.2 Simpson's Rule

Another method of numerical integration is called *Simpson's Rule*. This method uses a parabola to approximate the sections of the curve. It is amazingly efficient!

```
> simpson(f(x), x=a..b, n);
```

$$\frac{1}{3}(b-a)\left(f(a)+f(b)+4\left(\sum_{i=1}^{1/2\,n}f\left(a+\frac{(2\,i-1)(b-a)}{n}\right)\right)\right.$$
$$\left.+2\left(\sum_{i=1}^{1/2\,n-1}f\left(a+2\frac{i(b-a)}{n}\right)\right)\right)\Big/n$$

```
> f:=x->x^3+2*x^2;
```
$$f := x \rightarrow x^3 + 2\,x^2$$

Let's use Simpson's Rule to estimate the area under $f(x)$ between $x = 1$ and $x = 2$ using 6 sections.

```
> subs({a=1,b=2,n=6},simpson(f(x), x=a..b, n));
```

$$\frac{19}{18}+\frac{2}{9}\left(\sum_{i=1}^{3}\left(\left(\frac{5}{6}+\frac{1}{3}i\right)^3+2\left(\frac{5}{6}+\frac{1}{3}i\right)^2\right)\right)$$
$$+\frac{1}{9}\left(\sum_{i=1}^{2}\left(\left(1+\frac{1}{3}i\right)^3+2\left(1+\frac{1}{3}i\right)^2\right)\right)$$

```
> evalf(");
```
$$8.416666665$$

Compare this result to that obtained by integrating the function.

```
> Int(f(x),x)=evalf(int(f(x),x=1..2));
```
$$\int x^3 + 2\,x^2\,dx = 8.416666667$$

Exercise 20.8   Approximate the area under the curve $y(x) = -\frac{1}{3}x^2 + 4x + 9, -13 \leq x \leq 1$

(a) using the Trapeziod rule with $n = 20$

(b) using Simpson's rule with $n = 10$

Experiment 20.7   Integrate the function $f_n(x) = 1 + x + \frac{x^2}{2!} + \frac{x^3}{3!} + \ldots + \frac{x^n}{n!}$ for $n = 1, 2, \ldots$ What do you expect to be the limiting relationship with $f_n(x)$ as n approaches infinity?

Chapter 21

Trigonometry

Commands used in this chapter

- arccos
- arcsin
- arctan
- combine
- combine[trig]
- cos
- csc
- diff
- Diff
- dsolve
- evalf
- expand
- int
- Int

- limit
- Limit
- plot
- plot[options]
- sec
- series
- simplify
- sin
- student
- student[slope]
- subs
- tan
- value
- with

21.1 Compound Angles

From their graphs, the trigonometric functions are not linear, so we are not surprised by the following. The sine (or any other trigonometric function) of a sum of angles is not equal to the sum of the sines. For instance, consider the sine of $\frac{\pi}{6} + \frac{\pi}{4}$.

```
> evalf(sin(Pi/6+Pi/4));
```
$$.9659258263$$

```
> evalf(sin(Pi/6)+sin(Pi/4));
```
$$1.207106781$$

The addition formulas for sine are listed below.

```
> sin(theta+phi)=expand(sin(theta+phi));
```
$$\sin(\theta + \phi) = \sin(\theta)\cos(\phi) + \cos(\theta)\sin(\phi)$$

```
> sin(theta-phi)=expand(sin(theta-phi));
```
$$-\sin(-\theta + \phi) = \sin(\theta)\cos(\phi) - \cos(\theta)\sin(\phi)$$

Using these addition formulas, we can also express $\sin(2\theta)$.

> sin(2*theta)=expand(sin(2*theta));
$$\sin(2\theta) = 2\sin(\theta)\cos(\theta)$$

The addition formulas for cosine are as follows,

> cos(theta+phi)=expand(cos(theta+phi));
$$\cos(\theta + \phi) = \cos(\theta)\cos(\phi) - \sin(\theta)\sin(\phi)$$

> cos(theta-phi)=expand(cos(theta-phi));
$$\cos(-\theta + \phi) = \cos(\theta)\cos(\phi) + \sin(\theta)\sin(\phi)$$

> cos(2*theta)=expand(cos(2*theta));
$$\cos(2\theta) = 2\cos(\theta)^2 - 1$$

21.2   Graphs of Trigonometric Functions

We have already plotted the sine and cosine functions and their periodicity may
be illustrated by animations running through a range of functions such as $\sin(n\theta)$,
$-2\pi \leq \theta \leq 2\pi$ for $n = 1, 2, 3, \ldots$ The effect of *phase angle*, the intercept at
$x = 0$, can be illustrated by animations of $\sin(\theta + \frac{2\pi}{n})$, $-2\pi \leq \theta \leq 2\pi$ for
$n = 1, 2, 3, \ldots$ It turns out that $\cos(\theta) = \sin(\theta + \frac{\pi}{2})$. Having seen the power series
expansions of $\sin(\theta)$ and $\cos(\theta)$, we are not surprised that these functions have an
undulating appearance—what we would expect from polynomials of high degree
with alternating signs of its terms. The remarkable thing is that these functions
are *bounded* by ± 1. One of the most important indentities is $\cos^2\theta + \sin^2\theta = 1$.
It follows that if we put $x = \cos\theta$ and $y = \sin\theta$ then this equation becomes
$x^2 + y^2 = 1$ which represents a circle of $0 \leq \theta \leq 2\pi$. You obtain an ellipse if
you have $x^2 + \frac{y^2}{4} = 1$, which parametrically has the form $x = \cos\theta, y = 4\sin\theta$,
$0 \leq \theta \leq 2\pi$.

Exercise 21.1   Animate a family of ellipses $x^2 + \frac{y^2}{n} = 1$, for $n = 1, 2, 3, \ldots$
using *Maple's* functions plot and implicitplot. Here are the first two cases:

> plot([cos(t),sin(t),t=0..2*Pi]);

> with(plots):

> implicitplot(x^2+y^2=1,x=-1..1,y=-1..1,scaling=CONSTRAINED);

21.3 Derivative of Sine Function

Let's first approximate the derivative of the sine function by taking the derivative of its power series, since we already know how to differentiate polynomials.

```
> series(sin(x),x,9);
```

$$x - \frac{1}{6}x^3 + \frac{1}{120}x^5 - \frac{1}{5040}x^7 + O(x^9)$$

```
> diff(x-1/6*x^3+1/120*x^5-1/5040*x^7,x);
```

$$1 - \frac{1}{2}x^2 + \frac{1}{24}x^4 - \frac{1}{720}x^6$$

Notice that the derivative of the power series of the sine function is the same as the power series of the cosine function.

```
> series(cos(x),x,9);
```

$$1 - \frac{1}{2}x^2 + \frac{1}{24}x^4 - \frac{1}{720}x^6 + \frac{1}{40320}x^8 + O(x^{10})$$

Experiment 21.1 Let $s_n(x)$ be the power series expansion of $\sin x$ up to terms of order n. Differentiate $s_1(x), s_2(x), s_3(x), \ldots$ and check that the results are corresponding power series expansions of $\cos x$.

In order to differentiate the sine function from first principles, we must first examine the function $f(\theta) = \frac{\sin(\theta)}{\theta}$. This function will appear later in the proof so we must know how it behaves as $\theta \to 0$.

```
> f:=theta->sin(theta)/theta;
```

$$f := \theta \to \frac{\sin(\theta)}{\theta}$$

```
> plot(f(theta),theta=-5*Pi..5*Pi);
```

The plot of the function tells us that the limit $\lim\limits_{\theta \to 0} \dfrac{\sin(\theta)}{\theta} = 1$. Had we not seen the plot (Figure 21.1), we might have guessed that the limit was indeterminate.

```
> Limit(sin(theta)/theta,theta=0)=limit(sin(theta)/theta,theta=0);
```

$$\lim_{\theta \to 0} \frac{\sin(\theta)}{\theta} = 1$$

With this information, we can now find the derivative of sine.

```
> with(student):
> slope([x,sin(x)],[x+delta,sin(x+delta)]);
```

$$-\frac{\sin(x) - \sin(x + \delta)}{\delta}$$

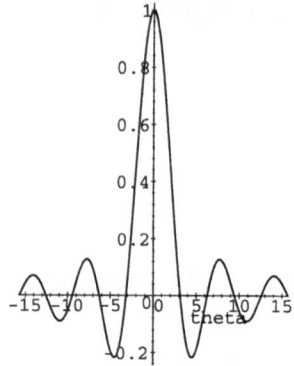

Figure 21.1. The graph of $f(x) = \dfrac{\sin(\theta)}{\theta}$, on $[-5\pi, 5\pi]$

```
> expand(");
```

$$-\frac{\sin(x)}{\delta} + \frac{\sin(x)\cos(\delta)}{\delta} + \frac{\cos(x)\sin(\delta)}{\delta}$$

```
> simplify(");
```

$$\frac{-\sin(x) + \sin(x)\cos(\delta) + \cos(x)\sin(\delta)}{\delta}$$

Notice that the first two terms will cancel out as $\delta \to 0$, leaving $\frac{\cos(x)\sin(\delta)}{\delta}$. As we just saw, $\frac{\sin(\delta)}{\delta} \to 1$ as $\delta \to 0$, leaving $\cos(x)$.

```
> limit(",delta=0);
```

$$\cos(x)$$

Exercise 21.2   Given that $\sin 30° = \frac{1}{2}$, use the approximation

$$\sin(x + h) \approx \sin x + h\frac{d\sin x}{dx}$$

with $x = \frac{\pi}{6}$ radians and $h = 0.0175$ radians to estimate $\sin 31°$.

21.4   Derivatives of Trigonometric Functions

The following is a list of the derivatives of the primary, secondary, and inverse trigonometric functions.

Primary Trigonometric Functions

```
> Diff(sin(x),x)=diff(sin(x),x);
```
$$\frac{\partial}{\partial x}\sin(x) = \cos(x)$$

```
> Diff(cos(x),x)=diff(cos(x),x);
```
$$\frac{\partial}{\partial x}\cos(x) = -\sin(x)$$

```
> Diff(tan(x),x)=diff(tan(x),x);
```
$$\frac{\partial}{\partial x}\tan(x) = 1 + \tan(x)^2$$

Secondary Trigonometric Functions

```
> Diff(csc(x),x)=diff(csc(x),x);
```
$$\frac{\partial}{\partial x}\csc(x) = -\csc(x)\cot(x)$$

```
> Diff(sec(x),x)=diff(sec(x),x);
```
$$\frac{\partial}{\partial x}\sec(x) = \sec(x)\tan(x)$$

```
> Diff(cot(x),x)=diff(cot(x),x);
```
$$\frac{\partial}{\partial x}\cot(x) = -1 - \cot(x)^2$$

Inverse Trigonometric Functions

```
> Diff(arcsin(x),x)=diff(arcsin(x),x);
```
$$\frac{\partial}{\partial x}\arcsin(x) = \frac{1}{\sqrt{1-x^2}}$$

```
> Diff(arccos(x),x)=diff(arccos(x),x);
```
$$\frac{\partial}{\partial x}\arccos(x) = -\frac{1}{\sqrt{1-x^2}}$$

```
> Diff(arctan(x),x)=diff(arctan(x),x);
```
$$\frac{\partial}{\partial x}\arctan(x) = \frac{1}{1+x^2}$$

Exercise 21.3 Differentiate the following trigonometric functions.

(a) $y = \sin(\frac{x}{2})$

(b) $y = \cos(3x)^2$

(c) $y = \cot(x)$

(d) $y = \sin(x^{-1})$

Exercise 21.4 Define, for constant a, b,

$$f(x) = \cos(a + x)\cos(b - x) - \sin(a + x)\sin(b - x)$$

and show that $f'(x) = 0$. Hence f must be a constant function and in particular, $f(0) = f(b)$. Hence deduce the formula for $\cos(a + b)$ and similarly derive that for $\sin(a + b)$.

Experiment 21.2 Solving equations like $\cos x = x$ is actually quite difficult other than by numerical iteration. Newton developed a general method to solve equations like $f(x) = 0$, so here you would put $f(x) = \cos x - x$. The steps are as follows

1. Guess the solution to be x_0.

2. Compute: $x_1 = x_0 - \dfrac{f(x_0)}{f'(x_0)}$

3. Continue: $x_{n+1} = x_n - \dfrac{f(x_n)}{f'(x_n)}$

Construct a table of values of n, x_n, and $(\cos(x_n) - n)$. Try solving some other equations, for example, $x^5 = 2$, and find an equation for which the method fails.

21.5 Maxima and Minima

Like polynomials, the local maxima and minima of trigonometric functions are found by setting the first derivative equal to zero. So, one of the extreme values of the sine function is given by,

```
> solve(diff(sin(x),x)=0);
```

$$\frac{1}{2}\pi$$

The function has a maximum value at the point $x = \frac{1}{2}\pi$, since the second derivative at that point is negative.

```
> evalf(subs(x=1/2*Pi,diff(sin(x),x$2)));
```
$$-1.$$

```
> sin(1/2*Pi);
```
$$1$$

As we know from its graph, the cosine function has a maximum value at the point $x = 0$.

```
> solve(diff(cos(x),x)=0);
```
$$0$$

```
> evalf(subs(x=0,diff(cos(x),x$2)));
```
$$-1.$$

```
> cos(0);
```
$$1$$

Experiment 21.3 Plot the graph of $f(x) = \sin(x \sin x), -\pi \leq x \leq \pi$ and by successively plotting on suitable decreasing intervals, locate and estimate the local maxima and minima. Remember that *Maple* plots show the coordinates of points over which you place the cursor and click the mouse.

Experiment 21.4 A particle is travelling round a circle centred at the origin with radius 1 m at a constant speed of 2π m/s; so it completes one orbit every second. A laser tracking device is located at the coordinates $(2,0)$ and measures the velocity of the particle's approach to the detector as a function of time. Find this function, plot it with its derivative and estimate the points on the orbit with extreme values of velocity or acceleration relative to the detector. Can you extend the results to the case where the detector is itself moving at some constant speed around a circle centred at the origin with radius 2?

Experiment 21.5 The curvature of a curve $y = f(x)$ is defined at each point by the function

$$K(x) = \frac{f''(x)}{(1 + (f'(x))^2)^{3/2}}.$$

Use this formula to plot graphs of the curvature of the curves $y = f(x)$ where

(a) $f(x) = x$

(b) $f(x) = x^2$

(c) $f(x) = x^3$

(d) $f(x) = e^x$

(e) $f(x) = e^{-x}$

(f) $f(x) = \sin x$

In each case plot $f(x)$ and $K(x)$ together. You should try to estimate by inspection of $K(x)$ its behaviour near $x = 0$.

21.6   Integrals of Trigonometric Functions

The integrals of the primary, secondary, and inverse trigonometric functions are listed below.

Primary Trigonometric Functions

```
> Int(sin(x),x)=int(sin(x),x);
```
$$\int \sin(x)\, dx = -\cos(x)$$

```
> Int(cos(x),x)=int(cos(x),x);
```
$$\int \cos(x)\, dx = \sin(x)$$

```
> Int(tan(x),x)=int(tan(x),x);
```
$$\int \tan(x)\, dx = -\ln(\cos(x))$$

Secondary Trigonometric Functions

```
> Int(csc(x),x)=int(csc(x),x);
```
$$\int \csc(x)\, dx = \ln(\csc(x) - \cot(x))$$

```
> Int(sec(x),x)=int(sec(x),x);
```
$$\int \sec(x)\, dx = \ln(\sec(x) + \tan(x))$$

```
> Int(cot(x),x)=int(cot(x),x);
```
$$\int \cot(x)\, dx = \ln(\sin(x))$$

Inverse Trigonometric Functions

```
> Int(arcsin(x),x)=int(arcsin(x),x);
```
$$\int \arcsin(x)\, dx = x\arcsin(x) + \sqrt{1-x^2}$$

```
> Int(arccos(x),x)=int(arccos(x),x);
```
$$\int \arccos(x)\, dx = x\arccos(x) - \sqrt{1-x^2}$$

```
> Int(arctan(x),x)=int(arctan(x),x);
```
$$\int \arctan(x)\, dx = x\arctan(x) - \frac{1}{2}\ln(1+x^2)$$

Exercise 21.5   Integrate the following trigonometric functions.

(a) $y = \sin(3x)$

(b) $y = \cos(x)^{3/2}\sin(x)$

(c) $y = x^2\cos(x^3+1)$

(d) $y = \sin(x) + \sin(2x)$

21.7   Areas Defined by Trigonometric Functions

As we saw in Chapter 20, The Fundamental Theorem of Calculus states that the area bounded by the graph of a function, the x-axis, and the lines $x = a$ and $x = b$, is equal to the definite integral of the function from $x = a$ to $x = b$. This theorem applies to trigonometric functions as well.

Example 21.1   What is the area bounded by the function $f(x) = 2\sin(4x)$, the x-axis, and the lines $x = 0$ and $x = \frac{\pi}{4}$? See Figure 21.2(a).

```
> f:=x->2*sin(4*x);
```
$$f := x \rightarrow 2\sin(4x)$$

```
> plot(f(x),x=0..Pi/4);
> int(f(x),x=0..Pi/4);
```
1

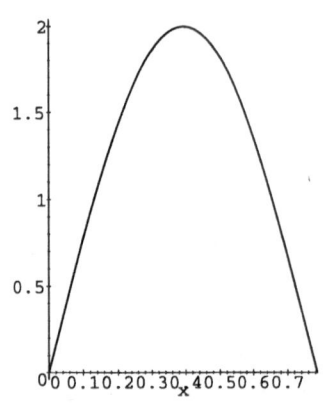
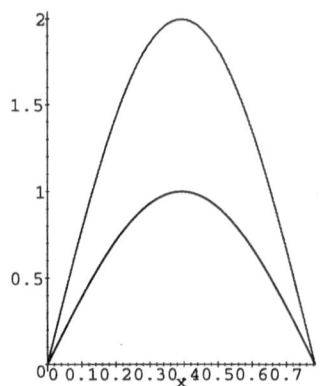

Figure 21.2. (a) The graph of $f(x) = 2\sin(4x)$, on $[0, \frac{\pi}{4}]$. (b) The graphs of
$f(x) = 2\sin(4x)$ and $g(x) = \sin(4x)$, on $[0, \frac{\pi}{4}]$

Example 21.2 What is the area between the points of intersection of the
curves $f(x) = 2\sin(4x)$ and $g(x) = \sin(4x)$, in the interval $0 \le x \le \frac{\pi}{4}$? See
Figure 21.2(b).

> g:=x->sin(4*x);

$$g := x \rightarrow \sin(4x)$$

> plot({g(x),f(x)},x=0..Pi/4);
> int(f(x)-g(x),x=0..Pi/4);

$$\frac{1}{2}$$

Chapter 22

Exponents and Logarithms

Commands used in this chapter

- `assign`
- `diff`
- `Diff`
- `dsolve`
- `evalf`
- `exp`
- `infinity`
- `int`
- `Int`
- `limit`
- `Limit`
- `ln`
- `log`
- `plot`
- `plot[options]`
- `solve`
- `student`
- `student[slope]`
- `with`

22.1   Base for Natural Logarithm

The base of the natural logarithm and exponential, e, is the limit as n approaches infinity of $(1+\frac{1}{n})^n$, shown in Figure 22.1. Like π, e is an irrational number. This seems a most unnatural choice at first, but later studies show that e is indeed the best base. It gives rise to the most important function in all mathematics.

```
> Limit((1+1/n)^n,n=infinity)=limit((1+1/n)^n,n=infinity);
```
$$\lim_{n \to \infty} \left(1 + \frac{1}{n}\right)^n = e$$

```
> plot((1+1/n)^n,n=0..100000);
> evalf(E,20);
```
$$2.7182818284590452354$$

The following expressions are equivalent.

```
> evalf(log[E](2));
```
$$.6931471806$$

```
> evalf(log(2));
```
$$.6931471806$$

```
> evalf(ln(2));
```
$$.6931471806$$

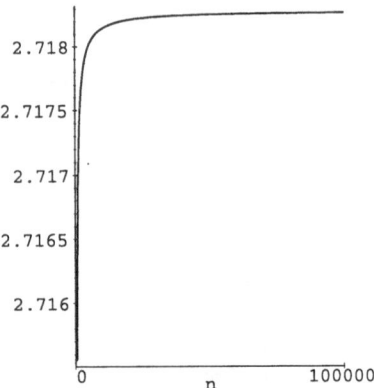

Figure 22.1. The graph of $(1 + \frac{1}{n})^n$, on $[0, 100000]$

22.2   Exponential Growth and Decay

Many phenomena can be modeled using a mathematical expression of the form 'the instantaneous change in a quantity with time is equal to a constant times the current value'. Some common examples of this are radioactive decay and simple population growth. If we translate the above word expression into mathematical symbols, we get the following differential equation.

> `> Diff(y(t),t)=k*y(t);`

$$\frac{\partial}{\partial t} y(t) = k\, y(t)$$

The solution of this differential equation is presented below. The initial value of $y(t)$ is set equal to a constant, $y(0) = y_0$.

> `> dsolve({diff(y(t),t)=k*y(t),y(0)=y0},y(t));`

$$y(t) = e^{(kt)}\, y0$$

If $k < 0$, the graph of the function will decay exponentially. If $k > 0$, the graph of the function will grow exponentially.

You may remember studying exponential growth and decay in the past, and in Chapter 6, before you had studied calculus. At that time, the equation for $y(t)$ was expressed in terms of half-life or doubling-period, rather than time. With that form of the equation, it is not necessary to use differential equations and integration.

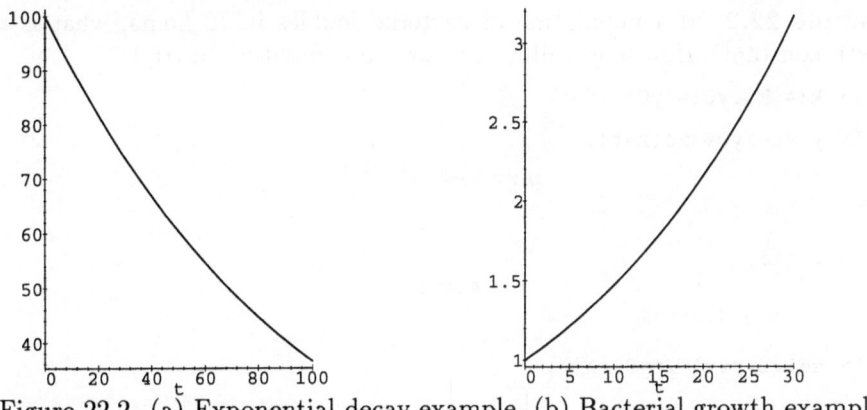

Figure 22.2. (a) Exponential decay example. (b) Bacterial growth example

Example 22.1 A 100 kg sample of uranium-232 decays to 80 kg after 22.18 years. What is the decay constant? How many years will it take for the sample to decay to 20 kg?

```
> y:=t->y0*exp(k*t);
```
$$y := t \rightarrow y0\, e^{(kt)}$$

```
> y0:=100;
```
$$y0 := 100$$

```
> solve(y(22.18)=80,{k});
```
$$\{\, k = -.01006057490 \,\}$$

```
> assign(");
> solve(y(t)=20,{t});
```
$$\{\, t = 159.9747458 \,\}$$

The decay curve is shown in Figure 22.2(a).

```
> plot(y(t),t=0..100);
```

Exercise 22.1 Advertising of a certain product has been discontinued; the company plans to resume advertising when sales have declined to 75% of their initial rate. (This phenomenon actually occurred when the Sony Corporation first introduced home videotape recorders in the United States in 1976.) If after 1 week without advertising, sales have declined to 95% of their initial rate, when should the company expect to resume advertising? (p395, Edwards and Penney [6])

Example 22.2 If a population of bacteria doubles in 18 hours, what is the growth constant? How long will it take for the population to triple?

```
> k:='k':y0:='y0':
> y:=t->y0*exp(k*t);
```
$$y := t \rightarrow y0\, e^{(k\,t)}$$

```
> t:=18;
```
$$t := 18$$

```
> solve(y(t)=2*y0,{k});
```
$$\left\{ k = \frac{1}{18}\ln(2)\right\}$$

```
> evalf(");
```
$$\{\, k = .03850817670 \,\}$$

```
> assign(");
> t:='t':
> solve(y(t)=3*y0,{t});
```
$$\{\, t = 28.52932502 \,\}$$

```
> y0:=1;
```
$$y0 := 1$$

The growth of the population is shown in Figure 22.2(b).

```
> plot(y(t),t=0..30);
```

22.3   Derivatives

We can differentiate the logarithmic function from first principles as we have with many other functions.

```
> with(student):
> f:=x->ln(x);
```
$$f := \ln$$

```
> slope([x,f(x)],[x+delta,f(x+delta)]);
```
$$-\frac{\ln(x) - \ln(x+\delta)}{\delta}$$

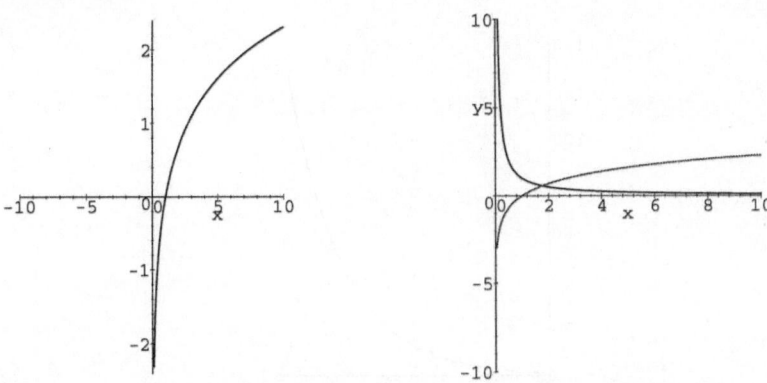

Figure 22.3. (a) The graph of $y = \log(x)$, on $[0, 10]$. (b) The graphs of the logarithmic function and its derivative, on $[0, 10]$

```
> limit(",delta=0);
```

$$\frac{1}{x}$$

```
> Diff(log(x),x)=diff(log(x),x);
```

$$\frac{\partial}{\partial x} \ln(x) = \frac{1}{x}$$

The following two plots can be seen in Figure 22.3.

```
> plot(log(x),x);
> plot({log(x),diff(log(x),x)},x=0..10,y=-10..10);
```

The exponential function has an unusual property. The derivative of the exponential function is equal to the exponential function. See Figure 22.4.

```
> Diff(exp(x),x)=diff(exp(x),x);
```

$$\frac{\partial}{\partial x} e^x = e^x$$

```
> Diff(exp(a*x),x)=diff(exp(a*x),x);
```

$$\frac{\partial}{\partial x} e^{(a x)} = a\, e^{(a x)}$$

```
> plot({exp(x),diff(exp(x),x)},x=0..5);
```

Exercise 22.2 Differentiate the following logarithmic functions.

(a) $y = \log(3x)$

(b) $y = \log(\sqrt{2 - x})$

(c) $y = \sin(\log(x))$

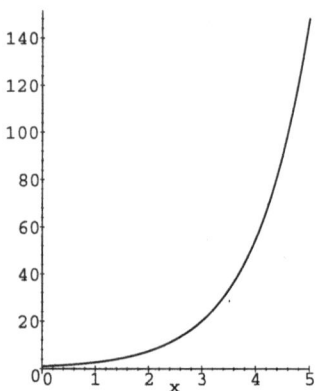

Figure 22.4. The graphs of the exponential function and its derivative, on $[0, 5]$

Exercise 22.3   Differentiate the following exponential functions.

(a) $y = e^{2x - x^2}$

(b) $y = \frac{e^x}{x^2}$

(c) $y = e^{\sin(x)}$

Exercise 22.4   Use the *Maple* function **series** to obtain quadratic approximations to e^x and $\frac{1}{1+x}$. Hence deduce that the slope of the tangent to $\frac{e^x}{1+x}$ is approximately x when x is small.

Exercise 22.5   Use **series** to show that for small x the slope of the tangent to $\left(\frac{1+x}{1-x}\right)^{1/3}$ is $\frac{2}{3}$.

Experiment 22.1   Let $f(x)$ satisfy the condition that $f'(x) = f(x)$. Now differentiate $e^{-x} f(x)$ to show that $f(x)$ is a constant multiple of the exponential function.

22.4   Integrals

The integral of the logarithmic function is,

```
> int(log(x),x);
```

$$x \ln(\,x\,) - x$$

But more often you will find that you need to integrate a function of the form $y = \frac{a}{x+b}$, the solution of which is,

```
> int(a/(x+b),x);
```

$$\ln(x+b)\,a$$

Because the derivative of the exponential function is equal to the exponential function, the integral of the exponential function is also equal to the exponential function.

```
> int(exp(x),x);
```

$$e^x$$

```
> int(2*x*exp(x^2),x);
```

$$e^{(x^2)}$$

Exercise 22.6 Integrate the following functions.

(a) $y = \frac{2}{x+3}$

(b) $y = \frac{x}{x^2-1}$

(c) $y = e^{-x+2}$

(d) $y = x^3 e^{x^4+1}$

Experiment 22.2 For positive numbers a, we define $a^x = e^{x \log a}$, for all real x and $x^r = e^{r \log x}$ for all real x, r. Plot $y = f(x)$ for $f = 2^x, e^x, 3^x$ over $-1 \leq x \leq 1$. What is $f'(x)$ for these functions? What is $\int f(x)dx$? Plot $g(x) = x^{1/x}$ for $0 \leq x \leq 10$, first showing that $g(x) \to 0$ as $x \to 0$. You see that g has a maximum near $x = 3$; try to locate it using the cursor in your plot. By considering the derivative of $\log(g(x))$ you can show that the maximum ocurs at $x = e$.

Chapter 23

Polar Coordinates

Commands used in this chapter
- assign
- cos
- factor
- Int
- lhs
- map
- plot
- plot[options]
- plots
- plots[implicitplot]

- rhs
- sin
- solve
- student
- student[completesquare]
- student[isolate]
- subs
- value
- with

23.1   Converting Coordinates

A point, P, can be described in *Cartesian coordinates* as $P(x, y)$, or in *polar coordinates* as $P(r, \theta)$, where r is the radius and θ is the angle from the terminal arm (the positive section of the x-axis). From Chapter 8, the expressions which relate x and y to r and θ are $x = r\cos(\theta)$ and $y = r\sin(\theta)$.

```
> x=r*cos(theta);
```
$$x = r\cos(\theta)$$

```
> solve(",{cos(theta)});
```
$$\left\{ \cos(\theta) = \frac{x}{r} \right\}$$

```
> assign(");
> y=r*sin(theta);
```
$$y = r\sin(\theta)$$

```
> solve(",{sin(theta)});
```
$$\left\{ \sin(\theta) = \frac{y}{r} \right\}$$

```
> assign(");
> cos(theta)^2+sin(theta)^2=1;
```

$$\frac{x^2}{r^2} + \frac{y^2}{r^2} = 1$$

```
> with(student):
> isolate("",r^2);
```
$$r^2 = x^2 + y^2$$

```
> isolate(",r);
```
$$r = \text{RootOf}(\_Z^2 - x^2 - y^2)$$

```
> allvalues(");
```
$$r = \sqrt{x^2 + y^2}, r = -\sqrt{x^2 + y^2}$$

```
> assign(");
```

With this relation, we can perform the transformation from Cartesian (x, y) to polar (r, θ) coordinates.

```
> P:=[1,1];
```
$$P := [1, 1]$$

```
> x:=P[1];
```
$$x := 1$$

```
> y:=P[2];
```
$$y := 1$$

```
> r;
```
$$-\sqrt{2}$$

```
> theta:=arctan(y/x);
```
$$\theta := \frac{1}{4}\pi$$

```
> evalf(");
```
$$.7853981635$$

```
> evalf(convert(",degrees));
```
$$44.99999998 \ degrees$$

We can also convert functions in terms of x and y into polar equations.

```
> y=x^2+3;
```
$$y = x^2 + 3$$

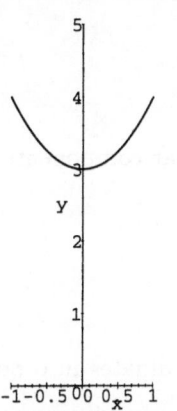
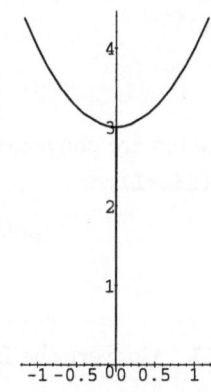

Figure 23.1. (a) The graph of $y = x^2 + 3$, on $[-1, 1]$. (b) The graph of polar equation $r \sin(t) = r^2 \cos(t) + 3$, on $[0, \pi]$

```
> subs({x=r*cos(t),y=r*sin(t)},");
```
$$r \sin(t) = r^2 \cos(t)^2 + 3$$

```
> isolate(",r);
```
$$r = -\frac{1}{2} \frac{-\sin(t) + \sqrt{\sin(t)^2 - 12 \cos(t)^2}}{\cos(t)^2}$$

```
> assign(");
```

These plots, Figure 23.1, show that the two forms of the equation have the same graph.

```
> plot(x^2+3,x=-1..1,y=0..5,scaling=CONSTRAINED);
> plot([r,t,t=0..Pi],coords=polar,scaling=CONSTRAINED);
```

Maple has a function called `polar` which can convert a given radius and angle into polar coordinates. Notice how it deals with negative radial values.

```
> readlib(polar):
> polar(3,Pi/2);
```
$$\text{polar}\left(3, \frac{1}{2}\pi\right)$$

```
> polar(-3, Pi/2);
```
$$\text{polar}\left(3, \frac{3}{2}\pi\right)$$

```
> polar(-1,Pi);
```
$$\text{polar}(1, 2\pi)$$

```
> polar(1);
```

$$\text{polar}(1, 0)$$

Complex numbers are converted into radial and angular components.

```
> polar(2+3*I);
```

$$\text{polar}\left(\sqrt{13}, \arctan\left(\frac{3}{2}\right)\right)$$

Exercise 23.1   Convert the following Cartesian coordinates into polar coordinates.

(a) $(2, -5)$

(b) $(-1, -3)$

(c) $(0, 6)$

Exercise 23.2   Convert the following polar coordinates into Cartesian coordinates.

(a) $(3, \frac{\pi}{4})$

(b) $(\sqrt{2}, -\frac{\pi}{6})$

(c) $(1, 50°)$

Exercise 23.3   Convert the following functions into polar form.

(a) $x + y + 1 = 0$

(b) $x^2 + y^2 - 4 = 0$

(c) $x^2 + xy + 8 = 0$

23.2   Circles in Polar Coordinates

We know that the equation of a circle in Cartesian coordinates is expressed as $(x - a)^2 + (y - b)^2 = r^2$, where a, b, and r are constants. How is the equation of a circle expressed in polar coordinates?

The polar equation $r = a$ gives a circle of radius a centred at the origin.

```
> subs({r=(x^2+y^2)^(1/2),theta=arctan(y/x)},r=a);
```

$$\sqrt{x^2 + y^2} = a$$

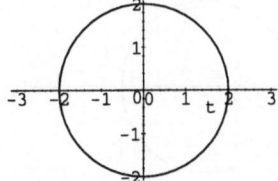

Figure 23.2. The graph of polar equation $r = 2$, on $[-3, 3]$

```
> map(x->x^2,");
```
$$x^2 + y^2 = a^2$$

So for the polar equation $r = 2$, the graph is shown in Figure 23.2.

```
> plot(2,t=-Pi..Pi,coords=polar);
```

A polar equation of the form $r = a\cos(t)$ gives a circle centred at $(\frac{a}{2}, 0)$ with radius $\frac{a}{2}$. The plot of the relation $r = 2\cos(t)$ is shown in Figure 23.3. Table 23.1 lists r values at various points along the circle.

t	$r(t)$
0	2
$\frac{\pi}{4}$	$\sqrt{2}$
$\frac{\pi}{2}$	0
$\frac{3\pi}{4}$	$-\sqrt{2}$
π	-2

Table 23.1. Values of $r(t) = 2\cos(t)$

```
> r:=t->2*cos(t);
```
$$r := t \rightarrow 2\cos(t)$$

We can convert the polar equation into the more familiar Cartesian form.

```
> subs(cos(theta)=x/r,r=a*cos(theta));
```
$$r = \frac{a\,x}{r}$$

```
> isolate(",x);
```
$$x = \frac{r^2}{a}$$

```
> isolate(",r^2);
```
$$r^2 = x\,a$$

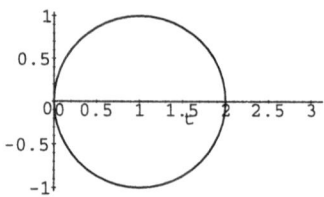

Figure 23.3. The graph of polar equation $r = 2\cos(t)$, on $[0, \pi]$

```
> subs(r=(x^2+y^2)^(1/2),");
```
$$x^2 + y^2 = x\,a$$

```
> lhs(")+rhs(")=0;
```
$$x^2 + y^2 + x\,a = 0$$

```
> completesquare(",x);
```
$$\left(x + \frac{1}{2}\,a\right)^2 - \frac{1}{4}\,a^2 + y^2 = 0$$

```
> plot(2*cos(t),t=0..Pi,coords=polar);
```

A polar equation of the form $r = a\sin(t)$, shown in Figure 23.4, gives a circle centred at $(0, \frac{a}{2})$ with radius $\frac{a}{2}$. Table 23.2 lists r values at various points along the circle.

t	$r(t)$
0	0
$\frac{\pi}{4}$	$\sqrt{2}$
$\frac{\pi}{2}$	2
$\frac{3\pi}{4}$	$\sqrt{2}$
π	0

Table 23.2. Values of $r(t) = 2\sin(t)$

```
> r:=t->2*sin(t);
```
$$r := t \rightarrow 2\sin(t)$$

```
> plot(2*sin(t),t=-Pi/2..Pi/2,coords=polar,scaling=CONSTRAINED);
```

A circle centred at (a, b), where a and b are not zero, can be expressed in polar coordinates, but the equation for the radius is much more complicated.

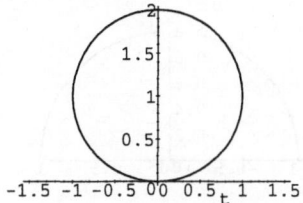

Figure 23.4. The graph of polar equation $r = 2\sin(t)$, on $[-\frac{\pi}{2}, \frac{\pi}{2}]$

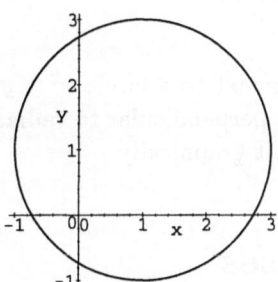 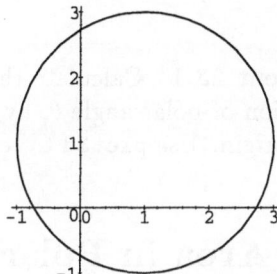

Figure 23.5. Circle of radius 2 and centred at $(1,1)$ in (a) Cartesian coordinates, (b) polar coordinates

```
> (x-1)^2+(y-1)^2=4;
```
$$(x-1)^2 + (y-1)^2 = 4$$

The following implicit plot is shown in Figure 23.5(a).
```
> with(plots):
> implicitplot((x-1)^2+(y-1)^2=4,x=-1..3,y=-1..3);
> subs({x=r*cos(theta),y=r*sin(theta)},(x-1)^2+(y-1)^2=4);
```
$$(r\cos(\theta) - 1)^2 + (r\sin(\theta) - 1)^2 = 4$$

```
> isolate(",r);
```
$$r = \frac{1}{2}\left(\right.$$
$$2\cos(\theta) + 2\sin(\theta) + 2\sqrt{3\cos(\theta)^2 + 2\cos(\theta)\sin(\theta) + 3\sin(\theta)^2}$$
$$\left.\right)/\left(\cos(\theta)^2 + \sin(\theta)^2\right)$$

```
> assign(");
```
The polar plot below is shown in Figure 23.5(b).

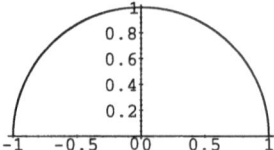

Figure 23.6. Half circle

```
> plot([r,theta,theta=-Pi..Pi],coords=polar);
```

Experiment 23.1   Calculate the slope of the tangent to a circle $x^2 + y^2 = 1$ as a function of polar angle θ, by considering lines perpendicular to radial lines from the origin. Use `plot3d` to represent your result graphically.

23.3   Area in Polar Coordinates

23.3.1   Sector of a Circle

The area of a sector of a circle is simply the fraction of the total area that the given angle is of 2π.

```
> area:=(theta/(2*Pi))*Pi*(r^2);
```

$$area := \frac{1}{2}\,\theta\,r^2$$

The area of a half-circle ($\theta = \pi$) of radius $r = 1$, shown in Figure 23.6, is $\frac{\pi}{2}$.

```
> subs(theta=Pi,r=1,area);
```

$$\frac{1}{2}\,\pi$$

```
> plot([1,theta, theta=0..Pi],coords=polar,scaling=CONSTRAINED);
```

Experiment 23.2   Consider parallel lines, separated by distance d, ruled in a large square. Drop a line segment (like dropping a needle on a plank floor) at random onto the square. Investigate the probability that a line segment, of length $L < d$ intersects one of the ruled lines. Do this by a simulation using *Maple's* random number generator `rand`, as explained in Chapter 12.5, to create random coordinates for one end of the segment and then compute the fraction of the circle of radius L which intersects ruled lines. This needs to be repeated many times to obtain a good estimate of the required probability. You might try to show analytically that the probability density function for intersection of a line at an angle θ is

$$p(\theta) = \frac{2}{\pi}\frac{L\cos\theta}{d} \text{ for } 0 \le \theta \le \frac{\pi}{2}$$

Then the expected probability at any angle is

$$\int_0^{\frac{\pi}{2}} p(\theta)d\theta = \frac{2L}{\pi d}$$

So if you fix L and d you could use your simulation to calculate π. This is the famous 'Needle Problem' of the Comte de Buffon, 1777.

23.3.2   Region Enclosed by a Curve

Consider a sector with a very small angle $d\theta$, like that shown in Figure 23.7.

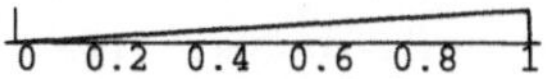

Figure 23.7. Sector of a circle of angle $d\theta$

Its area is approximately $\frac{r}{2}(rd\theta) = \frac{r^2}{2}d\theta$, half the area of a rectangle r by $rd\theta$. For a large sector, we just have to sum contributions like that, which is done by integration.

```
> Int(1/2*r^2,theta);
```

$$\int \frac{1}{2}r^2\,d\theta$$

Example 23.1   Find the area enclosed by a cardioid (Figure 23.8(a)), which has the polar equation $r = 2 + 2\cos(\theta)$ from 0 to 2π.

```
> Int(1/2*(2+2*cos(theta))^2,theta=0..2*Pi);
```

$$\int_0^{2\pi} \frac{1}{2}\left(2 + 2\cos(\theta)\right)^2 d\theta$$

```
> value(");
```

$$6\,\pi$$

```
> plot([2+2*cos(theta),theta,theta=0..2*Pi],coords=polar);
```

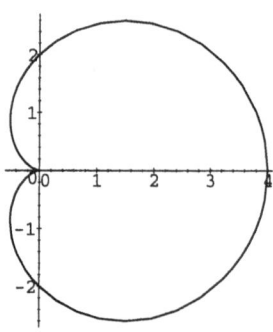 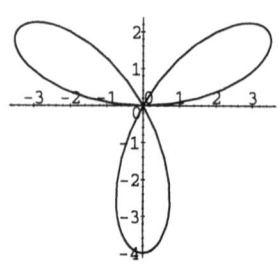

Figure 23.8. (a) Cardioid. (b) Three-leaved rose

Example 23.2   Find the area enclosed by a three-leaved rose with radius function $r = 4\sin(3\theta)$ from 0 to 2π as shown in Figure 23.8(b).

```
> Int(1/2*(4*sin(3*theta))^2,theta=0..2*Pi);
```
$$\int_0^{2\pi} 8\sin(3\,\theta)^2\,d\theta$$

```
> value(");
```
$$8\,\pi$$

```
> plot([4*sin(3*theta),theta,theta=0..2*Pi],coords=polar);
```

Now find the area enclosed by a three-leaved rose with radius function $r = 4\sin(3\theta)$ from 0 to $\frac{\pi}{3}$ as shown in Figure 23.8(b).

```
> Int(1/2*(4*sin(3*theta))^2,theta=0..Pi/3);
```
$$\int_0^{1/3\,\pi} 8\sin(3\,\theta)^2\,d\theta$$

```
> value(");
```
$$\frac{4}{3}\,\pi$$

```
> plot([4*sin(3*theta),theta,theta=0..Pi/3],coords=polar);
```

Exercise 23.4   In each of the following cases, find the area enclosed by the loop of the graph.

(a) $r = 3\sin(2\theta)$ Hint: $0 \le \theta \le \frac{\pi}{2}$

(b) $r = 5\cos(3\theta)$

(c) $r = \cos(3\theta/2)$

Exercise 23.5   Find the area of the circle $r = \sin(x) + \cos(x)$ by integration in polar coordinates. Check the answer by writing the equation of the circle in rectangular coordinates, finding its radius, and then using the familiar area formula. (p510, Edwards and Penney [6])

Exercise 23.6

(a) Show that the area enclosed by the cardioid $r = a(1 + \cos\theta)$ is
$$\frac{1}{2}\int_0^{2\pi} r^2 d\theta = \frac{3\pi a^2}{2}.$$

(b) Show that the area enclosed by one loop of the curve $r = a\cos 3\theta$ is
$$\frac{\pi a^2}{2}.$$

(c) Show that the area enclosed by the loops of the curve $r^2 = a^2 \sin 2\theta$
is $\dfrac{1}{2}\int_0^{\frac{\pi}{2}} r^2 d\theta + \dfrac{1}{2}\int_0^{\frac{3\pi}{2}} r^2 d\theta = a^2.$

Experiment 23.3   A curve given parametrically by $(x(t), y(t))$ has curvature given by
$$K(t) = \frac{x'(t)y''(t) - x''(t)y'(t)}{((x'(t))^2 + (y'(t))^2)^{3/2}}$$

Plot the figure of eight curve $(\sin t, \sin t \cos t)$, which you may have used for your first model railway, with $0 \leq t \leq 2\pi$. Now calculate its curvature and plot $K(t)$ for $0 \leq t \leq 2\pi$. Try other curves in parametric form, including circles, ellipses and cardioids.

Appendix A

Solutions to Part I Exercises

A.1   Functions

Exercise 2.1   Which of the following relations are functions and why?

(a) $f = \{(1,1),(3,4),(5,5),(5,6)\}$

```
> plot([1,1,3,4,5,5,5,6]);
```

No. The last two points are connected by a vertical line. On a vertical line there are an infinite number of values in the range for each value in the domain.

(b) $\frac{x^2}{4} + \frac{y^2}{2} = 1$

```
> with(plots):
> implicitplot((x/4)^2+(y/2)^2=1,x=-4..4,y=-2..2,scaling=CONSTRAINED);
```

No. An ellipse, like a circle, has two values of y for each value of x.

(c) $y = x^3 - 2x^2$

```
> plot(x^3-2*x^2,x=-2..4);
```

Yes. Each value in the domain maps onto a unique value in the range.

(d) $\frac{x^2}{4} - \frac{y^2}{2} = 1$

```
> implicitplot((x/4)^2-(y/2)^2=1,x=-10..10,y=-10..10);
```

No. A hyperbola also has two values of y for each value of x.

(e) $y = 1/x$

```
> plot(1/x,x=-2..2,y=-10..10);
```

Yes. Each value in the domain maps onto a unique value in the range. The function has an asymptote at $x = 0$.

```
> plot([1,1,3,4,5,5,5,6]);
```

No. The last two points are connected by a vertical line. On a vertical line there are an infinite number of values in the range for each value in the domain.

Exercise 2.2   Given $f(x) = 2x^2 - 4x + 9$, plot $f(x)$ and $3f(x)$, $-10 \le x \le 10$, on the same graph.

```
> f:=x->2*x^2-4*x+9;
```
$$f := x \rightarrow 2\,x^2 - 4\,x + 9$$

```
> plot({f(x),3*f(x)},x=-10..10);
```

Exercise 2.3   Perfect gases expand when heated according to the Ideal Gas Law $PV = nRT$ where P is the pressure, V is the volume, n is proportional to the amount of gas present, R is a constant, and T is the Kelvin temperature. Engineers who design cars use this formula to determine the *explosion point*, that is, the condition when increase in temperature will cause the windshield of a car to pop out. (p5, Goldenberg and Greenwald [7])

(a) Assume the interior of a locked car is airtight, the contents are at one atmosphere, and the initial temperature is 293 Kelvin (about 20°C). Suppose the temperature inside the car increases 15 Kelvin for each hour the car remains in the sun. The car manufacturer has designed the windshield so that it will pop out when the interior pressure is 1.6 times the exterior pressure. How long would these conditions have to be maintained before the windshield pops out?

```
> eq1:=P1*V=n*R*T1;
```
$$eq1 := P1\,V = n\,R\,T1$$

```
> eq2:=P2*V=n*R*T2;
```
$$eq2 := P2\,V = n\,R\,T2$$

```
> T1:=293;
```
$$T1 := 293$$

```
> dT:=15;
```
$$dT := 15$$

```
> P2:=1.6*P1;
```
$$P2 := 1.6\,P1$$

```
> lhs(eq1)/lhs(eq2)=rhs(eq1)/rhs(eq2);
```
$$.6250000000 = 293\,\frac{1}{T2}$$

```
> solve(",{T2});
```
$$\{\,T2 = 468.8000000\,\}$$

```
> assign(");
> hours:=(T2-T1)/dT;
```
$$hours := 11.72000000$$

(b) A car with a sun roof will heat up faster than a car with a solid roof. Suppose the temperature of the interior of a closed car with a sun roof increases at a rate of 20 Kelvin per hour. Suppose the same pressure difference as described in the example will cause the windshield to pop out. How long will it take to reach this pressure?

```
> dT:=20;
```
$$dT := 20$$

```
> hours:=(T2-T1)/dT;
```
$$hours := 8.790000000$$

(c) Older cars do not have windshields that pop out. If an older car has a sun roof and the interior temperature increases by 20 Kelvin per hour, what will the interior temperature be after 1.5 hours in the sun? (Assume the initial temperature is 293 Kelvin.)

```
> T2:=T1+1.5*dT;
```
$$T2 := 323.0$$

```
> P1:=1;
```
$$P1 := 1$$

```
> P2:='P2';
```
$$P2 := P2$$

```
> lhs(eq1)/lhs(eq2)=rhs(eq1)/rhs(eq2);
```
$$\frac{1}{P2} = .9071207430$$

```
> solve(",P2);
```
$$1.102389079$$

(d) A half-full bottle of sunscreen is left in the car. One window is partially open so the air pressure in the car does not build up, but the temperature does. The temperature in the bottle increases from 293 Kelvin to 353 Kelvin. Find the pressure within the bottle of sunscreen. Assume there is no change in volume and there is no increase in vapour pressure from the heated lotion.

```
> T1:=293;
```
$$T1 := 293$$

```
> T2:=353;
```
$$T2 := 353$$

```
> P1:=1;
```
$$P1 := 1$$

```
> lhs(eq1)/lhs(eq2)=rhs(eq1)/rhs(eq2);
```
$$\frac{1}{P2} = \frac{293}{353}$$

```
> fsolve(",P2);
```
$$1.204778157$$

(e) The plastic bottle described above is empty and expands. The volume
    is 15% more when heated. What is the new pressure?

```
> eq1:=P1*V1=n*R*T1;
```
$$eq1 := V1 = 293\,n\,R$$

```
> eq2:=P2*V2=n*R*T2;
```
$$eq2 := P2\,V2 = 353\,n\,R$$

```
> V2:=1.15*V1;
```
$$V2 := 1.15\,V1$$

```
> lhs(eq1)/lhs(eq2)=rhs(eq1)/rhs(eq2);
```
$$.8695652174\,\frac{1}{P2} = \frac{293}{353}$$

```
> fsolve(",P2);
```
$$1.047633180$$

Exercise 2.4   Given $f(x) = x^3 - 8$ and $g(x) = x + 4$, plot $f(x)$, $g(x)$, and
$f(x) + g(x)$, $-2 \le x \le 2$, on the same graph.

```
> f:=x->x^3-8;
```
$$f := x \to x^3 - 8$$

```
> g:=x->x+4;
```
$$g := x \to x + 4$$

```
> plot({f(x),g(x),f(x)+g(x)},x=-2..2);
```

Exercise 2.5   A function f is called *even* if $f(-x) = f(x)$ for all x and it is
called *odd* if $f(-x) = -f(x)$ for all x. Give examples of even functions and of
odd functions. Can you find a function that is odd and even?

```
> f:=x->x^2+2;
```
$$f := x \to x^2 + 2$$

```
> f(-x);
```
$$x^2 + 2$$

```
> f:=x->x;
```
$$f := x \rightarrow x$$

```
> f(-x);
```
$$-x$$

The zero function, $f(x) = 0$ for all x, is odd and even.

Exercise 2.6   Use the identity $f(x) = \frac{1}{2}(f(x) + f(-x)) + \frac{1}{2}(f(x) - f(-x))$ to show that every function is expressible as a sum of an even function and an odd function.

```
> 1/2*(f(x)+f(-x))+1/2*(f(x)-f(-x));
```
$$x$$

```
> f:=x->2*x^2-3*x+1;
```
$$f := x \rightarrow 2x^2 - 3x + 1$$

```
> 1/2*(f(x)+f(-x))+1/2*(f(x)-f(-x));
```
$$2x^2 - 3x + 1$$

```
> f:=x->1/x;
```
$$f := x \rightarrow \frac{1}{x}$$

```
> 1/2*(f(x)+f(-x))+1/2*(f(x)-f(-x));
```
$$\frac{1}{x}$$

```
> f:=x->sin(x);
```
$$f := \sin$$

```
> 1/2*(f(x)+f(-x))+1/2*(f(x)-f(-x));
```
$$\sin(x)$$

Exercise 2.7   Let $[[x]]$ be the greatest integer less than or equal to x. Plot the graphs of $[[x]]$, $x - [[x]]$, $\sqrt{x - [[x]]}$, $x + \sqrt{x - [[x]]}$, and $[[x]] + \sqrt{x - [[x]]}$ using *Maple's* trunc function to calculate $[[x]]$, ie trunc(x) = $[[x]]$.

```
> f:=x->trunc(x);
```
$$f := trunc$$

```
> plot(f(x),x=-2..2);
> f:=x->x-trunc(x);
```
$$f := x \rightarrow x - \text{trunc}(x)$$

```
> plot(f(x),x,-2..2);
> f:=x->sqrt(x-trunc(x));
```
$$f := x \rightarrow \text{sqrt}(x - \text{trunc}(x))$$

```
> plot(f(x),x,-2..2);
> f:=x->x+sqrt(x-trunc(x));
```
$$f := x \rightarrow x + \text{sqrt}(x - \text{trunc}(x))$$

```
> plot(f(x),x,0..5);
> f:=x->trunc(x)+sqrt(x-trunc(x));
```
$$f := x \rightarrow \text{trunc}(x) + \text{sqrt}(x - \text{trunc}(x))$$

```
> plot(f(x),x=0..5);
```

Exercise 2.8 Given $f(x) = x^2 - 1$ and $g(x) = x$, plot $f(x)$, $g(x)$, and $f(x) \times g(x)$, $-2 \leq x \leq 2$, on the same graph.

```
> f:=x->x^2-1;
```
$$f := x \rightarrow x^2 - 1$$

```
> g:=x->x;
```
$$g := x \rightarrow x$$

```
> plot({f(x),g(x),f(x)*g(x)},x=-2..2);
> f(x)*g(x);
```
$$(x^2 - 1)x$$

```
> expand(");
```
$$x^3 - x$$

Exercise 2.9 Given $f(x) = 8x - 7$ and $g(x) = x^2 + x$, find $f(g(x))$, $g(f(x))$, and $f(f(g(x)))$.

```
> f:=x->8*x-7;
```
$$f := x \rightarrow 8x - 7$$

```
> g:=x->x^2+x;
```
$$g := x \rightarrow x^2 + x$$

```
> (f@g)(x);
```
$$8x^2 + 8x - 7$$

```
> (g@f)(x);
```
$$(8x - 7)^2 + 8x - 7$$

```
> (f@(f@g))(x);
```
$$64x^2 + 64x - 63$$

Exercise 2.10   Let $f(x) = x^{1/3}$ and for $-1 \le x \le 1$. Consider the iterated functions $f_1(x) = f(x)$, $f_2(x) = f(f_1(x))$, $f_3(x) = f(f_2(x))$, ... Show that each of these new functions maps the interval $[-1, 1]$ to itself and plot their graphs for $n = 1, 2, \ldots$ Animate your sequence of plots for n up to 10. What do you expect the iterations of $g(x) = x^{1/5}$ to look like?

```
> f:=x->x^(1/3);
```
$$f := x \rightarrow x^{1/3}$$

```
> f1:=f(x);
```
$$f1 := x^{1/3}$$

```
> f2:=f(f1);
```
$$f2 := x^{1/9}$$

```
> f3:=f(f2);
```
$$f3 := x^{1/27}$$

```
> plot(f(x),x);
> plot(f1,x);
> plot(f2,x);
> plot(f3,x);
> display([seq(plot((f@@n)(x),x=0..20),n=1..10)],insequence=true);
```

Exercise 2.11   By considering the results of iterations of $f(x) = x^{1/3}$ in Exercise 2.10, deduce the form of iterations of $h(x) = x^3$.

```
> h:=x->x^3;
```
$$h := x \rightarrow x^3$$

```
> h1:=h(x);
```

$$h1 := x^3$$

```
> h2:=h(h1);
```

$$h2 := x^9$$

```
> h3:=h(h2);
```

$$h3 := x^{27}$$

Exercise 2.12 Given $f(x) = 5x - 9$, find the inverse of this function using simultaneous substitutions and plot these relations.

```
> y=5*x-9;
```

$$y = 5x - 9$$

```
> subs({x=y,y=x},y=5*x-9);
```

$$x = 5y - 9$$

```
> implicitplot({y=5*x-9,x=5*y-9},x=-10..10,y=-10..10);
```

The other way to plot these equations is to isolate y.

```
> with(student):
> isolate(x=5*y-9,y);
```

$$y = \frac{1}{5}x + \frac{9}{5}$$

```
> plot({5*x-9,1/5*x+9/5},x=-10..10);
```

A.2   Quadratics

Exercise 3.1   Given $f(x) = x^2$, plot $f(x)$, $2f(x)$, $\frac{1}{2}f(x)$, and $-2f(x)$, $-4 \leq x \leq 4$, on the same graph.

```
> f:=x->x^2;
```
$$f := x \rightarrow x^2$$

```
> plot({f(x),2*f(x),1/2*f(x),-2*f(x)},x=-4..4);
```

Exercise 3.2   Given $f(x) = 2x^2 + c$, plot $f(x)$ for $c = 1, 2$, and -2, $-4 \leq x \leq 4$, on the same graph.

```
> f:=x->2*x^2;
```
$$f := x \rightarrow 2\,x^2$$

```
> plot({f(x)+1,f(x)+2, f(x)-2},x=-4..4);
```

Exercise 3.3   Given $f(x) = 2x^2 - 6x + 7$, plot the function, $-4 \leq x \leq 4$, and find the vertex of the parabola, the axis of symmetry and the direction of opening.

```
> f:=x->2*x^2-6*x+7;
```
$$f := x \rightarrow 2\,x^2 - 6\,x + 7$$

```
> plot(f(x),x=-4..4);
> with(student):
> completesquare(f(x));
```
$$2\left(x - \frac{3}{2}\right)^2 + \frac{5}{2}$$

Vertex at $(\frac{3}{2}, \frac{5}{2})$. Axis of symmetry $x = \frac{3}{2}$. Opens upward.

Exercise 3.4   Given $f(x) = 5x^2$, find the equation of the parabola which has its vertex two units to the right and three units above it. Plot the parabolas over the interval $-4 \leq x \leq 4$.

```
> f:=x->5*x^2;
```
$$f := x \rightarrow 5\,x^2$$

```
> plot(f(x),x=-4..4);
> eq:=f(x-2)+3;
```
$$eq := 5\,(x - 2)^2 + 3$$

```
> plot({f(x),eq},x=-4..4);
```

Alternatively, the equation of the parabola can be determined from the coordinates of the vertex like this,

```
> a:=5;
```
$$a := 5$$

```
> p:=2;
```
$$p := 2$$

```
> q:=3;
```
$$q := 3$$

```
> y:=a*(x-p)^2+q;
```
$$y := 5\,(x-2)^2 + 3$$

Exercise 3.5 Given $f(x) = -3x^2 - 7x + 1$, find the equation of the parabola which has its vertex one unit to the left and four units below it. Plot the parabolas over the interval $-4 \le x \le 4$.

```
> f:=x->-3*x^2-7*x+1;
```
$$f := x \to -3\,x^2 - 7\,x + 1$$

```
> plot(f(x),x=-4..4);
> eq:=f(x+1)-4;
```
$$eq := -3\,(x+1)^2 - 7\,x - 10$$

```
> plot({f(x),eq},x=-4..4);
```

Or,

```
> completesquare(f(x));
```
$$-3\left(x + \frac{7}{6}\right)^2 + \frac{61}{12}$$

```
> p1:=-7/6;
```
$$p1 := \frac{-7}{6}$$

```
> q1:=61/12;
```
$$q1 := \frac{61}{12}$$

```
> a:=-3;
```
$$a := -3$$

```
> p2:=p1-1;
```
$$p2 := \frac{-13}{6}$$

```
> q2:=q1-4;
```
$$q2 := \frac{13}{12}$$

```
> y:=a*(x-p2)^2+q2;
```
$$y := -3 \left(x + \frac{13}{6} \right)^2 + \frac{13}{12}$$

```
> plot({y,f(x)},x=-4..4);
> x:='x':y:='y':a:='a':
```

Exercise 3.6 A golf ball is hit and in the air follows a parabolic trajectory starting at $x = 0$, $y = 0$ with maximum altitude $y = 20$ m; it strikes the ground at $x = 120$ m. Find an equation for the parabola.

```
>   f:=x->a*(x-p)^2+q;
```
$$f := x \rightarrow a\,(x - p)^2 + q$$

```
> p:=60;
```
$$p := 60$$

```
> q:=20;
```
$$q := 20$$

```
> f(0)=0;
```
$$3600\,a + 20 = 0$$

```
> solve(",{a});
```
$$\left\{ a = \frac{-1}{180} \right\}$$

```
> assign(");
> f(x);
```
$$-\frac{1}{180}\,(x - 60)^2 + 20$$

```
> plot(f(x),x=0..120);
```

A.3 Solving Quadratics

Exercise 4.1 Determine the roots of the function $y = 2x^2 - x - 1$ graphically.

```
> y:=2*x^2-x-1;
```
$$y := 2\,x^2 - x - 1$$

```
> plot(y,x=-2..2);
> root1:=-0.5;
```
$$root1 := -.5$$

```
> root2:=1;
```
$$root2 := 1$$

```
> y:=(x-root1)*(x-root2);
```
$$y := (\,x + .5\,)(\,x - 1\,)$$

```
> expand(");
```
$$x^2 - .5\,x - .5$$

```
> 2*";
```
$$2\,x^2 - 1.0\,x - 1.0$$

Exercise 4.2 Determine the roots of the following functions using any method.

(a) $y = 2x^2 + 3x + 9$
```
> y:=2*x^2+3*x+9;
```
$$y := 2\,x^2 + 3\,x + 9$$

```
> plot(y,x=-5..5);
```
As we can see on the graph, there are no real roots.
```
> x:=[(-b+sqrt(b^2-4*a*c))/(2*a),(-b-sqrt(b^2-4*a*c))/(2*a)];
```
$$x := \left[\frac{1}{2}\frac{-b + \sqrt{b^2 - 4\,a\,c}}{a}, \frac{1}{2}\frac{-b - \sqrt{b^2 - 4\,a\,c}}{a}\right]$$

```
> a:=2;
```
$$a := 2$$

```
> b:=3;
```
$$b := 3$$

```
> c:=9;
```
$$c := 9$$

```
> x;
```
$$\left[-\frac{3}{4} + \frac{1}{4}\sqrt{-63}, -\frac{3}{4} - \frac{1}{4}\sqrt{-63} \right]$$

```
> simplify(");
```
$$\left[-\frac{3}{4} + \frac{3}{4}I\sqrt{7}, -\frac{3}{4} - \frac{3}{4}I\sqrt{7} \right]$$

```
> x:='x':
> solve(y,{x});
```
$$\left\{ x = -\frac{3}{4} + \frac{3}{4}I\sqrt{7} \right\}, \left\{ x = -\frac{3}{4} - \frac{3}{4}I\sqrt{7} \right\}$$

(b) $y = x^2 - 9$

```
> y:=x^2-9;
```
$$y := x^2 - 9$$

```
> plot(y,x=-5..5);
```
From the graph, the roots of this function are located at $x = -3, 3$.

```
> root1:=-3;
```
$$root1 := -3$$

```
> root2:=3;
```
$$root2 := 3$$

```
> 'y'=(x-root1)*(x-root2);
```
$$y = (x + 3)(x - 3)$$

```
> a:=1;
```
$$a := 1$$

```
> b:=0;
```
$$b := 0$$

```
> c:=-9;
```
$$c := -9$$

```
> x:=[(-b+sqrt(b^2-4*a*c))/(2*a),(-b-sqrt(b^2-4*a*c))/(2*a)];
```
$$x := [3, -3]$$

```
> x:='x':
> solve(y,{x});
```
$$\{\,x = 3\,\},\{\,x = -3\,\}$$

(c) $y = x^2 - 4x + 1$
```
> y:=x^2-4*x+1;
```
$$y := x^2 - 4\,x + 1$$

```
> plot(y,x=-5..5);
```
From the graph, the roots of this function are located at $x = 0.25, 3.75$.

```
> root1:=0.25;
```
$$root1 := .25$$

```
> root2:=3.75;
```
$$root2 := 3.75$$

```
> 'y'=(x-root1)*(x-root2);
```
$$y = (\,x - .25\,)(\,x - 3.75\,)$$

```
> a:=1;
```
$$a := 1$$

```
> b:=-4;
```
$$b := -4$$

```
> c:=1;
```
$$c := 1$$

```
> x:=[(-b+sqrt(b^2-4*a*c))/(2*a),(-b-sqrt(b^2-4*a*c))/(2*a)];
```
$$x := \left[2 + \sqrt{3}, 2 - \sqrt{3}\right]$$

```
> evalf(");
```
$$[\,3.732050808, .267949192\,]$$

```
> x:='x':
> solve(y,{x});
```
$$\left\{x = 2 + \sqrt{3}\right\}, \left\{x = 2 - \sqrt{3}\right\}$$

The graphical estimate was a little off.

Exercise 4.3   The roots of $x^2 + bx + c$ are u, v; express $(u^2 + v^2)$ in terms of b, c.

```
> a:='a';b:='b':c:='c':
```
$$a := a$$

```
> u:=(-b+sqrt(b^2-4*c))/2;
```
$$u := -\frac{1}{2}b + \frac{1}{2}\sqrt{b^2 - 4c}$$

```
> v:=(-b-sqrt(b^2-4*c))/2;
```
$$v := -\frac{1}{2}b - \frac{1}{2}\sqrt{b^2 - 4c}$$

```
> u^2+v^2;
```
$$\left(-\frac{1}{2}b + \frac{1}{2}\sqrt{b^2 - 4c}\right)^2 + \left(-\frac{1}{2}b - \frac{1}{2}\sqrt{b^2 - 4c}\right)^2$$

```
> simplify(");
```
$$b^2 - 2c$$

Exercise 4.4   One root of $ax^2 + bx + c = 0$ is twice the other root; prove that $2b^2 = 9ac$.

```
> r1:=(-b+sqrt(b^2-4*a*c))/(2*a);
```
$$r1 := \frac{1}{2}\frac{-b + \sqrt{b^2 - 4ac}}{a}$$

```
> r2:=(-b-sqrt(b^2-4*a*c))/(2*a);
```
$$r2 := \frac{1}{2}\frac{-b - \sqrt{b^2 - 4ac}}{a}$$

```
> r1=2*r2;
```
$$\frac{1}{2}\frac{-b + \sqrt{b^2 - 4ac}}{a} = \frac{-b - \sqrt{b^2 - 4ac}}{a}$$

```
> with(student):
> isolate("",b^2);
```
$$b^2 = \frac{1}{9}b^2 + 4ac$$

```
> isolate(",b^2);
```
$$b^2 = \frac{9}{2}ac$$

Exercise 4.5   Determine the nature of the roots of the following functions.

(a) $y = 3x^2 - 1$

```
> a:='a':b:='b':c:='c':
> discriminant:=b^2-4*a*c;
```
$$discriminant := b^2 - 4\,a\,c$$

```
> y:=3*x^2-1;
```
$$y := 3\,x^2 - 1$$

```
> plot(y,x=-5..5);
> a:=3;
```
$$a := 3$$

```
> b:=0;
```
$$b := 0$$

```
> c:=-1;
```
$$c := -1$$

```
> discriminant;
```
$$12$$

Two real roots.

(b) $y = x^2 - 3x + 1$

```
> y:=x^2-3*x+1;
```
$$y := x^2 - 3\,x + 1$$

```
> plot(y,x=-5..5);
> a:=1;
```
$$a := 1$$

```
> b:=-3;
```
$$b := -3$$

```
> c:=1;
```
$$c := 1$$

```
> discriminant;
```
$$5$$

Two real roots.

(c) $y = 8x^2 - x + 1$

```
> y:=8*x^2-x+1;
```
$$y := 8\,x^2 - x + 1$$

```
> plot(y,x=-5..5);
> a:=8;
```
$$a := 8$$

```
> b:=-1;
```
$$b := -1$$

```
> c:=1;
```
$$c := 1$$

```
> discriminant;
```
$$-31$$

No real roots.

Exercise 4.6   For which value of k do the following equations have one distinct real root?

(a) $y = kx^2 - 3x + 2$

```
> a:='a':b:='b':c:='c':
> y:=k*x^2-3*x+2;
```
$$y := k\,x^2 - 3\,x + 2$$

```
> discriminant:=b^2-4*a*c;
```
$$discriminant := b^2 - 4\,a\,c$$

```
> a:=k;
```
$$a := k$$

```
> b:=-3;
```
$$b := -3$$

```
> c:=2;
```
$$c := 2$$

```
> k:=solve(discriminant);
```

$$k := \frac{9}{8}$$

```
> y;
```

$$\frac{9}{8} x^2 - 3x + 2$$

```
> plot(y,x=-5..5);
> k:='k':
```

(b) $y = x^2 - 2x + k$

```
> y:=x^2-2*x+k;
```

$$y := x^2 - 2x + k$$

```
> a:=1;
```

$$a := 1$$

```
> b:=-2;
```

$$b := -2$$

```
> c:=k;
```

$$c := k$$

```
> k:=solve(discriminant);
```

$$k := 1$$

```
> y;
```

$$x^2 - 2x + 1$$

```
> plot(y,x=-5..5);
> k:='k':
```

(c) $y = x^2 + kx + 5$.

```
> y:=x^2+k*x+5;
```

$$y := x^2 + kx + 5$$

```
> a:=1;
```

$$a := 1$$

```
> b:=k;
```

$$b := k$$

```
> c:=5;
```
$$c := 5$$

```
> k:=[solve(discriminant)];
```
$$k := \left[2\sqrt{5}, -2\sqrt{5}\right]$$

```
> k1:=op(1,k);
```
$$k1 := 2\sqrt{5}$$

```
> k2:=op(2,k);
```
$$k2 := -2\sqrt{5}$$

```
> y:=x^2+k1*x+5;
```
$$y := x^2 + 2\sqrt{5}\,x + 5$$

```
> plot(y,x=-5..5);
> y:=x^2+k2*x+5;
```
$$y := x^2 - 2\sqrt{5}\,x + 5$$

```
> plot(y,x=-5..5);
```

Exercise 4.7 Find the equation of the tangent at the point $(-1, 0)$ on the parabola $y = (x + 1)^2$.

```
> x:='x':y:='y':
> f:=x->(x+1)^2;
```
$$f := x \rightarrow (x + 1)^2$$

```
> (f(x)-f(a))/(x-a);
```
$$\frac{(x + 1)^2 - 4}{x - 1}$$

```
> simplify(");
```
$$x + 3$$

```
> subs(x=a,");
```
$$4$$

```
> subs(a=-1,");
```
$$4$$

```
> 0=(y-0)/(x-(-1));
```

$$0 = \frac{y}{x+1}$$

```
> isolate(",y);
```

$$y = 0$$

Exercise 4.8 Find the equation of the tangent at $x = 1$ on the parabola $f(x) = x^2 + 3x - 1$.

```
> f:=x->x^2+3*x-1;
```

$$f := x \rightarrow x^2 + 3x - 1$$

```
> (f(x)-f(a))/(x-a);
```

$$\frac{x^2 + 3x - 4}{x - 1}$$

```
> simplify(");
```

$$x + 4$$

```
> subs(x=a,");
```

$$5$$

```
> subs(a=1,");
```

$$5$$

```
> f(1);
```

$$3$$

```
> 5=(y-3)/(x-1);
```

$$5 = \frac{y - 3}{x - 1}$$

```
> isolate(",y);
```

$$y = 5x - 2$$

Exercise 4.9 A line perpendicular to a tangent is called a *normal* line; its slope is equal to the negative inverse of the slope of the tangent. Find the equations of the tangent and normal lines to the parabola $y = 4x^2$ at the points $x = 0, \pm 0.2, \pm 0.4, \ldots, \pm 1$. Construct two plots of the parabola, showing on one

the tangent lines and on the other the normal lines. Overlay these plots using
the `display` command.

```
> f:=x->4*x^2;
```
$$f := x \rightarrow 4\,x^2$$

```
> slope([x,f(x)],[a,f(a)]);
```
$$\frac{4\,x^2 - 4}{x - 1}$$

```
> simplify(");
```
$$4\,x + 4$$

```
> subs(x=a,");
```
$$8$$

```
> tangent_slope:=";
```
$$tangent\_slope := 8$$

```
> normal_slope:=-1/";
```
$$normal\_slope := \frac{-1}{8}$$

```
> a:=0;
```
$$a := 0$$

```
> tangent_slope=(y-f(a))/(x-a);
```
$$8 = \frac{y}{x}$$

```
> isolate(",y);
```
$$y = 8\,x$$

```
> tangent_eq:=0;
```
$$tangent\_eq := 0$$

```
> normal_slope=(y-f(a))/(x-a);
```
$$\frac{-1}{8} = \frac{y}{x}$$

As we should have spotted, if the tangent is horizontal then the normal will be
vertical.

```
> plot({f(x),tangent_eq},x=-2..2,y=-4..4,scaling=CONSTRAINED);
> a:=0.2;
```
$$a := .2$$

```
> tangent_slope=(y-f(a))/(x-a);
```

$$8 = \frac{y - .16}{x - .2}$$

```
> isolate(",y);
```

$$y = 8x - 1.44$$

```
> tangent_eq:=1.6*x-.16;
```

$$tangent\_eq := 1.6\,x - .16$$

```
> normal_slope=(y-f(a))/(x-a);
```

$$\frac{-1}{8} = \frac{y - .16}{x - .2}$$

```
> isolate(",y);
```

$$y = -\frac{1}{8}x + .1850000000$$

```
> normal_eq:=-0.625*x+.285;
```

$$normal\_eq := -.625\,x + .285$$

```
> plot({f(x),tangent_eq,normal_eq},x=-2..2,y=-4..4,
> scaling=CONSTRAINED);
```

A.4   Polynomials

Exercise 5.1   Divide the polynomial $y = 2x^3 + x^2 - 5x + 2$ by $x - 1$.

```
> y:=2*x^3+x^2-5*x+2;
```
$$y := 2\,x^3 + x^2 - 5\,x + 2$$

```
> divisor:=x-1;
```
$$divisor := x - 1$$

```
> divide(y,divisor);
```
$$true$$

```
> y/divisor;
```
$$\frac{2\,x^3 + x^2 - 5\,x + 2}{x - 1}$$

```
> simplify(");
```
$$2\,x^2 + 3\,x - 2$$

Exercise 5.2   Is the polynomial $y = x^3 - 4x^2 + 2x - 1$ divisible by $x - 3$? If not, find the quotient and remainder.

```
> y:=x^3-4*x^2+2*x-1;
```
$$y := x^3 - 4\,x^2 + 2\,x - 1$$

```
> divisor:=x-3;
```
$$divisor := x - 3$$

```
> divide(y,divisor);
```
$$false$$

```
> remainder:=rem(y,divisor,x,quotient);
```
$$remainder := -4$$

```
> quotient;
```
$$x^2 - x - 1$$

```
> expand(quotient*divisor+remainder);
```
$$x^3 - 4\,x^2 + 2\,x - 1$$

Exercise 5.3   Find the remainders on dividing $x^5 - 5x^4 + x^3 - 7$ by $(x-1)$ and by $(x-1)^2$.

```
> rem(x^5-5*x^4+x^3-7,x-1,x);
```
$$-10$$

```
> rem(x^5-5*x^4+x^3-7,(x-1)^2,x);
```
$$2 - 12\,x$$

Exercise 5.4   Find quadratic equations with the following roots.

(a) (3,6)
```
> y:=x^2-S*x+P;
```
$$y := x^2 - S\,x + P$$

```
> S:=3+6;
```
$$S := 9$$

```
> P:=3*6;
```
$$P := 18$$

```
> y;
```
$$x^2 - 9\,x + 18$$

```
> solve(y);
```
$$6, 3$$

(b) (2,-1)
```
> S:=2+(-1);
```
$$S := 1$$

```
> P:=2*(-1);
```
$$P := -2$$

```
> y;
```
$$x^2 - x - 2$$

```
> solve(y);
```
$$2, -1$$

(c) (9,0)

```
> S:=9+0;
```

$$S := 9$$

```
> P:=9*0;
```

$$P := 0$$

```
> y;
```

$$x^2 - 9\,x$$

```
> solve(y);
```

$$0, 9$$

Exercise 5.5   Show that if $y = f(x) = \frac{x+b}{ax-1}$ then $x = f(y)$. Can you find other functions with this property?

```
> x:='x':y:='y':
> y=(x+b)/(a*x-1);
```

$$y = \frac{x+b}{a\,x - 1}$$

```
> with(student):
> isolate("",x);
```

$$x = -\frac{b+y}{1-y\,a}$$

$y = f(x) = \frac{1}{x}$ has a similar property.

Exercise 5.6   Show that $x^2 + 5x + 6 = (x+2)(x+3)$ and so

$$\frac{2x^2 + 15x + 25}{x^2 + 5x + 6} = 2 + \frac{3}{x+2} + \frac{2}{x+3}$$

```
> divide(2*x^2+15*x+25,x^2+5*x+6);
```
$$\textit{false}$$

```
> rem(2*x^2+15*x+25,x^2+5*x+6,x,quot);
```
$$5\,x + 13$$

```
> quot;
```
$$2$$

```
> 2*x^2+15*x+25=2+(5*x+13)/(x^2+5*x+6);
```
$$2\,x^2 + 15\,x + 25 = 2 + \frac{5\,x + 13}{x^2 + 5\,x + 6}$$

```
> (5*x+13)/(x^2+5*x+6)=A/(x+2)+B/(x+3);
```
$$\frac{5\,x + 13}{x^2 + 5\,x + 6} = \frac{A}{x+2} + \frac{B}{x+3}$$

```
> (5*x+13)=A*(x+3)+B*(x+2);
```
$$5\,x + 13 = A\,(x+3) + B\,(x+2)$$

```
> expand(");
```
$$5\,x + 13 = A\,x + 3\,A + B\,x + 2\,B$$

```
> eq1:=5*x=A*x+B*x;
```
$$eq1 := 5\,x = A\,x + B\,x$$

```
> eq2:=13=3*A+2*B;
```
$$eq2 := 13 = 3\,A + 2\,B$$

```
> solve({eq1,eq2},{A,B});
```
$$\{\,A = 3, B = 2\,\}$$

Exercise 5.7 Use the Binomial Theorem in the form

$$(1 - x)^{-1} = 1 - x + x^2 + \cdots$$

to express $\frac{1}{1-x}$ in terms of such a series of powers of x, for $\mid x \mid < 1$. Hence deduce
that $2 = 1 + \frac{1}{2} + \frac{1}{2^2} + \frac{1}{2^3} + \cdots$

```
> 1/(1-x)=2;
```
$$\frac{1}{1 - x} = 2$$

```
> solve(",x);
```
$$\frac{1}{2}$$

```
> 1+(1/2)+(1/2)^2+(1/2)^3+(1/2)^4+(1/2)^5+(1/2)^6+(1/2)^7+(1/2)^8;
```
$$\frac{511}{256}$$

Etc. Try plotting the sum of the first n terms as a function of n.

Exercise 5.8   Find a when $(x - 1)$ is a factor of $7x^3 + ax^2 - 2x + 1$.

```
> f:=x-> 7*x^3+a*x^2-2*x+1;
```
$$f := x \rightarrow 7x^3 + ax^2 - 2x + 1$$

```
> solve(f(1)=0,a);
```
$$-6$$

Exercise 5.9   Find the sum and product of the roots of the following equations.

(a) $y = x^2 + 4x + 9$

```
> y:=x^2+4*x+9;
```
$$y := x^2 + 4x + 9$$

Knowing that the general form of the second order polymonial is $y = x^2 - \text{sum}x + \text{product}$, we can find the sum and product of the roots by inspection.

```
> S:=-4;
```
$$S := -4$$

```
> P:=9;
```
$$P := 9$$

We can also solve for the roots of the equations and then add and multiply them.

```
> x:=[solve(y)];
```
$$x := \left[-2 + I\sqrt{5}, -2 - I\sqrt{5} \right]$$

```
> 'sum'=op(1,x)+op(2,x);
```
$$sum = -4$$

```
> 'product'=expand(op(1,x)*op(2,x));
```
$$product = 9$$

```
> x:='x':
```

(b) $y = 2x^2 - 5x + 9$

```
> y:=2*x^2-5*x+9;
```
$$y := 2x^2 - 5x + 9$$

```
> y/2;
```
$$x^2 - \frac{5}{2}x + \frac{9}{2}$$

```
> S:=5/2;
```
$$S := \frac{5}{2}$$

```
> P:=9/2;
```
$$P := \frac{9}{2}$$

```
> x:=[solve(y)];
```
$$x := \left[\frac{5}{4} + \frac{1}{4}I\sqrt{47}, \frac{5}{4} - \frac{1}{4}I\sqrt{47}\right]$$

```
> 'sum'=op(1,x)+op(2,x);
```
$$sum = \frac{5}{2}$$

```
> 'product'=expand(op(1,x)*op(2,x));
```
$$product = \frac{9}{2}$$

```
> x:='x':
```

Exercise 5.10 Find the second root for each of the following equations.

(a) $y = x^2 + 4x - c$, root $= 1$
```
> y:=x^2+4*x-c;
```
$$y := x^2 + 4x - c$$

```
> root:=1;
```
$$root := 1$$

```
> eq1:=root2+root=-4;
```
$$eq1 := root2 + 1 = -4$$

```
> eq2:=root2*root=-c;
```
$$eq2 := root2 = -c$$

```
> solve({eq1,eq2});
```
$$\{c = 5, root2 = -5\}$$

   (b) $y = x^2 - bx + 3,\;\; \text{root} = 2$

```
> y:=x^2-d*x+3;
```

$$y := x^2 - d\,x + 3$$

```
> root:=2;
```

$$root := 2$$

```
> eq1:=root2+root=d;
```

$$eq1 := root2 + 2 = d$$

```
> eq2:=root2*root=3;
```

$$eq2 := 2\,root2 = 3$$

```
> solve({eq1,eq2});
```

$$\left\{ d = \frac{7}{2}, root2 = \frac{3}{2} \right\}$$

```
> S:='S':P:='P':
```

Exercise 5.11   What is the equation of a polynomial whose roots are two less than those of $y = x^2 - x - 3$?

```
> y:=x^2-x-3;
```

$$y := x^2 - x - 3$$

```
> eqn1:=S=r1+r2;
```

$$eqn1 := S = r1 + r2$$

```
> eqn2:=S=1;
```

$$eqn2 := S = 1$$

```
> eqn3:=P=r1*r2;
```

$$eqn3 := P = r1\; r2$$

```
> eqn4:=P=-3;
```

$$eqn4 := P = -3$$

```
> eqn5:=R1=r1-2;
```

$$eqn5 := R1 = r1 - 2$$

```
> eqn6:=R2=r2-2;
```

$$eqn6 := R2 = r2 - 2$$

```
> eqn7:=S2=R1+R2;
```
$$eqn7 := S2 = R1 + R2$$

```
> eqn8:=P2=R1*R2;
```
$$eqn8 := P2 = R1\ R2$$

```
> s:=solve({eqn1,eqn2,eqn3,eqn4,eqn5,eqn6,eqn7,eqn8});
```
$$s := \{S = 1, P = -3, P2 = -1, r2 = \%1, S2 = -3, R1 = -1 - \%1,$$
$$r1 = 1 - \%1, R2 = \%1 - 2\}$$
$$\%1 := \text{RootOf}(-3 - \_Z + \_Z^2\,)$$

```
> assign(s);
> 'y'=x^2-S2*x+P2;
```
$$y = x^2 + 3\,x - 1$$

Exercise 5.12   The roots of $x^3 + bx^2 + cx + d = 0$ are u, v, w. Express b, c, d in terms of u, v, w.

```
> eq1:=u^3+b*u^2+c*u+d=0;
```
$$eq1 := u^3 + b\,u^2 + c\,u + d = 0$$

```
> eq2:=v^3+b*v^2+c*v+d=0;
```
$$eq2 := v^3 + b\,v^2 + c\,v + d = 0$$

```
> eq3:=w^3+b*w^2+c*w+d=0;
```
$$eq3 := w^3 + b\,w^2 + c\,w + d = 0$$

```
> solve({eq1,eq2,eq3},{b,c,d});
```
$$\{b = -v - u - w, c = u\,v + u\,w + w\,v, d = -w\,u\,v\}$$

A.5   Exponential Functions

Exercise 6.1   Solve the following equations for x.

(a) $2^x = 32$

```
> fsolve(2^x=32,{x});
```
$$\{\, x = 5.000000000 \,\}$$

(b) $5^{3x-1} = 7^x$

```
> fsolve(5^(3*x-1)=7^x,{x});
```
$$\{\, x = .5583666073 \,\}$$

(c) $7^x = 49^4$

```
> fsolve(7^x=49^4,{x});
```
$$\{\, x = 8.000000000 \,\}$$

Exercise 6.2   A certain colony of bacteria has an initial population of 2500. If the population after 3 days is 8000, what is the doubling period? In how many days will the population be 20,000?

```
> y:=t->y_0*2^(t/k);
```
$$y := t \rightarrow y\_0\, 2^{\left(\frac{t}{k}\right)}$$

```
> y_0:=2500;
```
$$y\_0 := 2500$$

```
> solve(y(3)=8000,{k});
```
$$\left\{\, k = 3\,\frac{\ln(2)}{\ln\left(\dfrac{16}{5}\right)} \,\right\}$$

```
> assign(");
> solve(y(t)=20000,{t});
```
$$\left\{\, t = 3\,\frac{\ln(8)}{\ln\left(\dfrac{16}{5}\right)} \,\right\}$$

```
> evalf(");
```
$$\{\, t = 5.363298183 \,\}$$

Exercise 6.3 The half-life of one of the isotopes of polonium, Po^{216}, is 0.15 s. How long would it take for a sample of Po^{216} to decay to 1% of its original size?

```
> y_0:='y_0';
```
$$y\_0 := y\_0$$

```
> y:=t->y_0*2^(t/k);
```
$$y := t \rightarrow y\_0 \, 2^{\left(\frac{t}{k}\right)}$$

```
> k:=-0.15;
```
$$k := -.15$$

```
> solve(y(t)=0.01*y_0,{t});
```
$$\{\, t = .9965784284 \,\}$$

A.6   Logarithmic Functions

Exercise 7.1   Solve the following logarithmic equations.

(a) $\log(x) = 9$, base 3
```
> fsolve(log[3](x)=9,{x});
```
$$\{\, x = 19683.00000 \,\}$$

(b) $7\log(4) = 2x$, base 16
```
> fsolve(7*log[16](4)=2*x,{x});
```
$$\{\, x = 1.750000000 \,\}$$

(c) $\log(10) = 5$, base x
```
> fsolve(log[x](10)=5,{x});
```
$$\{\, x = 1.584893192 \,\}$$

Exercise 7.2   Solve the following equalities.

(a) $\log(x) + \log(2x) = 1$, base 3
```
> fsolve(log[3](x)+log[3](2*x)=1,{x});
```
$$\{\, x = 1.224744871 \,\}$$

(b) $\log(x^2) - \log(x) = \log(3)$, base 10
```
> fsolve(log10(x^2)-log10(x)=log10(3),{x});
```
$$\{\, x = 3.000000000 \,\}$$

(c) $\log(x + 7) + \log(x^2) = \log(9)$, base 5
```
> fsolve(log[5](x+7)+log[5](x^2)=log[5](9),{x});
```
$$\{\, x = 1.056907697 \,\}$$

Exercise 7.3   A scale for measuring the magnitude of earthquakes was developed in 1935 by Charles F. Richter of the California Institute of Technology. The so-called Richter Scale allows the 'size' of earthquakes to be compared. The Richter formula computes the magnitude of a quake from the logarithm of the amplitude of waves recorded by seismographs. Included in the formula is an adjustment to compensate for the variation in the distance of the seismograph and the earthquake's epicentre. The Richter Scale formula for magnitude M is given by $M = \log 10(\frac{x}{c})$ where x is the amplitude of the largest seismic wave as measured 100 kilometres from the epicentre and c is the amplitude of a reference earthquake of amplitude 1 micrometre on a standard graph at the same distance from the epicentre. (p63, Goldenberg and Greenwald [7])

(a) In 1989, San Francisco was struck by an earthquake of magnitude 7.1. As destructive as this earthquake was, it was not nearly as powerful as the 1906 San Francisco earthquake, which measured 8.3. The relative strengths of two earthquakes may be compared by looking at the ratio of the amplitudes. What is the ratio of amplitiude for these two San Francisco earthquakes?

```
> readlib(log10):
> solve({7.1=log10(x1/c),8.3=log10(x2/c)},{x1,x2});
```
$$\{\, x2 = .1995262311\,10^9\, c, x1 = .1258925411\,10^8\, c\,\}$$

```
> assign(");
> x2/x1;
```
$$15.84893190$$

(b) The largest earthquake magnitude ever measured was 8.9 for an earthquake in Japan in 1933. The largest earthquake magnitude ever recorded in the United States was 8.5 for the Alaskan earthquake of 1964. Determine the ratio of amplitude for these earthquakes.

```
> x1:='x1':x2:='x2':
> solve({8.9=log10(x1/c),8.5=log10(x2/c)},{x1,x2});
```
$$\{\, x1 = .7943282366\,10^9\, c, x2 = .3162277659\,10^9\, c\,\}$$

```
> assign(");
> x1/x2;
```
$$2.511886438$$

(c) When the amplitude of an earthquake is tripled, by how much does the magnitude increase?

```
> solve({M1=log10(x/c),M2=log10(3*x/c)},{M1.M2});
> M2=log10(3*x/c);
```
$$M2 = \frac{\ln\left(3\,\dfrac{x}{c}\right)}{\ln(10)}$$

```
> expand(");
```
$$M2 = \frac{\ln(3)}{\ln(10)} + \frac{\ln(x)}{\ln(10)} - \frac{\ln(c)}{\ln(10)}$$

(d) Assume that there are two earthquakes of unequal strength. If the ratio of amplitudes is 1.5 and the weaker earthquake has magnitude 5.6, determine the magnitude of the stronger earthquake.

```
> log10(x/c)=5.6;
```

$$\frac{\ln\left(\dfrac{x}{c}\right)}{\ln(10)} = 5.6$$

```
> eq1:=subs(x/c=a,");
```

$$eq1 := \frac{\ln(a)}{\ln(10)} = 5.6$$

```
> M=log10(1.5*x/c);
```

$$M = \frac{\ln\left(1.5\,\dfrac{x}{c}\right)}{\ln(10)}$$

```
> eq2:=subs(x/c=a,");
```

$$eq2 := M = \frac{\ln(1.5\,a)}{\ln(10)}$$

```
> solve({eq1,eq2});
```

$$\{\, M = 5.776091259, a = 398107.1706 \,\}$$

```
> x1:='x1':c:='c':x2:='x2':
```

(e) If the difference in magnitudes of two earthquakes is 0.5, determine the ratio of amplitudes for the two earthquakes.

```
> M=log10(x1/c);
```

$$M = \frac{\ln\left(\dfrac{x1}{c}\right)}{\ln(10)}$$

```
> with(student):
> isolate("",x1/c);
```

$$\frac{x1}{c} = e^{(M\ln(10))}$$

```
> isolate(",x1);
```

$$x1 = e^{(M\ln(10))}\,c$$

```
> assign(");
> M-0.5=log10(x2/c);
```

$$M - .5 = \frac{\ln\left(\dfrac{x2}{c}\right)}{\ln(10)}$$

```
> isolate(",x2/c);
```

$$\frac{x2}{c} = e^{(2.302585093\,M - 1.151292547)}$$

```
> isolate(",x2);
```

$$x2 = e^{(2.302585093\,M - 1.151292547)}\,c$$

```
> assign(");
> x1/x2;
```

$$\frac{e^{(M\ln(10))}}{e^{(2.302585093\,M - 1.151292547)}}$$

```
> simplify(");
```

$$3.162277662$$

Exercise 7.4   What is the concentration of H^+ ions in a cola, the pH of which is 3.4?

```
> pH:=H->-log[10](H);
```

$$pH := H \to -\log_{10}(H)$$

```
> solve(pH(H)=3.4,{H});
```

$$\{\,H = .0003981071706\,\}$$

Exercise 7.5   Find the simple interest of each of the following.

(a) $500 for 2 years at 4% pa

```
> eq:=A=P*(1+i*n);
```

$$eq := A = P(1 + in)$$

```
> P:=500;
```

$$P := 500$$

```
> i:=0.04;
```

$$i := .04$$

```
> n:=2;
```

$$n := 2$$

```
> solve(eq,{A});
```

$$\{\,A = 540.\,\}$$

(b) $800 for 4 years at 3% pa

```
> P:=800;
```
$$P := 800$$

```
> i:=0.03;
```
$$i := .03$$

```
> n:=4;
```
$$n := 4$$

```
> solve(eq,{A});
```
$$\{ A = 896. \}$$

Exercise 7.6   Calculate the amount for

(a) $500 for 2 years at 4% pa compounded annually

```
> A:='A':P:='P':i:='i':n:='n':
> eq:=A=P*(1+i)^n;
```
$$eq := A = P(1+i)^n$$

```
> P:=500;
```
$$P := 500$$

```
> i:=0.04;
```
$$i := .04$$

```
> n:=2;
```
$$n := 2$$

```
> solve(eq,{A});
```
$$\{ A = 540.8000000 \}$$

(b) $500 for 2 years at 4% pa compounded daily

```
> i:=0.04/(365);
```
$$i := .0001095890411$$

```
> n:=2*365;
```
$$n := 730$$

```
> A:='A':
> solve(eq,{A});
```
$$\{\, A = 541.6411435 \,\}$$

(c) \$800 for 4 years at 3% pa compounded semi-annually.
```
> P:=800;
```
$$P := 800$$

```
> i:=0.03/2;
```
$$i := .01500000000$$

```
> n:=2*4;
```
$$n := 8$$

```
> A:='A':
> solve(eq,{A});
```
$$\{\, A = 901.1940696 \,\}$$

A.7   Circular Functions

Exercise 8.1   Plot the following circular functions from -4π to 4π and determine the period and amplitude of each.

(a) $y = \sin(x)$

```
> plot(sin(x),x=-4*Pi..4*Pi);
> sin(0);
```
$$0$$

```
> sin(Pi);
```
$$0$$

```
> sin(2*Pi);
```
$$0$$

$$\text{Period} \; = 2\pi, \text{Amplitude} = 1$$

(b) $y = \cos(x)$

```
> plot(cos(x),x=-4*Pi..4*Pi);
> cos(0);
```
$$1$$

```
> cos(Pi);
```
$$-1$$

```
> cos(2*Pi);
```
$$1$$

$$\text{Period} \; = 2\pi, \text{Amplitude} = 1$$

(c) $y = 3\cos(2x)$

```
> f:=x->3*cos(2*x);
```
$$f := x \rightarrow 3\cos(2\,x)$$

```
> plot(f(x),x=-4*Pi..4*Pi);
> f(0);
```
$$3$$

```
> f(Pi/2);
```
$$-3$$

```
> f(Pi);
```
$$3$$

$$\text{Period} = \pi, \text{Amplitude} = 3$$

Exercise 8.2 Compare the plots of the functions $f(x) = |\sin(x)|$ and $g(x) = \sin(|x|)$.

```
> f:=x->abs(sin(x));
```
$$f := x \rightarrow |\sin(x)|$$

```
> g:=x->sin(abs(x));
```
$$g := x \rightarrow \sin(|x|)$$

```
> plot({f(x),g(x)},x=-4*Pi..4*Pi);
```

Exercise 8.3 In a certain city the average temperature $f(t)$ on the tth day of the year (with $t = 1$ on January 1) is given by a formula of the form $f(t) = A + B\sin(k(t - \alpha))$. (The phase shift, α, can be thought of as the distance between the y-axis and the closest point at which the function intersects the x-axis.) (p438, Edwards and Penney [6])

(a) Plot the function using the values $A = 15$, $B = -7$, $k = 1/50$, and $\alpha = -\pi/4$ to get an idea of the shape of the curve. What must k be for the period of $f(t)$ to be 365 days (where $period = \frac{2\pi}{k}$)?

```
> 15-7*sin(1/50*(x+Pi/4));
```
$$15 - 7\sin\left(\frac{1}{50}x + \frac{1}{200}\pi\right)$$

```
> plot(15-7*sin(1/50*(x+Pi/4)),x=0..365);
```

(b) The minimum daily average temperature of $10°C$ occurs on January 20 and the maximum of $25°C$ occurs half a year later. What are the values of A, B, and α?

```
> f:=x->A+B*sin(k*(x-t));
```
$$f := x \rightarrow A + B\sin(k(x - t))$$

```
> period:=365;
```
$$period := 365$$

```
> period=2*Pi/k;
```
$$365 = 2\frac{\pi}{k}$$

```
> solve(",{k});
```

$$\left\{ k = \frac{2}{365}\,\pi \right\}$$

```
> assign(");
> t:=20-365/4;
```

$$t := \frac{-285}{4}$$

```
> solve({f(20)=10,f(20+365/2)=25},{A,B});
```

$$\left\{ B = \frac{-15}{2}, A = \frac{35}{2} \right\}$$

```
> assign(");
> f(x);
```

$$\frac{35}{2} - \frac{15}{2} \sin\left(\frac{2}{365}\,\pi\,\left(x + \frac{285}{4} \right) \right)$$

    (c) On what days of the year is the average temperature 15°C? Repeat
       for 20°C?

```
> plot(f(x),x=0..365);
```

Exercise 8.4   One French horn emits a 200 Hz note and a second horn emits a 400 Hz note. If the amplitudes of the wave forms that represent each note are equal, show that the resultant of the combined sounds differs from each original sound and has a frequency of 200 Hz.

```
> t:='t':
> V1:=t->sin(2*200*Pi*t);
```

$$V1 := t \rightarrow \sin(400\,\pi\,t)$$

```
> V2:=t->sin(2*400*Pi*t);
```

$$V2 := t \rightarrow \sin(800\,\pi\,t)$$

```
> V1(t)+V2(t);
```

$$\sin(400\,\pi\,t) + \sin(800\,\pi\,t)$$

```
> plot({V1(t),V2(t),V1(t)+V2(t)},t=0..0.01);
> readlib(trigsubs):
> trigsubs(sin(2*x));
```

$$\left[\sin(2x), \sin(2x), 2\sin(x)\cos(x), \frac{1}{\csc(2x)}, \frac{1}{\csc(2x)}, 2\frac{\tan(x)}{1+\tan(x)^2},\right.$$

$$\left.-\frac{1}{2}I\left(e^{(2Ix)} - e^{(-2Ix)}\right)\right]$$

Notice that $\sin(2x) = 2\sin(x)\cos(x)$, therefore, the resultant sound has a frequency of 200 Hz.

A.8   Trigonometry

Exercise 9.1   Using this information and definitions of the circular functions, make a table of $\cos(x)$, $\sin(x)$, and $\tan(x)$ for $x = 30°, 45°, 60°$.

```
> cos(30*Pi/180);cos(45*Pi/180);cos(60*Pi/180);
```

$$\frac{1}{2}\sqrt{3}$$

$$\frac{1}{2}\sqrt{2}$$

$$\frac{1}{2}$$

```
> sin(30*Pi/180);sin(45*Pi/180);sin(60*Pi/180);
```

$$\frac{1}{2}$$

$$\frac{1}{2}\sqrt{2}$$

$$\frac{1}{2}\sqrt{3}$$

```
> tan(30*Pi/180);tan(45*Pi/180);tan(60*Pi/180);
```

$$\frac{1}{3}\sqrt{3}$$

$$1$$

$$\sqrt{3}$$

Exercise 9.2   Express the following in radians.

(a) 30°
```
> convert(30*degrees,radians);
```

$$\frac{1}{6}\pi$$

(b) 270°
```
> convert(270*degrees,radians);
```

$$\frac{3}{2}\pi$$

(c) −210°
```
> convert(-210*degrees,radians);
```

$$-\frac{7}{6}\pi$$

Exercise 9.3   Express the following in degrees.

(a) $\frac{\pi}{4}$ rad

```
> convert(Pi/4,degrees);
```
$$45 \; degrees$$

(b) $-\frac{9\pi}{4}$ rad

```
> convert(-9*Pi/4,degrees);
```
$$-405 \; degrees$$

(c) 4π rad

```
> convert(4*Pi,degrees);
```
$$720 \; degrees$$

Exercise 9.4   Express the following in terms of sine and cosine.

(a) $\csc(x)$

```
> convert(csc(x),sincos);
```
$$\frac{1}{\sin(x)}$$

(b) $\sec(x)^2$

```
> convert(sec(x)^2,sincos);
```
$$\frac{1}{\cos(x)^2}$$

(c) $\tan(x)\sin(x)$

```
> convert(tan(x)*sin(x),sincos);
```
$$\frac{\sin(x)^2}{\cos(x)}$$

Exercise 9.5   Find an equivalent form of the following.

(a) $\csc(x)^2$

```
> simplify(csc(x)^2,trig);
```
$$-\frac{1}{\cos(x)^2 - 1}$$

(b) $-\tan(x)^2$

```
> simplify(-tan(x)^2,trig);
```

$$\frac{\cos(x)^2 - 1}{\cos(x)^2}$$

(c) $1 + \cot(x)^2$

```
> simplify(1+cot(x)^2,trig);
```

$$-\frac{1}{\cos(x)^2 - 1}$$

Exercise 9.6 Solve the following trigonometric equations.

(a) $\sin(x) = \frac{1}{2}$

```
> solve(sin(x)=1/2,{x});
```

$$\left\{x = \frac{1}{6}\pi\right\}$$

(b) $\sin(x)^2 + 2\sin(x) - 3 = 0$

```
> solve(sin(x)^2+2*sin(x)-3=0,{x});
```

$$\left\{x = \frac{1}{2}\pi\right\}, \{x = -\arcsin(3)\}$$

(c) $2\sin(x) = \cos(x)$

```
> solve(2*sin(x)=cos(x),{x});
```

$$\left\{x = \arctan\left(\frac{1}{2}\right)\right\}$$

Exercise 9.7 Show that the general solution of $\cos(4x) = 0$ is $x = (2n+1)\frac{\pi}{8}$, and the general solution of $\sin(x) = 0$ is $x = n\pi$, for $n = 0, 1, 2, \ldots$ Hence deduce the general solution of $\sin(5x) - \sin(3x) = 0$.

```
> cos(4*x)=0;
```

$$\cos(4x) = 0$$

```
> solve(",x);
```

$$\frac{1}{8}\pi$$

```
> (2*0+1)*Pi/8;
```

$$\frac{1}{8}\pi$$

```
> (2*1+1)*Pi/8;
```
$$\frac{3}{8}\pi$$

```
> cos(4*3*Pi/8);
```
$$0$$

```
> sin(x)=0;
```
$$\sin(x) = 0$$

```
> solve(",x);
```
$$0$$

```
> sin(2*Pi);
```
$$0$$

```
> sin(4*Pi);
```
$$0$$

```
> sin(5*x)-sin(3*x);
```
$$\sin(5 x) - \sin(3 x)$$

```
> simplify(");
```
$$16 \sin(x) \cos(x)^4 - 16 \sin(x) \cos(x)^2 + 2 \sin(x)$$

Maple is unwilling to give us the trigonometric expression in the form we want, which is $2 \cos(4x) \sin(x)$.

Exercise 9.8   Show that

$$3 \cos x + \sin x = b(\cos x \cos a + \sin x \sin a) = b \cos(x - a)$$

where $b \cos a = 3$ and $b \sin a = 1$, and so $\tan a = \frac{1}{3}$ and $b = \sqrt{10}$. Hence solve the equation $3 \cos x + \sin x = 1$ for values of x between 0 and 4π.

```
> b*(cos(x)*cos(a)+sin(x)*sin(a));
```
$$b(\cos(x) \cos(a) + \sin(x) \sin(a))$$

```
> expand(");
```
$$b \cos(x) \cos(a) + b \sin(x) \sin(a)$$

```
> subs({cos(a)=3/b,sin(a)=1/b},");
```
$$3 \cos(x) + \sin(x)$$

```
> sin(a)/cos(a)=(1/b)/(3/b);
```

$$\frac{\sin(a)}{\cos(a)} = \frac{1}{3}$$

So we have a right-angled triangle with sides $1, 3$ and $\sqrt{10}$; angle a being opposite the side of length 1. Also, $\tan(a) = 1/3$ in $\sqrt{10}\cos(x - a) = 1$ so $x - a$ is the other angle in the triangle and $x = \frac{\pi}{2}$.

Exercise 9.9   Use the approximations $\sin\theta \doteq \theta$ and $\cos\theta \doteq 1 - \frac{\theta^2}{2}$ to obtain an approximation to $\frac{\sin 2\theta}{4 - \cos 3\theta}$. Over what range is this approximation accurate to three significant figures?

```
> f:=theta->sin(2*theta)/(4-cos(3*theta));
```

$$f := \theta \rightarrow \frac{\sin(2\theta)}{4 - \cos(3\theta)}$$

```
> subs({sin(2*theta)=2*theta,cos(3*theta)=1-(3*theta)^2/2},f(theta));
```

$$2\,\frac{\theta}{3 + \dfrac{9}{2}\theta^2}$$

```
> approx:=simplify(");
```

$$approx := \frac{4}{3}\,\frac{\theta}{2 + 3\theta^2}$$

```
> f(0);
```

$$0$$

```
> subs(theta=0,approx);
```

$$0$$

```
> evalf(f(Pi/50));
```

$$.04153252613$$

```
> evalf(subs(theta=Pi/50,approx));
```

$$.04164131209$$

Exercise 9.10   Solve the right-angled triangle $\triangle QRS$ given that $\angle R = 90°$, $s = 5$, and $q = 7$.

```
> with(geometry):
> triangle(QRS,[5,angle=Pi/2,7]);
```

$$QRS$$

```
> sides(QRS);
```
$$\left[5, 7, \sqrt{74}\right]$$

```
> evalf(");
```
$$[5., 7., 8.602325267]$$

```
> tan(S)=5/7;
```
$$\tan(S) = \frac{5}{7}$$

```
> solve(",{S});
```
$$\left\{S = \arctan\left(\frac{5}{7}\right)\right\}$$

```
> assign(");
> evalf(S);
```
$$.6202494860$$

```
> convert(",degrees);
```
$$111.6449075 \, \frac{degrees}{\pi}$$

```
> evalf(");
```
$$35.53767779 \, degrees$$

```
> tan(Q)=7/5;
```
$$\tan(Q) = \frac{7}{5}$$

```
> solve(",{Q});
```
$$\left\{Q = \arctan\left(\frac{7}{5}\right)\right\}$$

```
> assign(");
> evalf(Q);
```
$$.9505468408$$

```
> convert(",degrees);
```
$$171.0984313 \, \frac{degrees}{\pi}$$

```
> evalf(");
```
$$54.46232218 \, degrees$$

```
> Pi/2+Q+S;
```
$$\frac{1}{2}\pi + \arctan\left(\frac{7}{5}\right) + \arctan\left(\frac{5}{7}\right)$$

```
> convert(",degrees);
```
$$180\frac{\left(\frac{1}{2}\pi + \arctan\left(\frac{7}{5}\right) + \arctan\left(\frac{5}{7}\right)\right)}{\pi} \; degrees$$

```
> evalf(");
```
$$180.0000000 \; degrees$$

Exercise 9.11 On a clear day in a flat desert how far away is the horizon seen by a person whose eyes are 1.5 m above the ground? You may take the radius of the Earth to be 6400 km, and you will need to know Pythagoras' theorem for a right-angled triangle.

```
> x^2=r^2+(r+h)^2;
```
$$x^2 = r^2 + (r+h)^2$$

```
> expand(");
```
$$x^2 = 2r^2 + 2rh + h^2$$

```
> subs({r=6400,h=0.0015},");
```
$$x^2 = .8192001920 \, 10^8$$

```
> solve(",x);
```
$$9050.967856, -9050.967856$$

Exercise 9.12 Solve the following triangles using the Sine Law.

(a) $\triangle ABC$, given $a = 10$, $b = 15$, $\angle C = \frac{3\pi}{4}$

```
> triangle(ABC,[10,angle=3/4*Pi,15]);
```
$$ABC$$

```
> sides(ABC);
```
$$\left[10, 15, 5\sqrt{13 + 6\sqrt{2}}\right]$$

```
> 5*sqrt(13+6*sqrt(2))/sin(3/4*Pi)=10/sin(A);
```
$$5\sqrt{13 + 6\sqrt{2}}\sqrt{2} = 10\frac{1}{\sin(A)}$$

```
> solve(",{A});
```

$$\left\{ A = \arcsin\left(\frac{13}{97}\sqrt{13+6\sqrt{2}}\,\sqrt{2} - \frac{12}{97}\sqrt{13+6\sqrt{2}} \right) \right\}$$

```
> assign(");
> convert(A,degrees);
```

$$180\,\frac{\arcsin\left(\dfrac{13}{97}\sqrt{13+6\sqrt{2}}\,\sqrt{2} - \dfrac{12}{97}\sqrt{13+6\sqrt{2}} \right)\,degrees}{\pi}$$

```
> evalf(");
```

$$17.76427608\ degrees$$

```
> 5*sqrt(13+6*sqrt(2))/sin(3/4*Pi)=15/sin(B);
```

$$5\sqrt{13+6\sqrt{2}}\,\sqrt{2} = 15\,\frac{1}{\sin(B)}$$

```
> solve(",{B});
```

$$\left\{ B = \arcsin\left(\frac{39}{194}\sqrt{13+6\sqrt{2}}\,\sqrt{2} - \frac{18}{97}\sqrt{13+6\sqrt{2}} \right) \right\}$$

```
> assign(");
> convert(B,degrees);
```

$$180\,\frac{\arcsin\left(\dfrac{39}{194}\sqrt{13+6\sqrt{2}}\,\sqrt{2} - \dfrac{18}{97}\sqrt{13+6\sqrt{2}} \right)\,degrees}{\pi}$$

```
> convert(A+B+3/4*Pi,degrees);;
```

$$180\left(\arcsin\left(\frac{13}{97}\sqrt{13+6\sqrt{2}}\,\sqrt{2} - \frac{12}{97}\sqrt{13+6\sqrt{2}} \right) \right.$$
$$\left. + \arcsin\left(\frac{39}{194}\sqrt{13+6\sqrt{2}}\,\sqrt{2} - \frac{18}{97}\sqrt{13+6\sqrt{2}} \right) + \frac{3}{4}\pi \right)\,degrees \Big/$$
$$\pi$$

```
> evalf(");
```

$$180.0000000\ degrees$$

(b) $\triangle XYZ$, given $\angle X = 50°$, $z = 5$, $y = 3.5$

```
> triangle(XYZ,[5,angle=50*Pi/180,3.5]);
```

$$XYZ$$

```
> sides(XYZ);
```

$$\left[5, 3.5, \sqrt{37.25 - 35.0\cos\left(\frac{5}{18}\pi \right)} \right]$$

```
> evalf(");
```
$$[5., 3.5, 3.840889698]$$

```
> 3.840889698/sin(50*Pi/180)=5/sin(Z);
```
$$3.840889698 \; \frac{1}{\sin\left(\frac{5}{18}\,\pi\right)} = 5\,\frac{1}{\sin(Z)}$$

```
> solve(",{Z});
```
$$\{ Z = 1.496249254 \}$$

```
> assign(");
> convert(Z,degrees);
```
$$269.3248657 \; \frac{degrees}{\pi}$$

```
> evalf(");
```
$$85.72876732 \; degrees$$

```
> 3.840889698/sin(50*Pi/180)=3.5/sin(Y);
```
$$3.840889698 \; \frac{1}{\sin\left(\frac{5}{18}\,\pi\right)} = 3.5\,\frac{1}{\sin(Y)}$$

```
> solve(",{Y});
```
$$\{ Y = .7726787718 \}$$

```
> assign(");
> convert(Y,degrees);
```
$$139.0821789 \; \frac{degrees}{\pi}$$

```
> evalf(");
```
$$44.27123252 \; degrees$$

```
> 50*Pi/180+Y+Z;
```
$$\frac{5}{18}\,\pi + 2.268928026$$

```
> convert(",degrees);
```
$$180 \; \frac{\left(\frac{5}{18}\,\pi + 2.268928026\right) \; degrees}{\pi}$$

```
> evalf(");
```
$$179.9999999 \ degrees$$

Exercise 9.13   Solve the following triangles.

   (a) $\triangle JKL$, given $\angle J = 130°$, $\angle L = 30°$, $k = 10$
```
> J:=130*Pi/180;L:=30*Pi/180;k:=10;
```
$$J := \frac{13}{18} \pi$$
$$L := \frac{1}{6} \pi$$
$$k := 10$$

```
> K:=Pi-J-L;
```
$$K := \frac{1}{9} \pi$$

```
> eq1:=j^2=k^2+l^2-2*k*l*cos(J);
```
$$eq1 := j^2 = 100 + l^2 + 20\, l \cos\left(\frac{5}{18} \pi\right)$$

```
> eq3:=l^2=j^2+k^2-2*j*k*cos(L);
```
$$eq3 := l^2 = j^2 + 100 - 10\, j\, \sqrt{3}$$

```
> solve({eq1,eq3},{j,l});
```
   Maple cannot solve this system of equations, so use the Sine Law.
```
> solve(j/sin(J)=k/sin(K),{j});
```
$$\left\{ j = 10\, \frac{\sin\left(\dfrac{5}{18} \pi\right)}{\sin\left(\dfrac{1}{9} \pi\right)} \right\}$$

```
> evalf(");
```
$$\{\, j = 22.39764114 \,\}$$

```
> solve(l/sin(L)=k/sin(K),{l});
```
$$\left\{ l = 5\, \frac{1}{\sin\left(\dfrac{1}{9} \pi\right)} \right\}$$

```
> evalf(");
```
$$\{\, l = 14.61902200 \,\}$$

(b) $\triangle PQR$, given $p = 4$, $q = 6$, $\angle R = \frac{\pi}{4}$

```
> p:=4;q:=6;R:=Pi/4;
```
$$p := 4$$
$$q := 6$$
$$R := \frac{1}{4}\pi$$

```
> r^2=p^2+q^2-2*p*q*cos(R);
```
$$r^2 = 52 - 24\sqrt{2}$$

```
> [solve(",r)];
```
$$\left[2\sqrt{13 - 6\sqrt{2}},\, -2\sqrt{13 - 6\sqrt{2}} \right]$$

```
> r:=op(1,");
```
$$r := 2\sqrt{13 - 6\sqrt{2}}$$

```
> p^2=r^2+q^2-2*r*q*cos(P);
```
$$16 = 88 - 24\sqrt{2} - 24\sqrt{13 - 6\sqrt{2}}\cos(P)$$

```
> solve(",{P});
```
$$\left\{ P = \arccos\left(\frac{27}{97}\sqrt{13 - 6\sqrt{2}} + \frac{5}{97}\sqrt{13 - 6\sqrt{2}}\sqrt{2} \right) \right\}$$

```
> assign(");
> convert(P,degrees);
```
$$180\,\frac{\arccos\left(\frac{27}{97}\sqrt{13 - 6\sqrt{2}} + \frac{5}{97}\sqrt{13 - 6\sqrt{2}}\sqrt{2} \right)}{\pi}\ degrees$$

```
> evalf(");
```
$$41.72676503\ degrees$$

```
> q^2=r^2+p^2-2*r*p*cos(Q);
```
$$36 = 68 - 24\sqrt{2} - 16\sqrt{13 - 6\sqrt{2}}\cos(Q)$$

```
> solve(",{Q});
```
$$\left\{ Q = \arccos\left(\frac{8}{97}\sqrt{13 - 6\sqrt{2}} - \frac{15}{194}\sqrt{13 - 6\sqrt{2}}\sqrt{2} \right) \right\}$$

```
> assign(");
> convert(Q,degrees);
```

$$180\,\frac{\arccos\left(\frac{8}{97}\sqrt{13-6\sqrt{2}}-\frac{15}{194}\sqrt{13-6\sqrt{2}}\sqrt{2}\right)}{\pi}\ degrees$$

```
> evalf(");
```

$$93.27323495\ degrees$$

```
> P+Q+R;
```

$$\arccos\left(\frac{27}{97}\sqrt{13-6\sqrt{2}}+\frac{5}{97}\sqrt{13-6\sqrt{2}}\sqrt{2}\right)$$
$$+\arccos\left(\frac{8}{97}\sqrt{13-6\sqrt{2}}-\frac{15}{194}\sqrt{13-6\sqrt{2}}\sqrt{2}\right)+\frac{1}{4}\pi$$

```
> evalf(");
```

$$3.141592654$$

```
> convert(",degrees);
```

$$565.4866777\,\frac{degrees}{\pi}$$

```
> evalf(");
```

$$180.0000000\ degrees$$

Exercise 9.14 If a 50 m tall structure casts an 80 m long shadow, what is the angle of the sun's elevation?

```
> tan(theta)=50/80;
```

$$\tan(\theta)=\frac{5}{8}$$

```
> solve(",theta);
```

$$\arctan\left(\frac{5}{8}\right)$$

```
> evalf(");
```

$$.5585993153$$

```
> convert(",degrees);
```

$$100.5478768\,\frac{degrees}{\pi}$$

```
> evalf(");
```
$$32.00538321 \; degrees$$

Exercise 9.15 Determine the angle enclosed by two vectors, $\vec{a}(6,7)$ and $\vec{b}(4,3)$.

```
> with(linalg):
> a:=vector(2,[6,7]);
```
$$a := [\, 6 \; 7 \,]$$

```
> b:=vector(2,[4,3]);
```
$$b := [\, 4 \; 3 \,]$$

```
> angle(a,b);
```
$$\arccos \left(\frac{9}{425} \sqrt{85} \sqrt{25} \right)$$

```
> convert(",degrees);
```
$$180 \, \frac{\arccos \left(\dfrac{9}{425} \sqrt{85} \sqrt{25} \right) \; degrees}{\pi}$$

```
> evalf(");
```
$$12.52880767 \; degrees$$

Exercise 9.16 The angle between vectors $\vec{c}(2,3)$ and $\vec{d}(x,1)$ is 35°. Find x.
```
> c:=vector(2,[2,3]);
```
$$c := [\, 2 \; 3 \,]$$

```
> d:=vector(2,[x,1]);
```
$$d := [\, x \; 1 \,]$$

```
> angle(c,d)=35*Pi/180;
```
$$\arccos \left(\frac{1}{13} \frac{(\, 2\,x + 3\,) \sqrt{13}}{\sqrt{x^2 + 1}} \right) = \frac{7}{36} \pi$$

```
> evalf(");
```
$$\arccos \left(.2773500981 \, \frac{2.\, x + 3.}{\sqrt{x^2 + 1.}} \right) = .6108652381$$

```
> solve(",x);
```
$$2.563554177$$

Exercise 9.17   Use the dot product to determine the angle between the vectors $\vec{a}(5,3)$ and $\vec{b}(7,-4)$.

> a:=vector(2,[5,3]);
$$a := [\,5\ 3\,]$$

> b:=vector(2,[7,-4]);
$$b := [\,7\ -4\,]$$

> length_a:=sqrt(a[1]^2+a[2]^2);
$$length\_a := \sqrt{34}$$

> length_b:=sqrt(b[1]^2+b[2]^2);
$$length\_b := \sqrt{65}$$

> dotprod(a,b)=length_a*length_b*cos(theta);
$$23 = \sqrt{34}\,\sqrt{65}\,\cos(\,\theta\,)$$

> solve(",theta);
$$\arccos\left(\frac{23}{2210}\,\sqrt{65}\,\sqrt{34}\right)$$

> evalf(");
$$1.059565614$$

> convert(",degrees);
$$190.7218105\,\frac{degrees}{\pi}$$

> evalf(");
$$60.70863778\ degrees$$

Exercise 9.18   What is the x-coordinate of the vector $\vec{a}(x,-1)$ so that it forms a 40° angle with the vector $\vec{g}(3,3)$.

> a:=vector(2,[x,-1]);
$$a := [\,x\ -1\,]$$

> g:=vector(2,[3,3]);
$$g := [\,3\ 3\,]$$

```
> length_a:=sqrt(a[1]^2+a[2]^2);
```
$$length\_a := \sqrt{x^2 + 1}$$

```
> length_g:=sqrt(g[1]^2+g[2]^2);
```
$$length\_g := 3\sqrt{2}$$

```
> dotprod(a,g)=length_a*length_g*cos(40*Pi/180);
```
$$3\,x - 3 = 3\,\sqrt{x^2 + 1}\,\sqrt{2}\,\cos\left(\frac{2}{9}\,\pi\right)$$

```
> lhs(")^2=rhs(")^2;
```
$$(\,3\,x - 3\,)^2 = 18\,(\,x^2 + 1\,)\cos\left(\frac{2}{9}\,\pi\right)^2$$

```
> expand(");
```
$$9\,x^2 - 18\,x + 9 = 18\cos\left(\frac{2}{9}\,\pi\right)^2 x^2 + 18\cos\left(\frac{2}{9}\,\pi\right)^2$$

```
> solve(",x);
```
$$\frac{1}{2}\,\frac{18 + 36\,\sqrt{\cos\left(\frac{2}{9}\,\pi\right)^2 - \cos\left(\frac{2}{9}\,\pi\right)^4}}{9 - 18\cos\left(\frac{2}{9}\,\pi\right)^2}\,,\,\frac{1}{2}\,\frac{18 - 36\,\sqrt{\cos\left(\frac{2}{9}\,\pi\right)^2 - \cos\left(\frac{2}{9}\,\pi\right)^4}}{9 - 18\cos\left(\frac{2}{9}\,\pi\right)^2}$$

```
> evalf(");
```
$$-11.43005230, \; -.08748866480$$

A.9 Similar Figures

Exercise 10.1   If a 50 m tall structure casts an 80 m shadow at a given time of day, what would be the length of the shadow cast by a 1.8 m tall man?

```
> 50/80=1.8/x;
```

$$\frac{5}{8} = 1.8\,\frac{1}{x}$$

```
> solve(",x);
```

$$2.880000000$$

Exercise 10.2   What is the shortest distance between the point $F(3,3)$ and the line $y = 2x + 6$ and what is the equation of the line joining them?

```
> with(geometry):
> point(F,(3,3));
```

$$F$$

```
> line(l,[2*x-y=-6]);
```

$$l$$

```
> distance(F,l);
```

$$\frac{9}{5}\sqrt{5}$$

```
> perpendicular(F,l,perp);
```

$$perp$$

```
> perp[equation];
```

$$-x - 2\,y + 9 = 0$$

A.10   Circles and Spheres

Exercise 11.1   Plot the relations $y - x + 1 = 0$ and $(x - 1)^2 + y^2 = 1$ to determine if they intersect. If so, calculate their point(s) of intersection.

```
> with(plots):
> p1:=implicitplot(y-x+1=0,x=-2..2,y=-2..2):
> p2:=implicitplot((x-1)^2+y^2=1,x=-2..2,y=-2..2):
> display({p1,p2},scaling=CONSTRAINED);
> solve({y-x+1=0,(x-1)^2+y^2=1},{x,y});
```
$$\{\, x = \mathrm{RootOf}(\, 2\,\_Z^2 - 4\,\_Z + 1\,), y = \mathrm{RootOf}(\, 2\,\_Z^2 - 4\,\_Z + 1\,) - 1 \,\}$$

```
> allvalues(");
```
$$\left\{\, y = \frac{1}{2}\sqrt{2}, x = 1 + \frac{1}{2}\sqrt{2} \,\right\}, \left\{\, y = -\frac{1}{2}\sqrt{2}, x = 1 + \frac{1}{2}\sqrt{2} \,\right\},$$
$$\left\{\, y = \frac{1}{2}\sqrt{2}, x = 1 - \frac{1}{2}\sqrt{2} \,\right\}, \left\{\, y = -\frac{1}{2}\sqrt{2}, x = 1 - \frac{1}{2}\sqrt{2} \,\right\}$$

```
> evalf(");
```
$$\{\, y = .7071067810, x = 1.707106781 \,\},$$
$$\{\, y = -.7071067810, x = 1.707106781 \,\},$$
$$\{\, x = .2928932190, y = .7071067810 \,\},$$
$$\{\, x = .2928932190, y = -.7071067810 \,\}$$

From the plot, the first and fourth solutions are the points of intersection.

Exercise 11.2   The height of a circular cylinder is equal to its radius. Express its total surface area A (including both ends) as a function of its volume. (p38, Edwards and Penney [6])

```
> h:=r;
```
$$h := r$$

```
> volume:=Pi*r^2*h;
```
$$volume := \pi r^3$$

```
> area:=2*Pi*r*h+2*Pi*r^2;
```
$$area := 4\pi r^2$$

```
> solve(area=volume*a,{a});
```
$$\left\{\, a = 4\frac{1}{r} \,\right\}$$

```
> assign(");
> surface_area='volume'*a;
```

$$surface\_area = 4\,\frac{volume}{r}$$

Exercise 11.3 A piece of wire 100 cm long is cut into two pieces of lengths x and $100 - x$. The first piece is bent into the shape of a square and the second is bent into the shape of a circle. Express as a function of x the sum A of the area of the square and the circle.

```
> A1:=(x/4)^2;
```

$$A1 := \frac{1}{16}\,x^2$$

```
> perimeter:=2*Pi*r;
```

$$perimeter := 2\,\pi\,r$$

```
> perimeter=100-x;
```

$$2\,\pi\,r = 100 - x$$

```
> solve(",{r});
```

$$\left\{ r = -\frac{1}{2}\,\frac{-100 + x}{\pi} \right\}$$

```
> assign(");
> A2:=r^2*Pi;
```

$$A2 := \frac{1}{4}\,\frac{(-100 + x)^2}{\pi}$$

```
> A=A1+A2;
```

$$A = \frac{1}{16}\,x^2 + \frac{1}{4}\,\frac{(-100 + x)^2}{\pi}$$

Exercise 11.4 A spherical cloud of toxic powder forms over a factory following a fire. The radius of the sphere is 100 m and contains a uniform dispersion of powder at a density of 0.001 kg$\cdot$m^{-3}. This cloud subsequently is deposited on the ground over a circular region of radius 100 m. Find the total mass of powder in the cloud and find also the average areal density in kg$\cdot$m^{-2} of powder deposited on the ground.

```
> r:=100;
```

$$r := 100$$

```
> v:=4/3*r^3*Pi;
```

$$v := \frac{4000000}{3} \pi$$

```
> evalf(v);
```

$$.4188790204 \, 10^7$$

```
> density:=v/100;
```

$$density := \frac{40000}{3} \pi$$

```
> evalf(density);
```

$$41887.90204$$

```
> r:='r':h:='h':
```

Exercise 11.5 A giant Sequoia tree, perhaps the largest tree in the world, stands 82.9 m tall and is about 10.4 m in diameter at its base. Its habitat is the central California mountain range, the Sierra Nevada, in the Sequoia National Park. The great age, size, and rapid growth of this tree have contributed to its fame. It has the heroic name General Sherman. (p4, Goldenberg and Greenwald [7])

(a) We can approximate the volume of a tree by disregarding the branch structure and considering the trunk as a cylinder. If the General Sherman has an average radius of 2.3 m, what is its volume?

```
> V:=Pi*r^2*h;
```

$$V := \pi r^2 h$$

```
> r:=2.3;
```

$$r := 2.3$$

```
> h:=82.9;
```

$$h := 82.9$$

```
> evalf(V);
```

$$1377.717184$$

(b) Assuming the General Sherman produces an annual growth ring of 0.00009 m, how much new wood is added to the tree (trunk) each year?

```
> delta_r:=0.00009+r;
```

$$delta\_r := 2.30009$$

```
> V:=Pi*(delta_r^2-r^2)*h;
```
$$V := .0343212632\,\pi$$

```
> evalf(V);
```
$$.1078234283$$

(c) Suppose the paper industry cultivated a fast-growing tree that could grow from a seedling to a height of 15.2 m in one year. If the increase in volume of this tree were to match that of the General Sherman, what would be the average radius of the tree?

```
> r:='r':delta_r:='delta_r':
> h:=15.2;
```
$$h := 15.2$$

```
> V=Pi*r^2*h;
```
$$.0343212632\,\pi = 15.2\,\pi\,r^2$$

```
> solve(",r);
```
$$-.04751818433, .04751818433$$

(d) The baobab tree, native to Australia and Africa, has a trunk that measures as much as 18.3 m in diameter. However, this tree only grows to a height of 12.2 m. Approximate the amount of new wood this tree would produce each year. Use a cylindrical model with a radius of 12.2 m, a height of 18.3 m, and an annual growth ring of 0.0009 m.

```
> h:=18.3;
```
$$h := 18.3$$

```
> r1:=12.2;
```
$$r1 := 12.2$$

```
> r2:=12.2+0.0009;
```
$$r2 := 12.2009$$

```
> V:=Pi*r2^2*h-Pi*r1^2*h;;
```
$$V := .401883\,\pi$$

(e) Model the baobab tree as a cone. Approximate the amount of new wood it would produce in one year if the increase in thickness is 0.0009 m.

```
> V:=1/3*Pi*r2^2*h-1/3*Pi*r1^2*h;
```

$$V := .1339610\,\pi$$

A.11 Loci

Exercise 12.1 Let a circle of diameter b have centre at $x = \frac{b}{2}$. Take a straight line $\overline{PQ}$ of any length $2a$ and move it such that its midpoint M moves round the given circle and the line $\overline{PQ}$ passes through the origin O. Then the point P and Q have as loci the curve $r = \overline{MP} + \overline{OP} = a + b\cos(\theta)$. This curve is called a *limaçon*; in particular it is called a *cardioid* if $a = b$. Plot a family of limaçons for a range of values of a and b.

```
> r:=theta->a+b*cos(theta);
```
$$r := \theta \rightarrow a + b\cos(\theta)$$

```
> a:=1;b:=1;
```
$$a := 1$$
$$b := 1$$

```
> with(plots):
> polarplot(r(theta),theta=-2*Pi..2*Pi);
> a:=2;b:=4;
```
$$a := 2$$
$$b := 4$$

```
> polarplot(r(theta),theta=-2*Pi..2*Pi);
```

Exercise 12.2 Let $\overline{AB}$ be the diameter of a circle and P a point on the upper semicircle. The tangent to the circle at B is a vertical line and suppose that the line $\overline{AP}$ extends to meet this vertical line at T. Now choose Q on $\overline{AT}$ such that $\overline{AQ} = \overline{PT}$ and the locus of Q is given by $y^2(2a - x) = x^3$ when A is the origin and $\overline{AB}$ lies along the $x-$axis. This is called a *cissoid* and has polar equation $r = \frac{2a\sin^2\theta}{\cos\theta}$. Plot a family of cissoids for a range of values of a.

```
> r:=theta->(2*a*sin(theta)^2)/(cos(theta));
```
$$r := \theta \rightarrow 2\frac{a\sin(\theta)^2}{\cos(\theta)}$$

```
> a:=2;
```
$$a := 2$$

```
> plot(r(theta),theta=-0.3..0.3,coords=polar);
> a:=1;
```
$$a := 1$$

```
> plot(r(theta),theta=-0.3..0.3,coords=polar);
```

Exercise 12.3   Try plotting the cycloid curves with the following parameterizations:

(a) $x(t) = t - \sin(t), y(t) = 1 - \cos(t)$ (cycloid)

```
> plot([t-sin(t),1-cos(t),t=-4*Pi..4*Pi],scaling=CONSTRAINED);
```

(b) $x(t) = t - 3\sin(t), y(t) = 1 - 3\cos(t)$ (prolate cycloid)

```
> plot([t-3*sin(t),1-3*cos(t),t=-4*Pi..4*Pi],scaling=CONSTRAINED);
```

(c) $x(t) = 2t - \sin(t), y(t) = 2 - \cos(t)$ (curate cycloid)

```
> plot([2*t-sin(t),2-cos(t),t=-4*Pi..4*Pi],scaling=CONSTRAINED);
```

A.12   Sequences and Series

Exercise 13.1   Calculate the first 4 terms of the following sequences.

(a) $f(n) = n - 1$
```
> seq(n-1,n=1..4);
```
$$0, 1, 2, 3$$

(b) $f(n) = \frac{1}{2}n + 7$
```
> seq((1/2)*n+7,n=1..4);
```
$$\frac{15}{2}, 8, \frac{17}{2}, 9$$

(c) $f(n) = -3n$
```
> seq(-3*n,n=1..4);
```
$$-3, -6, -9, -12$$

```
> n:='n':
```

Exercise 13.2   Determine the nth term of the following sequences.

(a) $5, 8, 11, 14, \ldots$
```
> 8-5;
```
$$3$$

```
> 11-8;
```
$$3$$

```
> d:=3;
```
$$d := 3$$

```
> a:=5-d;
```
$$a := 2$$

```
> a+n*d;
```
$$2 + 3n$$

(b) $\frac{5}{3}, \frac{4}{3}, 1, \frac{2}{3}, \frac{1}{3}, \ldots$
```
> 4/3-5/3;
```
$$\frac{-1}{3}$$

```
> 1-4/3;
```
$$\frac{-1}{3}$$

```
> d:=-1/3;
```
$$d := \frac{-1}{3}$$

```
> a:=5/3-d;
```
$$a := 2$$

```
> a+n*d;
```
$$2 - \frac{1}{3}\,n$$

(c) $5, 9, 13, 17, \ldots$

```
> 9-5;
```
$$4$$

```
> 13-9;
```
$$4$$

```
> d:=4;
```
$$d := 4$$

```
> a:=5-d;
```
$$a := 1$$

```
> a+n*d;
```
$$1 + 4\,n$$

```
> a:='a':d:='d':
```

Exercise 13.3   The third term of an arithmetic sequence is 14 and the seventh is 26. Find the ninth term.

```
> solve({a+3*d=14,a+7*d=26},{a,d});
```
$$\{\,d = 3, a = 5\,\}$$

```
> 5+9*3;
```
$$32$$

Exercise 13.4   Calculate the first four terms of the following sequences.

(a) $f(n) = 7^n$

```
> seq(7^n,n=1..4);
```
$$7, 49, 343, 2401$$

(b) $f(n) = \frac{1}{2}(3)^n$

```
> seq((1/2)*(3)^n,n=1..4);
```
$$\frac{3}{2}, \frac{9}{2}, \frac{27}{2}, \frac{81}{2}$$

(c) $f(n) = -5(2^n)$

```
> seq(-5*(2^n),n=1..4);
```
$$-10, -20, -40, -80$$

```
> n:='n':
```

Exercise 13.5   Calculate the nth term of the following sequences.

(a) $6, 18, 54, 162, \ldots$

```
> 18/6;
```
$$3$$

```
> 54/18;
```
$$3$$

```
> r:=3;
```
$$r := 3$$

```
> a:=6/r;
```
$$a := 2$$

```
> a*r^n;
```
$$2\,3^n$$

(b) $1, 4, 16, 64, \ldots$

```
> 4/1;
```
$$4$$

```
> 16/4;
```
$$4$$

```
> r:=4;
```
$$r := 4$$

```
> a:=1/r;
```
$$a := \frac{1}{4}$$

```
> a*r^n;
```
$$\frac{1}{4}\, 4^n$$

(c) $-6, 12, -24, 96. \ldots$

```
> 12/(-6);
```
$$-2$$

```
> -24/12;
```
$$-2$$

```
> r:=-2;
```
$$r := -2$$

```
> a:=-6/r;
```
$$a := 3$$

```
> a*r^n;
```
$$3\,(-2)^n$$

Exercise 13.6 For the geometric sequence $-8, 16, -32, \ldots, t_n = 256$, find n .

```
> 16/(-8);
```
$$-2$$

```
> -32/16;
```
$$-2$$

```
> r:=-2;
```
$$r := -2$$

```
> a:=-8/r;
```
$$a := 4$$

```
> a*r^6;
```
$$256$$

```
> a*r^6;
```
$$256$$

Exercise 13.7   Calculate the sum of the following arithmetic series.

(a) $5 + 11 + 17 + \ldots + 59$

```
> 11-5;
```
$$6$$

```
> 17-11;
```
$$6$$

```
> d:=6;
```
$$d := 6$$

```
> a:=5-d;
```
$$a := -1$$

```
> solve(a+n*d=59,{n});
```
$$\{ n = 10 \}$$

```
> sum(a+n*d,n=1..10);
```
$$320$$

(b) $-9 + (-11) + (-13) + \ldots + (-25)$

```
> -11-(-9);
```
$$-2$$

```
> -13-(-11);
```
$$-2$$

```
> d:=-2;
```
$$d := -2$$

```
> a:=-9-d;
```
$$a := -7$$

```
> solve(a+n*d=-25,{n});
```
$$\{ n = 9 \}$$

```
> sum(a+n*d,n=1..9);
```
$$-153$$

(c) $-2 + 1 + 4 + \ldots + 295$
```
> 1-(-2);
```
$$3$$

```
> 4-1;
```
$$3$$

```
> d:=3;
```
$$d := 3$$

```
> a:=-2-d;
```
$$a := -5$$

```
> solve(a+n*d=295,{n});
```
$$\{\, n = 100 \,\}$$

```
> sum(a+n*d,n=1..100);
```
$$14650$$

Exercise 13.8   Calculate the sum of the first 20 terms of the following arithmetic series.

(a) $4 + 5 + 6 + 7 + \ldots$
```
> 5-4;
```
$$1$$

```
> 6-5;
```
$$1$$

```
> d:=1;
```
$$d := 1$$

```
> a:=4-d;
```
$$a := 3$$

```
> sum(a+n*d,n=1..20);
```
$$270$$

(b) $5 + 9 + 13 + 17 + \ldots$

```
> 9-5;
```
$$4$$

```
> 13-9;
```
$$4$$

```
> d:=4;
```
$$d := 4$$

```
> a:=5-d;
```
$$a := 1$$

```
> sum(a+n*d,n=1..20);
```
$$860$$

(c) $-\frac{15}{2} - 7 - \frac{13}{2} - 6 - \ldots$

```
> -7-(-15/2);
```
$$\frac{1}{2}$$

```
> -13/2-(-7);
```
$$\frac{1}{2}$$

```
> d:=1/2;
```
$$d := \frac{1}{2}$$

```
> a:=-15/2-d;
```
$$a := -8$$

```
> sum(a+n*d,n=1..20);
```
$$-55$$

Exercise 13.9   Find the sum of the following geometric series.

(a) $6 + 12 + 24 + \ldots + 384$

```
> 12/6;
```
$$2$$

```
> 24/12;
```
$$2$$

```
> r:=2;
```
$$r := 2$$

```
> a:=6/r;
```
$$a := 3$$

```
> solve(a*r^n=384,n);
```
$$\frac{\ln(128)}{\ln(2)}$$

```
> evalf(");
```
$$7.000000000$$

```
> sum(a*r^n,n=1..7);
```
$$762$$

(b) $\frac{1}{4} + \frac{1}{16} + \frac{1}{64} + \cdots + \frac{1}{65536}$

```
> (1/16)/(1/4);
```
$$\frac{1}{4}$$

```
> (1/64)/(1/16);
```
$$\frac{1}{4}$$

```
> r:=1/4;
```
$$r := \frac{1}{4}$$

```
> a:=1/4/r;
```
$$a := 1$$

```
> solve(a*r^n=1/65536,n);
```
$$\frac{\ln(65536)}{\ln(4)}$$

```
> evalf(");
```
$$8.000000001$$

```
> sum(a*r^n,n=1..8);
```
$$\frac{21845}{65536}$$

(c) $-5 + 25 - 125 + \ldots + 15625$

```
> 25/(-5);
```
$$-5$$

```
> -125/25;
```
$$-5$$

```
> r:=-5;
```
$$r := -5$$

```
> a:=-5/r;
```
$$a := 1$$

```
> a*r^6;
```
$$15625$$

```
> sum(a*r^n,n=1..6);
```
$$13020$$

A.13   Statistics and Probability

Exercise 14.1   The mid-term and final exam marks for a given class are as follows.

Student	Mid-term Mark	Final Exam Mark
1	80	85
2	72	70
3	55	50
4	77	82
5	64	66
6	71	77
7	81	88
8	32	40
9	48	58
10	79	70
11	29	20
12	92	89
13	83	78
14	67	70
15	45	78

Create a scatterplot of the data and determine if the two data sets are associated. Using a pencil and paper, try to find the equation of the line which best fits this data.

```
> with(stats):
> midterm:=[80,72,55,77,64,71,81,32,48,79,29,92,83,67,45]:
> exam:=[85,70,50,82,66,77,88,40,58,70,20,89,78,70,78]:
> statplots[scatter2d](midterm,exam);
```

Exercise 14.2   The circumferences of 40 trees living on a plot of land, given in centimetres, are listed below.

43.2	80.3	71.8	35.1	95.0	60.0	97.9	30.4	102.3	91.1
62.6	81.3	73.5	66.1	56.4	42.2	42.7	69.4	46.5	70.0
43.9	76.7	18.8	75.6	24.5	41.4	69.6	56.5	41.1	87.9
15.0	41.4	54.3	70.0	75.3	94.5	41.4	63.5	63.2	94.7

(a) Create a histogram of this data using a class width of 10. What is the modal frequency?

```
> tree:=[43.2,80.3,71.8,35.1,95.0,60.0,97.9,30.4,102.3,91.1,62.6,
```

```
> 81.3,73.5,66.1,56.4,42.2,42.7,69.4,46.5,70.0,43.9,76.7,18.8,75.6,
> 24.5,41.4,69.6,56.5,41.1,87.9,15.0,41.4,54.3,70.0,75.3,94.5,41.4,
> 63.5,63.2,94.7]:
> tallied_tree:=transform[tallyinto](tree,[0..10,10..20,20..30,
> 30..40,40..50,50..60,60..70,70..80,80..90,90..100,100..110]);
```

$tallied\_tree := [\text{Weight}(0..10,0), \text{Weight}(10..20,2), 20..30,$
$\text{Weight}(40..50,9), \text{Weight}(50..60,3), \text{Weight}(60..70,7),$
$\text{Weight}(70..80,7), \text{Weight}(80..90,3), \text{Weight}(90..100,5),$
$100..110, \text{Weight}(30..40,2)]$

```
> statplots[histogram](tallied_tree);
> describe[mode](tree);
```
$$41.4$$

(b) What is the mean circumference of the trees?
```
> describe[mean](tree);
```
$$61.67750000$$

(c) What is the standard deviation?
```
> describe[standarddeviation](tree);
```
$$22.43890027$$

(d) What are the mean and standard deviation of the diameters of the trees?
```
> diameter:=[evalf(seq(tree[n]/Pi,n=1..40))];
```

$diameter := [13.75098708, 25.56028385, 22.85464982, 11.17267700,$
$30.23943918, 19.09859317, 31.16253785, 9.676620537,$
$32.56310135, 28.99803062, 19.92619887, 25.87859374,$
$23.39577663, 21.04028347, 17.95267758, 13.43267719,$
$13.59183214, 22.09070610, 14.80140970, 22.28169203,$
$13.97380400, 24.41436826, 5.984225859, 24.06422739,$
$7.798592209, 13.17802928, 22.15436807, 17.98450856,$
$13.08253632, 27.97943899, 4.774648292, 13.17802928,$
$17.28422682, 22.28169203, 23.96873442, 30.08028424,$
$13.17802928, 20.21267777, 20.11718480, 30.14394621]$

```
> describe[mean](diameter);
```
$$19.63255800$$

```
> describe[standarddeviation](diameter);
```
$$7.142523779$$

Exercise 14.3   Calculate the probability that a 13-card bridge hand contains

(a) 4 hearts, 3 clubs, 3 diamonds, 3 spades

```
> binomial(13,4)*binomial(13,3)*binomial(13,3)*binomial(13,3)/
> binomial(52,13);
```

$$\frac{418161601}{15875338990}$$

```
> evalf(");
```

$$.02634032579$$

(b) 5 hearts, 4 clubs, 2 diamonds, 2 spades

```
> binomial(13,5)*binomial(13,4)*binomial(13,2)*binomial(13,2)/
> binomial(52,13);
```

$$\frac{279926361}{31750677980}$$

```
> evalf(");
```

$$.008816390037$$

(c) 6 hearts, 3 clubs, 3 diamonds, 1 spade

```
> binomial(13,6)*binomial(13,3)*binomial(13,3)*binomial(13,1)/
> binomial(52,13);
```

$$\frac{114044073}{39688347475}$$

```
> evalf(");
```

$$.002873490086$$

Exercise 14.4   A computer operator has to invent a seven character password from the letters of the alphabet and the numbers $0, 1, 2, \ldots, 9$. At least one letter must be a capital letter and exactly one character must be a number. In how many ways can the password by chosen? Another computer operator uses a random character selector to create a password meeting the same conditions; what is the probability that this password will contain at least two capital letters?

```
> case1:=binomial(26,1)*binomial(10,1)*binomial(26,5);
```
$$case1 := 17102800$$

```
> case2:=binomial(26,2)*binomial(10,1)*binomial(26,4);
```
$$case2 := 48587500$$

```
> case3:=binomial(26,3)*binomial(10,1)*binomial(26,3);
                        case3 := 67600000


> case4:=binomial(26,4)*binomial(10,1)*binomial(26,2);
                        case4 := 48587500


> case5:=binomial(26,5)*binomial(10,1)*binomial(26,1);
                        case5 := 17102800


> case6:=binomial(26,6)*binomial(10,1);
                        case6 := 2302300


> case1+case2+case3+case4+case5+case6;
                        201282900
```

Total number of ways of creating the password is,

```
> binomial(26,1)*binomial(10,1)*binomial(52,5);
                        675729600


> twocapitals:=binomial(26,1)*binomial(10,1)*binomial(26,2)*
> binomial(26,3);
                   twocapitals := 219700000


> probability:=(")/("");
```

$$probability := \frac{1625}{4998}$$

```
> n:='n':
```

Exercise 14.5   In how many ways can 3 flags be flown on 2 flagpoles? What about r flags on n flagpoles?

```
> binomial(3,2);
```

$$3$$

```
> r!/(n!*(r-n)!);
```

$$\frac{r!}{n!\,(r-n)!}$$

Exercise 14.6   In a manufacturing process N computer chips are made per day. Of these, on average d are defective and $(N-d)$ are good. A sample of r chips is chosen at random from the production on a given day. What is the probability that this group will contain exactly k defectives where k varies from

0 to the smaller of d and r? For the case $N = 1000, r = 100, d = 10$, plot this probability for $k = 1, 2, \ldots, 10$.

The sample has $r - k$ good chips and these can be chosen in $\left(\frac{N-d}{r-k}\right)$ ways while the k defective chips can be chosen in $\left(\frac{d}{k}\right)$ ways. The ordering of chips is unimportant so we divide by the number of ways of choosing r from N to give the required probability $p(k) = \dfrac{\left(\frac{d}{k}\right)\left(\frac{N-d}{r-k}\right)}{\left(\frac{N}{R}\right)}$ and $p(k) = 0$ if $k > d$ or $k > r$.

```
> binomial(d,k)*binomial(N-d,r-k)/binomial(N,r);
```
$$\frac{\mathrm{binomial}(\,d,k\,)\,\mathrm{binomial}(\,N-d,r-k\,)}{\mathrm{binomial}(\,N,r\,)}$$

```
> f:=(k)->binomial(10,k)*binomial(1000-10,100-k)/
> binomial(1000,100);
```
$$f := k \rightarrow \frac{\mathrm{binomial}(\,10,k\,)\,\mathrm{binomial}(\,990,100-k\,)}{\mathrm{binomial}(\,1000,100\,)}$$

```
> t:=[[k,f(k)] $k=1..10];
```
$$
t := \left[\left[1, \frac{8271249852054420400}{21242706488868565551}\right], \left[2, \frac{1376996191737759450}{7080902162956188517}\right],\right.
$$
$$
\left[3, \frac{402973129646287200}{7080902162956188517}\right], \left[4, \frac{76515311809236300}{7080902162956188517}\right],
$$
$$
\left[5, \frac{9848674771423488}{7080902162956188517}\right], \left[6, \frac{870186107966175}{7080902162956188517}\right],
$$
$$
\left[7, \frac{52108612294200}{7080902162956188517}\right], \left[8, \frac{4047411701025}{14161804325912377034}\right],
$$
$$
\left.\left[9, \frac{46021737300}{7080902162956188517}\right], \left[10, \frac{13959926981}{21242706488868565510}\right]\right]\right]
$$

```
> plot(t,x=1..10, style=POINT);
```

Appendix B

Solutions to Part II Exercises

B.1   Secants and Tangents

Exercise 15.1   Find the slope of a secant to $y = 2x^2 - 3x$ which passes through the following points:

(a) $x = 1$ and $x = 4$

```
> with(student):
> f:=x->2*x^2-3*x;
```
$$f := x \rightarrow 2\,x^2 - 3\,x$$

```
> slope([1,f(1)],[4,f(4)]);
```
$$7$$

(b) $x = 0$ and $x = -1$

```
> slope([0,f(0)],[-1,f(-1)]);
```
$$-5$$

(c) $x = -2$ and $x = 2$

```
> slope([-2,f(-2)],[2,f(2)]);
```
$$-3$$

Exercise 15.2   Find the general equation of the slope of a secant passing through the points $(x, f(x))$ and $(a, f(a))$ to the following curves:

(a) x^2

```
> f:=x->x^2;
```
$$f := x \rightarrow x^2$$

```
> slope([x,f(x)],[a,f(a)]);
```
$$\frac{x^2 - a^2}{x - a}$$

```
> simplify(");
```
$$a + x$$

(b) $3x^2 - 5x + 8$

```
> f:=x->3*x^2-5*x+8;
```
$$f := x \rightarrow 3\,x^2 - 5\,x + 8$$

```
> slope([x,f(x)],[a,f(a)]);
```
$$\frac{3\,x^2 - 5\,x - 3\,a^2 + 5\,a}{x - a}$$

```
> simplify(");
```
$$3\,a - 5 + 3\,x$$

(c) $-7x^2 + 1$
```
> f:=x-> -7*x^2+1;
```
$$f := x \rightarrow -7\,x^2 + 1$$

```
> slope([x,f(x)],[a,f(a)]);
```
$$\frac{-7\,x^2 + 7\,a^2}{x - a}$$

```
> simplify(");
```
$$-7\,a - 7\,x$$

Exercise 15.3   Find the equation of the tangent at $x = 2$ to the following curves. Find also the equations of the normal lines to these curves.

(a) x^2
```
> f:=x->x^2;
```
$$f := x \rightarrow x^2$$

```
> slope([x,f(x)],[a,f(a)]);
```
$$\frac{x^2 - a^2}{x - a}$$

```
> simplify(");
```
$$a + x$$

```
> subs(x=a,");
```
$$2\,a$$

```
> m:=subs(a=2,");
```
$$m := 4$$

```
> m=(y-f(2))/(x-2);
```
$$4 = \frac{y - 4}{x - 2}$$

```
> isolate(",y);
```
$$y = 4x - 4$$

```
> -1/m=(y-f(2))/(x-2);
```
$$\frac{-1}{4} = \frac{y - 4}{x - 2}$$

```
> isolate(",y);
```
$$y = -\frac{1}{4}x + \frac{9}{2}$$

```
> m:='m':
```

(b) $3x^2 - 5x + 8$

```
> f:=x->3*x^2-5*x+8;
```
$$f := x \rightarrow 3x^2 - 5x + 8$$

```
> slope([x,f(x)],[a,f(a)]);
```
$$\frac{3x^2 - 5x - 3a^2 + 5a}{x - a}$$

```
> simplify(");
```
$$3a - 5 + 3x$$

```
> subs(x=a,");
```
$$6a - 5$$

```
> m:=subs(a=2,");
```
$$m := 7$$

```
> m=(y-f(2))/(x-2);
```
$$7 = \frac{y - 10}{x - 2}$$

```
> isolate(",y);
```
$$y = 7x - 4$$

```
> -1/m=(y-f(2))/(x-2);
```
$$\frac{-1}{7} = \frac{y - 10}{x - 2}$$

```
> isolate(",y);
```
$$y = -\frac{1}{7}x + \frac{72}{7}$$

```
> m:='m':
```

(c) $-7x^2 + 1$

```
> f:=x->-7*x^2+1;
```
$$f := x \rightarrow -7\,x^2 + 1$$

```
> slope([x,f(x)],[a,f(a)]);
```
$$\frac{-7\,x^2 + 7\,a^2}{x - a}$$

```
> simplify(");
```
$$-7\,a - 7\,x$$

```
> subs(x=a,");
```
$$-14\,a$$

```
> m:=subs(a=2,");
```
$$m := -28$$

```
> m=(y-f(2))/(x-2);
```
$$-28 = \frac{y + 27}{x - 2}$$

```
> isolate(",y);
```
$$y = -28\,x + 29$$

```
> -1/m=(y-f(2))/(x-2);
```
$$\frac{1}{28} = \frac{y + 27}{x - 2}$$

```
> isolate(",y);
```
$$y = \frac{1}{28}\,x - \frac{379}{14}$$

```
> m:='m':
```

Exercise 15.4 Find the point on the graph $y = 3x^2 + 1$ at which the slope is

(a) 2

```
> f:=x->3*x^2+1;
```
$$f := x \rightarrow 3\,x^2 + 1$$

```
> slope([x,f(x)],[a,f(a)]);
```
$$\frac{3\,x^2 - 3\,a^2}{x - a}$$

```
> simplify(");
```
$$3\,a + 3\,x$$

```
> subs(a=x,");
```
$$6\,x$$

```
> solve("=2,{x});
```
$$\left\{ x = \frac{1}{3} \right\}$$

```
> assign(");
> f(x);
```
$$\frac{4}{3}$$

```
> x:='x':
```
(b) $\frac{1}{5}$
```
> 6*1/5;
```
$$\frac{6}{5}$$

```
> f(");
```
$$\frac{133}{25}$$

(c) -4
```
> -4*6;
```
$$-24$$

```
> f(");
```
$$1729$$

Exercise 15.5 Determine the equations of the lines through the following points:

(a) $(0,1)$ and $(4,1)$
```
> m:=slope([0,1],[4,1]);
```
$$m := 0$$

```
> m=(y-1)/(x-0);
```

$$0 = \frac{y-1}{x}$$

```
> isolate(",y);
```

$$y = 1$$

```
> m:='m'
```

(b) $(7,0)$ and $(8,9)$

```
> m:=slope([7,0],[8,9]);
```

$$m := 9$$

```
> m=(y-1)/(x-0);
```

$$9 = \frac{y-1}{x}$$

```
> isolate(",y);
```

$$y = 9x + 1$$

```
> m:='m':
```

(c) $(3,8)$ and $(3,-1)$

Here the method fails, because the slope is infinite. However, the equation of the line is obvious: $x = 3$.

Exercise 15.6 Determine the equations of the lines with slope of 2 which pass through the following points:

(a) $(2,1)$

```
> 2=(y-1)/(x-2);
```

$$2 = \frac{y-1}{x-2}$$

```
> isolate(",y);
```

$$y = 2x - 3$$

(b) $(-8,-1)$

```
> 2=(y-(-1))/(x-(-8));
```

$$2 = \frac{y+1}{x+8}$$

```
> isolate(",y);
```

$$y = 2x + 15$$

(c) $(4,4)$

```
> 2=(y-4)/(x-4);
```

$$2 = \frac{y-4}{x-4}$$

```
> isolate(",y);
```

$$y = 2x - 4$$

Exercise 15.7 Determine the equations of the lines which pass through the point $(2,4)$ and have the following slopes. Find also the equations of the lines perpendicular to these.

(a) 3

```
> 3=(y-4)/(x-2);
```

$$3 = \frac{y-4}{x-2}$$

```
> isolate(",y);
```

$$y = 3x - 2$$

```
> -1/3=(y-4)/(x-2);
```

$$\frac{-1}{3} = \frac{y-4}{x-2}$$

```
> isolate(",y);
```

$$y = -\frac{1}{3}x + \frac{14}{3}$$

(b) $\frac{1}{2}$

```
> 1/2=(y-4)/(x-2);
```

$$\frac{1}{2} = \frac{y-4}{x-2}$$

```
> isolate(",y);
```

$$y = \frac{1}{2}x + 3$$

```
> -2=(y-4)/(x-2);
```

$$-2 = \frac{y-4}{x-2}$$

```
> isolate(",y);
```

$$y = -2x + 8$$

(c) $-\frac{3}{8}$

```
> -3/8=(y-4)/(x-2);
```

$$\frac{-3}{8} = \frac{y-4}{x-2}$$

```
> isolate(",y);
```

$$y = -\frac{3}{8}x + \frac{19}{4}$$

```
> 8/3=(y-4)/(x-2);
```

$$\frac{8}{3} = \frac{y-4}{x-2}$$

```
> isolate(",y);
```

$$y = \frac{8}{3}x - \frac{4}{3}$$

Exercise 15.8   The owner of a grocery store finds that he can sell 980 litres of milk at \$1.69/l and 1220 litres of milk each week at \$1.49/l. Assume a linear relationship between selling price and demand. How many litres could he sell weekly at \$1.56/l? (p22, Edwards and Penney [6])

```
> m:=slope([1.69,980],[1.49,1220]);
```

$$m := -1200.000000$$

```
> eq:m=(y-980)/(x-1.69);
```

$$-1200.000000 = \frac{y-980}{x-1.69}$$

```
> isolate(",y);
```

$$y = -1200.000000\,x + 3008.000000$$

```
> assign(");
> subs(x=1.56,y);
```

$$1136.000000$$

```
> y:='y':
```

Exercise 15.9   In hilly areas, reception for both television and radio is frequently poor. We consider a situation where an FM transmitter is located behind a hill, and a radio receiver is on the opposite side of the hill. How far from the base of the hill should the radio be located so that its reception is not obstructed? Obviously, the height of the transmitter, compared with the height of the hill is crucial, as is the specific positioning of the base of the transmitter. The radio can receive a clear signal if located far from the hill, provided the signal is strong enough. What is the closest that the radio can be to the hill so that reception is not obstructed? (p18, Goldenberg and Greenwald [7])

We consider an idealized situation: the contour of the hill is a semicircle of radius 1 unit. The transmitter is at the base of the hill and has a height of 2 units. The radio is on the side of the hill opposite the transmitter.

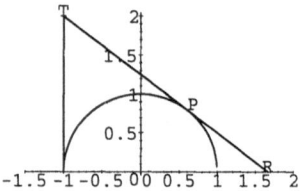

Figure B.1. Transmitter problem

Figure B.1 illustrates the situation. The position of the tangent line $\overline{TR}$ is the key element, where T is the top of the transmitter and R is the radio receiver. P is the point where the tangent line intersects the circle.

(a) We wish to find R. In other words, how far from the base of the hill should the radio be placed in order for it to receive an unobstructed signal from the transmitter?

```
> circle:=y=sqrt(1-x^2);
```
$$circle := y = \sqrt{1 - x^2}$$

```
> m:=diff(rhs(circle),x);
```
$$m := -\frac{x}{\sqrt{1 - x^2}}$$

```
> transmitter:=[-1,2];
```
$$transmitter := [-1, 2]$$

```
> tangent:=m=(transmitter[2]-y)/(transmitter[1]-x);
```
$$tangent := -\frac{x}{\sqrt{1 - x^2}} = \frac{2 - y}{-1 - x}$$

```
> solve({circle,tangent},{x,y});
```
$$\left\{ x = \frac{3}{5}, y = \frac{4}{5} \right\}$$

```
> slope([-1,2],[3/5,4/5]);
```
$$\frac{-3}{4}$$

```
> -(3/5)/(sqrt(1-(3/5)^2));
```
$$\frac{-3}{4}$$

```
> m:=";
```
$$m := \frac{-3}{4}$$

```
> tangent:=m=(transmitter[2]-y)/(transmitter[1]-x);
```
$$tangent := \frac{-3}{4} = \frac{2-y}{-1-x}$$

```
> isolate(tangent,y);
```
$$y = \frac{5}{4} - \frac{3}{4}x$$

```
> assign(");
> solve(y,{x});
```
$$\left\{ x = \frac{5}{3} \right\}$$

```
> evalf(");
```
$$\{\, x = 1.666666667 \,\}$$

(b) Solve the problem as given in the example if the transmitting tower
    (of height one unit) is placed on the top of the hill.

```
> y:='y':
> m:=diff(rhs(circle),x);
```
$$m := -\frac{x}{\sqrt{1-x^2}}$$

```
> transmitter:=[0,2];
```
$$transmitter := [\,0,2\,]$$

```
> tangent:=m=(transmitter[2]-y)/(transmitter[1]-x);
```
$$tangent := -\frac{x}{\sqrt{1-x^2}} = -\frac{2-y}{x}$$

```
> solve({circle,tangent},{x,y});
```
$$\left\{ x = -\frac{1}{2}\sqrt{3}, y = \frac{1}{2} \right\}, \left\{ x = \frac{1}{2}\sqrt{3}, y = \frac{1}{2} \right\}$$

```
> subs(x=1/2*sqrt(3),m);
```
$$-\frac{1}{2}\sqrt{4}\sqrt{3}$$

```
> m:=";
```
$$m := -\frac{1}{2}\sqrt{4}\sqrt{3}$$

```
> tangent:=m=(transmitter[2]-y)/(transmitter[1]-x);
```
$$tangent := -\frac{1}{2}\sqrt{4}\sqrt{3} = -\frac{2-y}{x}$$

```
> isolate(tangent,y);
```
$$y = -\frac{1}{2}\sqrt{4}\sqrt{3}\,x + 2$$

```
> assign(");
> solve(y,{x});
```
$$\left\{x = \frac{1}{3}\sqrt{4}\sqrt{3}\right\}$$

```
> evalf(");
```
$$\{\,x = 1.154700539\,\}$$

(c) Consider the problem as given in the example, but suppose the height of the transmitter is unknown. If it is known that the radio must be placed at least one unit from the base of the hill in order to receive an unobstructed signal, determine the height of the transmitter.

```
> m:='m':y:='y':
> m:=diff(rhs(circle),x);
```
$$m := -\frac{x}{\sqrt{1-x^2}}$$

```
> receiver:=[2,0];
```
$$receiver := [\,2,0\,]$$

```
> tangent:=m=(receiver[2]-y)/(receiver[1]-x);
```
$$tangent := -\frac{x}{\sqrt{1-x^2}} = -\frac{y}{2-x}$$

```
> solve({circle,tangent},{x,y});
```
$$\left\{x = \frac{1}{2}, y = \frac{1}{2}\sqrt{3}\right\}$$

```
> subs(x=1/2,m);
```
$$-\frac{1}{6}\sqrt{4}\sqrt{3}$$

```
> m:=";
```

$$m := -\frac{1}{6}\sqrt{4}\sqrt{3}$$

```
> tangent:=m=(receiver[2]-y)/(receiver[1]-x);
```

$$tangent := -\frac{1}{6}\sqrt{4}\sqrt{3} = -\frac{y}{2-x}$$

```
> isolate(tangent,x);
```

$$x = -\frac{1}{2}\sqrt{4}\sqrt{3}\,y + 2$$

```
> assign(");
> solve(x=-1,{y});
```

$$\left\{ y = \frac{1}{2}\sqrt{4}\sqrt{3} \right\}$$

```
> evalf(");
```

$$\{\, y = 1.732050808 \,\}$$

(d) Solve the problem as given in the example if the hill contour is described by the equation $y = x - x^2$.

```
> m:='m':x:='x':
> circle:=y=x-x^2;
```

$$circle := y = x - x^2$$

```
> m:=diff(rhs(circle),x);
```

$$m := -2\,x + 1$$

```
> tangent:=m=(transmitter[2]-y)/(transmitter[1]-x);
```

$$tangent := -2\,x + 1 = -\frac{2-y}{x}$$

```
> solve({circle,tangent},{x,y});
```

$$\{\, y = \mathrm{RootOf}(\_Z^2 - 2) - 2,\, x = \mathrm{RootOf}(\_Z^2 - 2)\,\}$$

```
> subs(x=sqrt(2),m);
```

$$-2\sqrt{2} + 1$$

```
> m:=";
```

$$m := -2\sqrt{2} + 1$$

```
> tangent:=m=(transmitter[2]-y)/(transmitter[1]-x);
```
$$tangent := -2\sqrt{2} + 1 = -\frac{2-y}{x}$$

```
> isolate(tangent,y);
```
$$y = -\left(2\sqrt{2} - 1\right)x + 2$$

```
> assign(");
> solve(y,{x});
```
$$\left\{x = -2\frac{1}{-2\sqrt{2} + 1}\right\}$$

```
> evalf(");
```
$$\{\,x = 1.093836322\,\}$$

(e) Solve the problem as given in the example, but suppose the transmitter is located at the point (-1,4).

```
> m:='m':y:='y':
> circle:=y=sqrt(1-x^2);
```
$$circle := y = \sqrt{1 - x^2}$$

```
> m:=diff(rhs(circle),x);
```
$$m := -\frac{x}{\sqrt{1 - x^2}}$$

```
> transmitter:=[-1,4];
```
$$transmitter := [-1, 4]$$

```
> tangent:=m=(transmitter[2]-y)/(transmitter[1]-x);
```
$$tangent := -\frac{x}{\sqrt{1 - x^2}} = \frac{4 - y}{-1 - x}$$

```
> solve({circle,tangent},{x,y});
```
$$\left\{x = \frac{15}{17}, y = \frac{8}{17}\right\}$$

```
> subs(x=15/17,m);
```
$$-\frac{15}{1088}\sqrt{64}\sqrt{289}$$

```
> m:=";
```
$$m := -\frac{15}{1088}\sqrt{64}\sqrt{289}$$

```
> tangent:=m=(transmitter[2]-y)/(transmitter[1]-x);
```

$$tangent := -\frac{15}{1088}\sqrt{64}\sqrt{289} = \frac{4-y}{-1-x}$$

```
> isolate(tangent,y);
```

$$y = \frac{15}{1088}\sqrt{64}\sqrt{289}\,(-1-x)+4$$

```
> assign(");
> solve(y,{x});
```

$$\left\{ x = \frac{1}{255}\left(-\frac{15}{1088}\sqrt{64}\sqrt{289}+4\right)\sqrt{64}\sqrt{289}\right\}$$

```
> evalf(");
```

$$\{\, x = 1.133333332\,\}$$

B.2   Sequences and Limits

Exercise 16.1   Find the first six terms of the sequence with the following nth terms:

(a) $f(n) = 2n - 1$, where $n \in \mathbb{N}$

```
> f:=n->2*n-1;
```
$$f := n \rightarrow 2n - 1$$

```
> seq(f(n),n=1..6);
```
$$1, 3, 5, 7, 9, 11$$

(b) $f(n) = 3^{n+1} + n$, where $n \in \mathbb{N}$

```
> f:=n->3^(n+1)+n;
```
$$f := n \rightarrow 3^{(n+1)} + n$$

```
> seq(f(n),n=1..6);
```
$$10, 29, 84, 247, 734, 2193$$

(c) $f(n) = 1 + \frac{7}{n}$, where $n \in \mathbb{N}$

```
> f:=n->1+7/n;
```
$$f := n \rightarrow 1 + 7\frac{1}{n}$$

```
> seq(f(n),n=1..6);
```
$$8, \frac{9}{2}, \frac{10}{3}, \frac{11}{4}, \frac{12}{5}, \frac{13}{6}$$

Exercise 16.2   Find the stated terms of sequences defined by the following functions:

(a) 7th term of $f(n) = -5^n$, where $n \in \mathbb{N}$

```
> f:=n->-5^n;
```
$$f := n \rightarrow -5^n$$

```
> f(7);
```
$$-78125$$

```
> seq(f(n),n=1..7);
```
$$-5, -25, -125, -625, -3125, -15625, -78125$$

(b) 11th term of $f(n) = 3n^2$, where $n \in \mathbf{N}$
```
> f:=n->3*n^2;
```
$$f := n \to 3\,n^2$$

```
> f(11);
```
$$363$$

(c) 100th term of $f(n) = \frac{3}{n^2}$, where $n \in \mathbf{N}$
```
> f:=n->3/n^2;
```
$$f := n \to 3\,\frac{1}{n^2}$$

```
> f(100);
```
$$\frac{3}{10000}$$

Exercise 16.3 Try to prove analytically that $\lim\limits_{x \to 0} \dfrac{\sqrt{4 - x^2} - 2}{x} = 0$. Later in this chapter we give a number of rules for limits but your intuition will probably guide you here since all of the rules are statements of what you would expect to be true! Here's a hint: on any exam about limits you can expect to need the formula for the difference of two squares. Try multiplying top and bottom of $\frac{\sqrt{4-x^2}-2}{x}$ by $\sqrt{4 - x^2} + 2$ and simplify. Then find the limit of the result as x tends to zero. The multiplication is quite legitimate because as $x \to 0$ so $(\sqrt{4 - x^2} + 2) \to 4$.

```
> f:=x->(sqrt(4-x^2)-2)/x;
```
$$f := x \to \frac{\mathrm{sqrt}(\,4 - x^2\,) - 2}{x}$$

```
> f(x)*(sqrt(4-x^2)+2);
```
$$\frac{\left(\sqrt{4 - x^2} - 2\right)\left(\sqrt{4 - x^2} + 2\right)}{x}$$

```
> expand(");
```
$$-x$$

```
> "*1/(sqrt(4-x^2)+2);
```
$$-\frac{x}{\sqrt{4 - x^2} + 2}$$

```
> subs(x=0,");
```
$$0$$

Exercise 16.4 Find the limiting value, as x increases without bound, of the following functions, or explain why no real limit exists.

(a) $f(x) = \frac{x^2-1}{x^2+1}$

```
> f:=x->(x^2-1)/(x^2+1);
```

$$f := x \rightarrow \frac{x^2 - 1}{x^2 + 1}$$

```
> limit(f(x),x=infinity);
```

$$1$$

(b) $f(x) = \frac{3+x}{x}$

```
> f:=x->(3+x)/x;
```

$$f := x \rightarrow \frac{3 + x}{x}$$

```
> limit(f(x),x=infinity);
```

$$1$$

(c) $f(x) = e^x$

```
> f:=x->exp(x);
```

$$f := \exp$$

```
> limit(f(x),x=infinity);
```

$$\infty$$

Exercise 16.5 Tabulate the sequence $(1 + \frac{1}{n})^n$ for $n = 1, 2, \ldots$ and estimate its limiting value correct to five decimal places.

```
> f:=n->(1+1/n)^n;
```

$$f := n \rightarrow \left(1 + \frac{1}{n}\right)^n$$

```
> n:='n':
> sum(f(n),n=1);
```

$$2$$

```
> evalf(sum(f(n),n=2));
```

$$2.250000000$$

```
> evalf(sum(f(n),n=10));
```

$$2.593742460$$

```
> evalf(sum(f(n),n=100));
```
$$2.704813829$$

```
> evalf(sum(f(n),n=1000));
```
$$2.716923932$$

```
> limit(f(n),n=infinity);
```
$$e$$

Exercise 16.6   Let $x = 1 + z$ and use the Binomial Theorem

$$(1+z)^n = 1 + nz + \frac{n(n-1)z^2}{2!} + \frac{n(n-1)(n-2)z^3}{3!} + \dots \quad \text{for } |x| < 1$$

to prove that $\displaystyle\lim_{x \to 1} \frac{x^n - 1}{x - 1} = n$.

We create a function to third order in z, since we are interested in z tending to zero.

```
> f:=z->1+n*z+(n*(n-1)*z^2)/(2!)+(n*(n-1)*(n-2)*z^3)/(3!);
```
$$f := z \to 1 + nz + \frac{1}{2}n(n-1)z^2 + \frac{1}{6}n(n-1)(n-2)z^3$$

```
> (f(z)-1)/z;
```
$$\frac{nz + \frac{1}{2}n(n-1)z^2 + \frac{1}{6}n(n-1)(n-2)z^3}{z}$$

```
> simplify(");
```
$$\frac{1}{6}n\left(6 + 3nz - 3z + z^2 n^2 - 3z^2 n + 2z^2\right)$$

```
> subs(z=0,");
```
$$n$$

Exercise 16.7   Use the Binomial Theorem to expand $(1 + \frac{1}{n})^n$ and deduce that

$$\lim_{n \to \infty} (1 + \frac{1}{n})^n = 1 + 1 + \frac{1}{2!} + \frac{1}{3!} + \dots$$

We create a function to third order in x, since we are interested in $x = \frac{1}{n}$ tending to zero.

```
> f:=x->1+n*x+(n*(n-1)*x^2)/(2!)+(n*(n-1)*(n-2)*x^3)/(3!);
```
$$f := x \to 1 + nx + \frac{1}{2}n(n-1)x^2 + \frac{1}{6}n(n-1)(n-2)x^3$$

```
> f(1/n);
```

$$2 + \frac{1}{2} \frac{n-1}{n} + \frac{1}{6} \frac{(n-1)(n-2)}{n^2}$$

As $n \to \infty$, the $\frac{n-1}{n}$ and $\frac{(n-1)(n-2)}{n^2}$ approach one, so the limit of $(1+x)^n$ is equal to the sum of the constants in the series. In fact, *Maple* can sum the whole series and gives the right answer, e.

```
> limit((1+1/n)^n,n=infinity);
```

$$e$$

Exercise 16.8   Use *Maple's* **series** function to investigate the limits of $\frac{\sin x}{x}$ and $\frac{\tan x}{x}$ as $x \to 0$.

```
> series(sin(x)/x,x,9);
```

$$1 - \frac{1}{6} x^2 + \frac{1}{120} x^4 - \frac{1}{5040} x^6 + O(x^8)$$

```
> series(tan(x)/x,x,9);
```

$$1 + \frac{1}{3} x^2 + \frac{2}{15} x^4 + \frac{17}{315} x^6 + O(x^8)$$

Exercise 16.9   Determine whether the following functions are continuous over the whole real line.

(a) $f(x) = \frac{x^3 + 3x^2 - 9x - 27}{x+3}$

```
> f:=x->(x^3+3*x^2-9*x-27)/(x+3);
```

$$f := x \to \frac{x^3 + 3x^2 - 9x - 27}{3 + x}$$

```
> readlib(iscont):
> iscont(f(x),x=-4..-2);
```

$$false$$

```
> limit(f(x),x=-3);
```

$$0$$

(b) $f(\theta) = \frac{\cos(\theta)}{\theta}$

```
> f:=theta->cos(theta)/theta;
```

$$f := \theta \to \frac{\cos(\theta)}{\theta}$$

```
> iscont(f(theta),theta=-1..1);
```
$$false$$

```
> limit(f(theta),theta=0);
```
$$undefined$$

(c) $f(x) = \frac{2}{x^2}$
```
> f:=x->2/x^2;
```
$$f := x \to 2\,\frac{1}{x^2}$$

```
> iscont(f(x),x=-1..1);
```
$$false$$

```
> limit(f(x),x=0);
```
$$\infty$$

Exercise 16.10   For which values of the real number x is the function $\log(\log x)$ defined?

```
> f:=x->log(log(x));
```
$$f := x \to \log(\log(x))$$

```
> plot(f(x),x);
```

Since $\log x$ is defined only for $x > 0$, $\log(\log x)$ is defined only for $\log(x) > 0$, which means $x > 1$.

B.3   Derivatives of Functions

Exercise 17.1   Differentiate the following functions from first principles using the concept of limits.

(a) $y = 3x^3 + 2x^2 + 1$

```
> f:=x->3*x^3+2*x^2+1;
```
$$f := x \rightarrow 3\,x^3 + 2\,x^2 + 1$$

```
> with(student):
> slope([x+delta,f(x+delta)],[x,f(x)]);
```
$$\frac{3\,(x + \delta)^3 + 2\,(x + \delta)^2 - 3\,x^3 - 2\,x^2}{\delta}$$

```
> limit(",delta=0);
```
$$9\,x^2 + 4\,x$$

(b) $y = \frac{\sin(x)}{x^2}$

```
> f:=x->sin(x)/x^2;
```
$$f := x \rightarrow \frac{\sin(x)}{x^2}$$

```
> slope([x+delta,f(x+delta)],[x,f(x)]);
```
$$\frac{\dfrac{\sin(x + \delta)}{(x + \delta)^2} - \dfrac{\sin(x)}{x^2}}{\delta}$$

```
> limit(",delta=0);
```
$$\frac{\cos(x)\,x - 2\sin(x)}{x^3}$$

(c) $y = \frac{x+2}{x^2+7}$

```
> f:=x->(x+2)/(x^2+7);
```
$$f := x \rightarrow \frac{x + 2}{x^2 + 7}$$

```
> slope([x+delta,f(x+delta)],[x,f(x)]);
```
$$\frac{\dfrac{x + \delta + 2}{(x + \delta)^2 + 7} - \dfrac{x + 2}{x^2 + 7}}{\delta}$$

```
> limit(",delta=0);
```
$$-\frac{x^2 - 7 + 4\,x}{(x^2 + 7)^2}$$

Exercise 17.2   Find the derivatives of the following functions.

(a) $y = x^2 + 2$

```
> f:=x->x^2+2;
```
$$f := x \rightarrow x^2 + 2$$

```
> diff(f(x),x);
```
$$2\,x$$

```
> D(f)(x);
```
$$2\,x$$

Can you see in what way the function $f(x) = 2x$ is the best linear approximation to $f(x) = x^2 + 2$ at point x? At $x = a$, the tangent to $x^2 + 2$ has equation $y = 2ax + 2 - a^2$ which has slope $2a$ and is the best linear approximation to $y = x^2 + 2$ near $x = a$. Obviously, it is the tangent line at $x = a$.

(b) $y = \frac{1}{2+3x}$

```
> f:=x->1/(2+3*x);
```
$$f := x \rightarrow \frac{1}{2 + 3\,x}$$

```
> diff(f(x),x);
```
$$-3\,\frac{1}{(2 + 3\,x)^2}$$

(c) $y = \frac{6}{x^3}$

```
> f:=x->6/x^3;
```
$$f := x \rightarrow 6\,\frac{1}{x^3}$$

```
> diff(f(x),x);
```
$$-18\,\frac{1}{x^4}$$

Exercise 17.3   Differentiate the product of the functions $f(x)$ and $g(x)$ using the product rule.

(a) $f(x) = (3x - 1)$ and $g(x) = (x^2 + 2)$

```
> f:=x->(3*x-1);
```
$$f := x \rightarrow 3\,x - 1$$

```
> g:=x->(x^2+2);
```
$$g := x \rightarrow x^2 + 2$$

```
> g(x)*diff(f(x),x)+f(x)*diff(g(x),x);
```
$$3\,x^2 + 6 + 2\,(\,3\,x - 1\,)\,x$$

```
> diff(f(x)*g(x),x);
```
$$3\,x^2 + 6 + 2\,(\,3\,x - 1\,)\,x$$

(b) $f(x) = \ln(x)$ and $g(x) = 2x^2$

```
> f:=x->ln(x);
```
$$f := \ln$$

```
> g:=x->2*x^2;
```
$$g := x \rightarrow 2\,x^2$$

```
> g(x)*diff(f(x),x)+f(x)*diff(g(x),x);
```
$$2\,x + 4\ln(\,x\,)\,x$$

```
> diff(f(x)*g(x),x);
```
$$2\,x + 4\ln(\,x\,)\,x$$

(c) $f(x) = \frac{1}{x}$ and $g(x) = (5x - 6)$

```
> f:=x->1/x;
```
$$f := x \rightarrow \frac{1}{x}$$

```
> g:=x->(5*x-6);
```
$$g := x \rightarrow 5\,x - 6$$

```
> g(x)*diff(f(x),x)+f(x)*diff(g(x),x);
```
$$-\frac{5\,x - 6}{x^2} + 5\,\frac{1}{x}$$

```
> diff(f(x)*g(x),x);
```
$$-\frac{5\,x - 6}{x^2} + 5\,\frac{1}{x}$$

Exercise 17.4   Differentiate the quotient of the functions $f(x)$ and $g(x)$ using
the quotient rule.

(a) $f(x) = x^2 - 1$ and $g(x) = 6x - 8$
```
> f:=x->x^2-1;
```
$$f := x \to x^2 - 1$$

```
> g:=x->6*x-8;
```
$$g := x \to 6x - 8$$

```
> (g(x)*diff(f(x),x)-f(x)*diff(g(x),x))/g(x)^2;
```
$$\frac{2(6x-8)x - 6x^2 + 6}{(6x-8)^2}$$

```
> simplify(");
```
$$\frac{1}{2}\frac{3x^2 - 8x + 3}{(3x-4)^2}$$

```
> diff(f(x)/g(x),x);
```
$$2\frac{x}{6x-8} - 6\frac{x^2 - 1}{(6x-8)^2}$$

```
> simplify(");
```
$$\frac{1}{2}\frac{3x^2 - 8x + 3}{(3x-4)^2}$$

(b) $f(x) = \sin(x)$ and $g(x) = x^3 + x$
```
> f:=x->sin(x);
```
$$f := \sin$$

```
> g:=x->x^3+x;
```
$$g := x \to x^3 + x$$

```
> (g(x)*diff(f(x),x)-f(x)*diff(g(x),x))/g(x)^2;
```
$$\frac{(x^3 + x)\cos(x) - \sin(x)(3x^2 + 1)}{(x^3 + x)^2}$$

```
> simplify(");
```
$$\frac{\cos(x)x^3 + \cos(x)x - 3\sin(x)x^2 - \sin(x)}{x^2(x^4 + 2x^2 + 1)}$$

```
> diff(f(x)/g(x),x);
```
$$\frac{\cos(x)}{x^3 + x} - \frac{\sin(x)(3x^2 + 1)}{(x^3 + x)^2}$$

```
> simplify(");
```
$$\frac{\cos(x)\,x^3 + \cos(x)\,x - 3\sin(x)\,x^2 - \sin(x)}{x^2\,(x^4 + 2\,x^2 + 1)}$$

(c) $f(x) = 11x + 9$ and $g(x) = e^x$

```
> f:=x->11*x+9;
```
$$f := x \rightarrow 11\,x + 9$$

```
> g:=x->exp(x);
```
$$g := \exp$$

```
> (g(x)*diff(f(x),x)-f(x)*diff(g(x),x))/g(x)^2;
```
$$\frac{11\,e^x - (11\,x + 9)\,e^x}{(e^x)^2}$$

```
> simplify(");
```
$$-e^{(-x)}\,(-2 + 11\,x)$$

```
> diff(f(x)/g(x),x);
```
$$11\,\frac{1}{e^x} - \frac{11\,x + 9}{e^x}$$

```
> simplify(");
```
$$2\,e^{(-x)} - 11\,e^{(-x)}\,x$$

Exercise 17.5 Find the first and second derivatives of the following functions and plot them on the same graph.

(a) $y = -2x$
```
> f:=x->-2*x;
```
$$f := x \rightarrow -2\,x$$

```
> diff(f(x),x);
```
$$-2$$

```
> diff(f(x),x$2);
```
$$0$$

```
> plot({f(x),diff(f(x),x),diff(f(x),x$2)},x);
```
(b) $y = x^3 - 8$
```
> f:=x->x^3-8;
```
$$f := x \rightarrow x^3 - 8$$

```
> diff(f(x),x);
```
$$3\,x^2$$

```
> diff(f(x),x$2);
```
$$6\,x$$

```
> plot({f(x),diff(f(x),x),diff(f(x),x$2)},x);
```
(c) $y = \frac{9}{x}$
```
> f:=x->9/x;
```
$$f := x \to 9\,\frac{1}{x}$$

```
> diff(f(x),x);
```
$$-9\,\frac{1}{x^2}$$

```
> diff(f(x),x$2);
```
$$18\,\frac{1}{x^3}$$

```
> plot({f(x),diff(f(x),x),diff(f(x),x$2)},x,y=-10..10);
```

Exercise 17.6 Differentiate the following equations implicitly and solve for $y(x)$.

(a) $xy(x) = 6$
```
> diff(x*y(x)=6,x);
```
$$\mathrm{y}(\,x\,) + x\,\left(\frac{\partial}{\partial x}\,\mathrm{y}(\,x\,)\right) = 0$$

```
> isolate(",diff(y(x),x));
```
$$\frac{\partial}{\partial x}\,\mathrm{y}(\,x\,) = -\,\frac{\mathrm{y}(\,x\,)}{x}$$

```
> lhs(")=subs(y(x)=6/x,rhs("));
```
$$\frac{\partial}{\partial x}\,\mathrm{y}(\,x\,) = -6\,\frac{1}{x^2}$$

(b) $xy(x)^2 = 12$
```
> diff(x*y(x)^2=12,x);
```
$$\mathrm{y}(\,x\,)^2 + 2\,x\,\mathrm{y}(\,x\,)\,\left(\frac{\partial}{\partial x}\,\mathrm{y}(\,x\,)\right) = 0$$

```
> isolate(",diff(y(x),x));
```

$$\frac{\partial}{\partial x} y(x) = -\frac{1}{2}\frac{y(x)}{x}$$

```
> lhs(")=subs(y(x)=sqrt(12/x),rhs("));
```

$$\frac{\partial}{\partial x} y(x) = -\frac{\sqrt{3}\sqrt{\frac{1}{x}}}{x}$$

(c) $x^2 y(x) = 18$

```
> diff(x^2*y(x)=18,x);
```

$$2x\,y(x) + x^2 \left(\frac{\partial}{\partial x} y(x)\right) = 0$$

```
> isolate(",diff(y(x),x));
```

$$\frac{\partial}{\partial x} y(x) = -2\frac{y(x)}{x}$$

```
> lhs(")=subs(y(x)=sqrt(18/x^2),rhs("));
```

$$\frac{\partial}{\partial x} y(x) = -6\frac{\sqrt{2}\sqrt{\frac{1}{x^2}}}{x}$$

Exercise 17.7 By now, pictures of astronauts and their equipment floating with their space vehicle are familiar. Here the concept of weight and weightlessness are quite visibly demonstrated. We know that an object at a distance from the earth has less weight than if measured on the earth's surface. But what of the rate of change of weight, as an object rises? Let's consider a rocket carrying a satellite of known weight on earth. To determine the rate of change of the weight of an object, we need to know the force, F, between the earth and the satellite as a function of the distance, r, between them. This is described by $F(r) = \frac{K}{r^2}$ where K is a constant. Using the Chain Rule $\frac{dF}{dt} = \frac{dF}{dr}\cdot\frac{dr}{dt}$ and calculating $\frac{dF}{dr}$ and $\frac{dr}{dt}$, we can determine $\frac{dF}{dt}$, the quantity we wish to find. (p22, Goldenberg and Greenwald [7])

(a) A satellite weighs 10 kg on the surface of the earth. A rocket carrying the satellite is leaving the earth at a rate of 2000 km/hr. Assume that the radius of the earth is approximately 6373 km and that the force describing the rocket's weight varies inversely as the square of the distance from the centre of the earth. Determine the rate of change of the weight of the satellite with respect ot time, when the satellite is 60 km above the surface of the earth.

```
> F:=r->K/r^2;
```

$$F := r \rightarrow \frac{K}{r^2}$$

```
> solve(F(6373)=10,{K});
```
$$\{ K = 406151290 \}$$

```
> assign(");
> F:='F':
> diff(F(t),t)=diff(F(r),r)*diff(r(t),t);
```
$$\frac{\partial}{\partial t} F(t) = \left(\frac{\partial}{\partial r} F(r) \right) \left(\frac{\partial}{\partial t} r(t) \right)$$

```
> subs(diff(r(t),t)=2000,F(r)=K/r^2,");
```
$$\frac{\partial}{\partial t} F(t) = 2000 \left(\frac{\partial}{\partial r} \left(406151290 \frac{1}{r^2} \right) \right)$$

```
> subs(r=60+6373,");
```
$$\frac{\partial}{\partial t} F(t) = \frac{-1624605160000}{266219984737}$$

```
> evalf(");
```
$$\frac{\partial}{\partial t} F(t) = -6.102491372$$

(b) Rework the example with the position of the satellite given by $r = -16t^2 + 21120000$ m. Find $\frac{dF}{dt}$ when $t = 1$.

```
> r:=t->(-16*t^2+21120000)/1000;
```
$$r := t \rightarrow -\frac{2}{125} t^2 + 21120$$

```
> diff(F(t),t)=diff(K/r^2,r)*diff(r(t),t);
```
$$\frac{\partial}{\partial t} F(t) = \frac{649842064}{25} \frac{t}{r^3}$$

```
> r(1);
```
$$\frac{2639998}{125}$$

```
> lhs("")=subs(r=",t=1,rhs(""));
```
$$\frac{\partial}{\partial t} F(t) = \frac{6346113906250}{2299962772803959999}$$

```
> evalf(");
```
$$\frac{\partial}{\partial t} F(t) = .2759224619\, 10^{-5}$$

(c) The satellite in the example weighs about 1.66 kg on the surface of the moon. If the radius of the moon is approximately 1733 km, and if the speed of the rocket is 2000 km/hour, find the rate of change of the weight of the satellite when the satellite is 60 km above the surface of the moon.

```
> F:=r->K2/r^2;
```

$$F := r \rightarrow \frac{K2}{r^2}$$

```
> solve(F(1733)=1.66,{K2});
```

$$\{ K2 = .4985459740\,10^7 \}$$

```
> assign(");
> F:='F':
> diff(F(t),t)=diff(F(r),r)*diff(r(t),t);
```

$$\frac{\partial}{\partial t}\,F(t) = -\frac{4}{125}\,\left(\frac{\partial}{\partial r}\,F(r)\right)\,t$$

```
> subs(diff(r(t),t)=2000,F(r)=K2/r^2,");
```

$$\frac{\partial}{\partial t}\,F(t) = -\frac{4}{125}\,\left(\frac{\partial}{\partial r}\,\left(.4985459740\,10^7\,\frac{1}{r^2}\right)\right)\,t$$

```
> subs(r=60+1733,");
```

$$\frac{\partial}{\partial t}\,F(t) = .00005535340217\,t$$

(d) The rocket in the example is rising at a constant velocity. If the change in the weight of the satellite is given by $\frac{dF}{dt} = -10$ kg/hr determine the rate at which the rocket is rising when the rocket is 160 km above the surface of the earth.

```
> F:='F':r:='r':
> diff(F(t),t)=diff(F(r),r)*diff(r(t),t);
```

$$\frac{\partial}{\partial t}\,F(t) = \left(\frac{\partial}{\partial r}\,F(r)\right)\,\left(\frac{\partial}{\partial t}\,r(t)\right)$$

```
> subs(diff(F(t),t)=-10,F(r)=K2/r^2,");
```

$$-10 = \left(\frac{\partial}{\partial r}\,\left(.4985459740\,10^7\,\frac{1}{r^2}\right)\right)\,\left(\frac{\partial}{\partial t}\,r(t)\right)$$

```
> isolate(",diff(r(t),t));
```

$$\frac{\partial}{\partial t}\,r(t) = .1002916533\,10^{-5}\,r^3$$

```
> lhs(")=subs(r=160+6373,rhs("));
```

$$\frac{\partial}{\partial t} r(t) = 279642.2354$$

(e) Consider the satellite in the example except assume that the weight of the satellite is unknown and the change in the weight of the satellite is given by $\frac{dF}{dt} = -20$ kg/hr when the satellite is 60 km above the surface of the earth, determine the weight of the satellite on the surface of the earth.

```
> F:='F':r:='r':K:='K':
> diff(F(t),t)=diff(F(r),r)*diff(r(t),t);
```

$$\frac{\partial}{\partial t} F(t) = \left(\frac{\partial}{\partial r} F(r) \right) \left(\frac{\partial}{\partial t} r(t) \right)$$

```
> subs(diff(F(t),t)=-20,diff(r(t),t)=2000,F(r)=K/r^2,");
```

$$-20 = 2000 \left(\frac{\partial}{\partial r} \frac{K}{r^2} \right)$$

```
> subs(r=60+6373,");
```

$$-20 = -\frac{4000}{266219984737} K$$

```
> solve(",{K});
```

$$\left\{ K = \frac{266219984737}{200} \right\}$$

```
> assign(");
> F:=K/r^2;
```

$$F := \frac{266219984737}{200} \frac{1}{r^2}$$

```
> subs(r=60+6373,F);
```

$$\frac{6433}{200}$$

B.4   Functions and Graphs

Exercise 18.1   Determine the $x-$ and $y-$intercepts for the following functions.

(a) $y = 2x^2 - 9$

```
> f:=x->2*x^2-9;
```

$$f := x \rightarrow 2x^2 - 9$$

```
> f(0);
```

$$-9$$

```
> solve(f(x),x);
```

$$\frac{3}{2}\sqrt{2}, -\frac{3}{2}\sqrt{2}$$

(b) $y = -x^3$

```
> f:=x->-x^3;
```

$$f := x \rightarrow -x^3$$

```
> f(0);
```

$$0$$

```
> solve(f(x),x);
```

$$0, 0, 0$$

(c) $y = \frac{x+11}{6x^2+5x+1}$

```
> f:=x->(x+11)/(6*x^2+5*x+1);
```

$$f := x \rightarrow \frac{x+11}{6x^2 + 5x + 1}$$

```
> f(0);
```

$$11$$

```
> solve(f(x),x);
```

$$-11$$

Exercise 18.2   Find the asymptotes of the following functions.

(a) $f(x) = \frac{x}{\sqrt{4x-1}}$

```
> f:=x->x/(4*x-1)^(1/2);
```

$$f := x \rightarrow \frac{x}{\sqrt{4x-1}}$$

```
> solve(denom(f(x))=0);
```

$$\frac{1}{4}$$

```
> limit(f(x),x=1/4);
```

$$undefined$$

```
> limit(f(x),x=infinity);
```

$$\infty$$

The function $f(x) = \frac{x}{\sqrt{4x-1}}$ has a vertical asymptote at $x = \frac{1}{4}$.

(b) $f(x) = \frac{3x^2+5x-6}{x+2}$

```
> f:=x->(3*x^2+5*x-6)/(x+2);
```

$$f := x \to \frac{3x^2 + 5x - 6}{x + 2}$$

```
> solve(denom(f(x))=0);
```

$$-2$$

```
> limit(f(x),x=-2);
```

$$undefined$$

```
> limit(f(x),x=infinity);
```

$$\infty$$

```
> solve(limit(diff(f(x),x)-m,x=infinity)=0,{m});
```

$$\{m = 3\}$$

```
> assign(");
> solve(limit(f(x)-(m*x+b),x=infinity)=0,{b});
```

$$\{b = -1\}$$

```
> m:='m':
```

The function $f(x) = \frac{3x^2+5x-6}{x+2}$ has a vertical asymptote at $x = -2$ and an oblique asymptote at $y = 3x - 1$.

(c) $f(x) = \frac{x^2}{7-x}$

```
> f:=x->x^2/(7-x);
```

$$f := x \to \frac{x^2}{7 - x}$$

```
> solve(denom(f(x))=0);
```

$$7$$

```
> limit(f(x),x=7);
```
$$undefined$$

```
> limit(f(x),x=infinity);
```
$$-\infty$$

```
> solve(limit(diff(f(x),x)-m,x=infinity)=0,{m});
```
$$\{m = -1\}$$

```
> assign(");
> solve(limit(f(x)-(m*x+b),x=infinity)=0,{b});
```
$$\{b = -7\}$$

```
> m:='m':
```
The function $f(x) = \frac{x^2}{7-x}$ has a vertical asymptote at $x = 7$ and an oblique asymptote at $y = -x - 7$.

Exercise 18.3 Determine whether the following functions are even, odd, or neither.

(a) $y = 3x^4 - x^2 + 2$
```
> f:=x->3*x^4-x^2+2;
```
$$f := x \rightarrow 3x^4 - x^2 + 2$$

```
> f(-x);
```
$$3x^4 - x^2 + 2$$

Even.

(b) $y = \frac{x^2-4}{x+2}$
```
> f:=x->(x^2-4)/(x+2);
```
$$f := x \rightarrow \frac{x^2 - 4}{x + 2}$$

```
> f(-x);
```
$$\frac{x^2 - 4}{-x + 2}$$

Neither.

(c) $y = \frac{x^3-x}{3}$
```
> f:=x->(x^3-x)/3;
```
$$f := x \rightarrow \frac{1}{3}x^3 - \frac{1}{3}x$$

```
> f(-x);
```
$$-\frac{1}{3}x^3 + \frac{1}{3}x$$

Odd.

(d) $y = \sin(x)$
```
> f:=x->sin(x);
```
$$f := \sin$$

```
> f(-x);
```
$$-\sin(x)$$

Odd.

Exercise 18.4 Calculate the intervals over which the following functions are increasing, decreasing, concave, and convex.

(a) $y = x^3 - 2x$
```
> f:=x->x^3-2*x;
```
$$f := x \rightarrow x^3 - 2x$$

```
> diff(f(x),x);
```
$$3x^2 - 2$$

```
> solve(">0,x);
```
$$\left\{ x < -\frac{1}{3}\sqrt{6} \right\}, \left\{ \frac{1}{3}\sqrt{6} < x \right\}$$

```
> solve(""<0,x);
```
$$\left\{ -\frac{1}{3}\sqrt{6} < x, x < \frac{1}{3}\sqrt{6} \right\}$$

```
> diff(f(x),x$2);
```
$$6x$$

```
> solve(">0);
```
$$\{ 0 < x \}$$

```
> solve(""<0);
```
$$\{ x < 0 \}$$

(b) $y = x^4 + 8x$

```
> f:=x->x^4+8*x;
```
$$f := x \rightarrow x^4 + 8x$$

```
> diff(f(x),x);
```
$$4x^3 + 8$$

```
> solve(">0,x);
```
$$\{-2^{1/3} < x\}$$

```
> solve(""<0,x);
```
$$\{x < -2^{1/3}\}$$

```
> diff(f(x),x$2);
```
$$12x^2$$

```
> solve(">0);
```
$$\{x \neq 0\}$$

(c) $y = \frac{1}{x} + x$

```
> f:=x->1/x+x;
```
$$f := x \rightarrow \frac{1}{x} + x$$

```
> diff(f(x),x);
```
$$-\frac{1}{x^2} + 1$$

```
> solve(">0,x);
```
$$\{x < -1\}, \{1 < x\}$$

```
> solve(""<0,x);
```
$$\{x \neq 0, -1 < x, x < 1\}$$

```
> diff(f(x),x$2);
```
$$2\frac{1}{x^3}$$

```
> solve(">0);
```
$$\{0 < x\}$$

```
> solve(""<0);
```
$$\{\,x < 0\,\}$$

Exercise 18.5 Find the local extrema of the following functions and determine if each is a maximum, minimum, or inflection point.

(a) $y = -5x^2 + 8x - 1$
```
> f:=x->-5*x^2+8*x-1;
```
$$f := x \rightarrow -5\,x^2 + 8\,x - 1$$

```
> diff(f(x),x);
```
$$-10\,x + 8$$

```
> solve(",x);
```
$$\frac{4}{5}$$

```
> f(");
```
$$\frac{11}{5}$$

```
> diff(f(x),x$2);
```
$$-10$$

The point $(\frac{4}{5}, \frac{11}{5})$ is a local maximum.

(b) $y = x^3 - 7x$
```
> f:=x->x^3-7*x;
```
$$f := x \rightarrow x^3 - 7\,x$$

```
> diff(f(x),x);
```
$$3\,x^2 - 7$$

```
> root:=solve(",x);
```
$$root := \frac{1}{3}\,\sqrt{21}, -\frac{1}{3}\,\sqrt{21}$$

```
> f(root[1]);
```
$$-\frac{14}{9}\,\sqrt{21}$$

```
> f(root[2]);
```
$$\frac{14}{9}\,\sqrt{21}$$

```
> diff(f(x),x$2);
```
$$6\,x$$

```
> subs(x=root[1],");
```
$$2\,\sqrt{21}$$

```
> subs(x=root[2],"");
```
$$-2\,\sqrt{21}$$

The point $(\frac{1}{3}\sqrt{21}, -\frac{14}{9}\sqrt{21})$ is a local minimum. The point $(-\frac{1}{3}\sqrt{21}, \frac{14}{9}\sqrt{21})$ is a local maximum, show this on a plot.

(c) $y = 2x^4 - x^3 + 6x^2 + 9$

```
> f:=x->2*x^4-x^3-6*x^2+9;
```
$$f := x \rightarrow 2\,x^4 - x^3 - 6\,x^2 + 9$$

```
> diff(f(x),x);
```
$$8\,x^3 - 3\,x^2 - 12\,x$$

```
> root:=solve(",x);
```
$$root := 0, \frac{3}{16} + \frac{1}{16}\,\sqrt{393}, \frac{3}{16} - \frac{1}{16}\,\sqrt{393}$$

```
> f(root[1]);
```
$$9$$

```
> evalf(f(root[2]));
```
$$2.169452776$$

```
> evalf(f(root[3]));
```
$$5.973613637$$

```
> subs(x=root[1],diff(f(x),x$2));
```
$$-12$$

```
> evalf(subs(x=root[2],diff(f(x),x$2)));
```
$$28.27954267$$

```
> evalf(subs(x=root[3],diff(f(x),x$2)));
                    20.84545731

> solve(diff(f(x),x$2));
```

$$\frac{1}{8} + \frac{1}{8}\sqrt{33}, \frac{1}{8} - \frac{1}{8}\sqrt{33}$$

B.5   Rates

Exercise 19.1   A vehicle travelling at 100 km/h applies its breaks, giving a constant deceleration, and stops after a distance of 50 m is covered in a time of 4 s. What was the constant deceleration?

First convert km/h to m/s to keep units consistent.

```
> 100*(1000/3600);
```
$$\frac{250}{9}$$

```
> evalf(");
```
$$27.77777778$$

Then divide by the time taken to stop to get deceleration.

```
> "/4;
```
$$6.944444445$$

Exercise 19.2   A vehicle travelling at 150 km/h passes a stationary police car. It takes 10 s for the police car to start moving and then it continues at constant acceleration until it is adjacent to the vehicle, which happens 20 s after the initial passing. How far has the police car travelled and what is its speed when it catches up with the other vehicle? What was the constant acceleration of the police car?

```
> 150*(1000/3600);
```
$$\frac{125}{3}$$

```
> "*20;
```
$$\frac{2500}{3}$$

```
> 1/2*a*(t-10)^2=";
```
$$\frac{1}{2}a(t-10)^2 = \frac{2500}{3}$$

```
> subs(t=20,");
```
$$50\,a = \frac{2500}{3}$$

Hence, $a = \frac{50}{3}$ and speed of police car at passing is $\frac{500}{3}$.

Exercise 19.3 In order to escape the earth's gravitational pull, a rocket must be launched with a certain initial velocity. This velocity is called the escape velocity. A rocket launched from the surface of the earth (approximate radius: 4000 miles) has velocity, v, given by $v = \sqrt{\frac{2GM}{r} + v_0^2 - \frac{2GM}{R}} \approx \sqrt{\frac{192000}{r} + v_0^2 - 48}$ mi/sec where v_0 represents the initial velocity, r represents the distance from the rocket to the centre of the earth, G is the gravitational constant, M is the mass of the earth, and R is the radius of the earth. (p13, Goldenberg and Greenwald [7])

(a) Find the value of v_0 for which we obtain an infinite limit for r as $v \to 0$. This value for v_0 is the escape velocity for earth.

```
> eq1:=v=sqrt(192000/r+v0^2-48);
```

$$eq1 := v = \sqrt{192000\,\frac{1}{r} + v0^2 - 48}$$

```
> with(student):
> isolate(eq1,r);
```

$$r = 192000\,\frac{1}{v^2 - v0^2 + 48}$$

```
> assign(");
> limit(r,v=0);
```

$$-192000\,\frac{1}{v0^2 - 48}$$

```
> solve(denom(")=0,{v0});
```

$$\left\{v0 = 4\sqrt{3}\right\}, \left\{v0 = -4\sqrt{3}\right\}$$

If $v_0 > \sqrt{48}$, show that the velocity v is always positive. In this case, the rocket will travel indefinitely.

```
> r:='r':
> eq1:=v=sqrt(192000/r+v0^2-48);
```

$$eq1 := v = \sqrt{192000\,\frac{1}{r} + v0^2 - 48}$$

```
> eq2:=v0>sqrt(48);
```

$$eq2 := 4\sqrt{3} < v0$$

(b) A rocket launched from the surface of the moon has velocity given by $v = \sqrt{\frac{1920}{r} + v_0^2 - 2.17}$ mi/sec. Find the escape velocity for the moon.

```
> eq3:=v=sqrt(1920/r+v0^2-2.17);
```

$$eq3 := v = \sqrt{1920\,\frac{1}{r} + v0^2 - 2.17}$$

```
> isolate(eq3,r);
```
$$r = 1 \Big/ (.0005208333333\, v^2 - .0005208333333\, v0^2 + .001130208333\,)$$

```
> assign(");
> limit(r,v=0);
```
$$-.3333333333\,10^{13}\; \frac{1}{.1736111111\,10^{10}\,v0^2 - .3767361110\,10^{10}}$$

```
> solve(denom(")=0,{v0});
```
$$\{\,v0 = 1.473091986\,\},\{\,v0 = -1.473091986\,\}$$

(c) A rocket launched from the surface of Mars has velocity given by
$v = \sqrt{\frac{20544}{r} + v_0^2 - 9.55}$ mi/sec. Find the escape velocity for Mars.

```
> r:='r':
> eq4:=v=sqrt(20544/r+v0^2-9.55);
```
$$eq4 := v = \sqrt{20544\,\frac{1}{r} + v0^2 - 9.55}$$

```
> isolate(eq4,r);
```
$$r = 1 \Big/ (.00004867601246\, v^2 - .00004867601246\, v0^2 + .0004648559190$$
$$)$$

```
> assign(");
> limit(r,v=0);
```
$$-.5000000000\,10^{14}\; \frac{1}{.2433800623\,10^{10}\,v0^2 - .2324279595\,10^{11}}$$

```
> solve(denom(")=0,{v0});
```
$$\{\,v0 = 3.090307429\,\},\{\,v0 = -3.090307429\,\}$$

(d) A rocket launched from the surface of Planet X has velocity given by
$v = \sqrt{\frac{10600}{r} + v_0^2 - 6.99}$ mi/sec. Find the escape velocity for Planet X.

```
> r:='r':
> eq5:=v=sqrt(10600/r+v0^2-6.99);
```
$$eq5 := v = \sqrt{10600\,\frac{1}{r} + v0^2 - 6.99}$$

```
> isolate(eq5,r);
```
$$r = 1 \Big/ (.00009433962264\, v^2 - .00009433962264\, v0^2 + .0006594339623$$
$$)$$

```
> assign(");
> limit(r,v=0);
```

$$-.5000000000\,10^{14}\;\frac{1}{.4716981132\,10^{10}\,v0^2 - .3297169812\,10^{11}}$$

```
> solve(denom(")=0,{v0});
```

$$\{\,v0 = 2.643860812\,\},\{\,v0 = -2.643860812\,\}$$

(e) Is the planet larger or smaller than earth? (Assume that the mean density of Planet X is the same as that of the earth.)

Exercise 19.4 The surface area of a sphere of radius r is $4\pi r^2$ and its volume is $\frac{4}{3}\pi r^3$. What is the rate of change of volume per unit change in area? What is the percentage change in area caused by a 5% increase in radius?

```
> r:='r':
> diff(A(t),t)=4*Pi*diff(r(t)^2,t);
```

$$\frac{\partial}{\partial t}\mathrm{A}(t) = 8\,\pi\,\mathrm{r}(t)\left(\frac{\partial}{\partial t}\mathrm{r}(t)\right)$$

```
> diff(V(t),t)=4/3*Pi*diff(r(t)^3,t);
```

$$\frac{\partial}{\partial t}\mathrm{V}(t) = 4\,\pi\,\mathrm{r}(t)^2\left(\frac{\partial}{\partial t}\mathrm{r}(t)\right)$$

```
> diff(V(t),t)/diff(A(t),t)=(4/3*Pi*diff(r(t)^3,t))/
> (4*Pi*diff(r(t)^2,t));
```

$$\frac{\frac{\partial}{\partial t}\mathrm{V}(t)}{\frac{\partial}{\partial t}\mathrm{A}(t)} = \frac{1}{2}\mathrm{r}(t)$$

```
> diff(A(t),t)=4*Pi*diff(r(t)^2,t);
```

$$\frac{\partial}{\partial t}\mathrm{A}(t) = 8\,\pi\,\mathrm{r}(t)\left(\frac{\partial}{\partial t}\mathrm{r}(t)\right)$$

```
> isolate(",diff(r(t),t));
```

$$\frac{\partial}{\partial t}\mathrm{r}(t) = \frac{1}{8}\frac{\frac{\partial}{\partial t}\mathrm{A}(t)}{\pi\,\mathrm{r}(t)}$$

```
> lhs(")=subs(r(t)=r(t)+.05*r(t),rhs("));
```

$$\frac{\partial}{\partial t}\mathrm{r}(t) = .1190476191\frac{\frac{\partial}{\partial t}\mathrm{A}(t)}{\pi\,\mathrm{r}(t)}$$

Exercise 19.5 An inverted cone has radius $r = \frac{z}{2}$ where z is the height above the x,y plane in metres. The top of the cone is at $z = 1$ m. Find the rate of increase of fluid height in the cone if fluid is pumped in at the apex at a rate of $0.1 \ m^3/s$. (Recall that the volume of a cone is $\frac{1}{3}\pi r^2 h$.)

```
> diff(v(t),t)=diff(1/3*Pi*(h(t)/2)^2*h(t),t);
```

$$\frac{\partial}{\partial t}v(t) = \frac{1}{4}\pi\,h(t)^2 \left(\frac{\partial}{\partial t}h(t)\right)$$

```
> subs(diff(v(t),t)=0.1,lhs("))=rhs(");
```

$$.1 = \frac{1}{4}\pi\,h(t)^2 \left(\frac{\partial}{\partial t}h(t)\right)$$

```
> isolate(",diff(h(t),t));
```

$$\frac{\partial}{\partial t}h(t) = .4\,\frac{1}{\pi\,h(t)^2}$$

```
> evalf(");
```

$$\frac{\partial}{\partial t}h(t) = .1273239544\,\frac{1}{h(t)^2}$$

Exercise 19.6 Air is being pumped into a spherical balloon at the constant rate of $10 \ cm^3/s$. At what rate is the surface area of the balloon increasing when its radius is 5 cm? (p170, Edwards and Penney [6])

```
> v:=t->4/3*Pi*r(t)^3;
```

$$v := t \rightarrow \frac{4}{3}\pi\,r(t)^3$$

```
> diff(v(t),t)=10;
```

$$4\pi\,r(t)^2 \left(\frac{\partial}{\partial t}r(t)\right) = 10$$

```
> solve(",{diff(r(t),t)});
```

$$\left\{\frac{\partial}{\partial t}r(t) = \frac{5}{2}\,\frac{1}{\pi\,r(t)^2}\right\}$$

```
> a:=t->4*Pi*r(t)^2;
```

$$a := t \rightarrow 4\pi\,r(t)^2$$

```
> diff(a(t),t);
```

$$8\pi\,r(t) \left(\frac{\partial}{\partial t}r(t)\right)$$

```
> subs("''",");
```

$$20\,\frac{1}{\mathrm{r}(t)}$$

```
> subs(r(t)=5,");
```

$$4$$

B.6   Integration

Exercise 20.1   Approximate the area under the following curves by dividing the region into 10 rectangular subregions.

(a) $y = -2x^2 + 6, -1 \le x \le 1$

```
> y:=x->-2*x^2+6;
```
$$y := x \rightarrow -2\,x^2 + 6$$

```
> plot(y(x),x=-1..1);
> with(student):
> leftbox(y(x),x=-1..1,10);
> leftsum(y(x),x=-1..1,10);
```
$$\frac{1}{5} \left(\sum_{i=0}^{9} \left(-2 \left(-1 + \frac{1}{5} i \right)^2 + 6 \right) \right)$$

```
> value(");
```
$$\frac{266}{25}$$

```
> evalf(");
```
$$10.64000000$$

The above is an approximation to the integration:

```
> int(y(x),x=-1..1);
```
$$\frac{32}{3}$$

(b) $y = x^3 + x + 10, -2 \le x \le 2$

```
> y:=x->x^3+x+10;
```
$$y := x \rightarrow x^3 + x + 10$$

```
> plot(y(x),x=-2..2);
> rightbox(y(x),x=-2..2,10);
> rightsum(y(x),x=-2..2,10);
```
$$\frac{2}{5} \left(\sum_{i=1}^{10} \left(\left(-2 + \frac{2}{5} i \right)^3 + 8 + \frac{2}{5} i \right) \right)$$

```
> value(");
```
$$44$$

Here the correct answer is:

```
> int(y(x),x=-2..2);
```
$$40$$

(c) $y = \sqrt{1-x^2}, -1 \le x \le 1$

```
> y:=x->(1-x^2)^(1/2);
```
$$y := x \to \sqrt{1-x^2}$$

```
> plot(sqrt(1-x^2),x=-1..1);
> middlebox(y(x),x=-1..1,10);
> middlesum(y(x),x=-1..1,10);
```
$$\frac{1}{5}\left(\sum_{i=0}^{9}\sqrt{1-\left(-\frac{9}{10}+\frac{1}{5}i\right)^2}\right)$$

```
> value(");
```
$$\frac{1}{250}\sqrt{19}\sqrt{100} + \frac{1}{250}\sqrt{51}\sqrt{100} + \frac{1}{10}\sqrt{3}\sqrt{4} + \frac{1}{250}\sqrt{91}\sqrt{100}$$
$$+ \frac{1}{250}\sqrt{99}\sqrt{100}$$

```
> evalf(");
```
$$1.585993912$$

The correct answer is:

```
> int(y(x),x=-1..1);
```
$$\frac{1}{2}\pi$$

Exercise 20.2 Integrate the following functions.

(a) $y = x^2 + 3x + 2, 0 \le x \le 2$

```
> y:=x->x^2+3*x+2;
```
$$y := x \to x^2 + 3x + 2$$

```
> plot(y(x),x=0..2);
> int(y(x),x=0..2);
```
$$\frac{38}{3}$$

(b) $y = x^5, -2 \le x \le 2$

```
> y:=x->x^5;
```

$$y := x \rightarrow x^5$$

```
> plot(y(x),x=-2..2);
> int(y(x),x=-2..2);
```

$$0$$

(c) $y = 2^x, -\pi \le x \le \pi$

```
> y:=x->2^x;
```

$$y := x \rightarrow 2^x$$

```
> plot(y(x),x=-1..1);
> int(y(x),x=-1..1);
```

$$\frac{3}{2} \frac{1}{\ln(2)}$$

Exercise 20.3   Integrate the following functions.

(a) $y = x + 6$

```
> y:=x->x+6;
```

$$y := x \rightarrow x + 6$$

```
> int(y(x),x);
```

$$\frac{1}{2} x^2 + 6 x$$

(b) $y = x^2 + \frac{x}{10}$

```
> y:=x->x^2+x/10;
```

$$y := x \rightarrow x^2 + \frac{1}{10} x$$

```
> int(y(x),x);
```

$$\frac{1}{3} x^3 + \frac{1}{20} x^2$$

(c) $y = 4e^{2x}$

```
> y:=x->4*exp(2*x);
```

$$y := x \rightarrow 4 e^{(2x)}$$

```
> int(y(x),x);
```

$$2\,e^{(2x)}$$

```
> y:='y':
```

Exercise 20.4   Determine the area enclosed by the following curves.

(a) $f(x) = 2x^2 - 4, g(x) = -2x^2$

```
> f:=x->2*x^2-4;
```

$$f := x \to 2\,x^2 - 4$$

```
> g:=x->-2*x^2;
```

$$g := x \to -2\,x^2$$

```
> plot({f(x),g(x)},x=-2..2);
> solve(f(x)=g(x),x);
```

$$1, -1$$

```
> int(g(x)-f(x),x=-1..1);
```

$$\frac{16}{3}$$

(b) $f(x) = x + 2, g(x) = x^2$

```
> f:=x->x+2;
```

$$f := x \to x + 2$$

```
> g:=x->x^2;
```

$$g := x \to x^2$$

```
> plot({f(x),g(x)},x=-2..3);
> solve(f(x)=g(x),x);
```

$$-1, 2$$

```
> int(f(x)-g(x),x=-1..2);
```

$$\frac{9}{2}$$

(c) $f(x) = \sqrt{x}, g(x) = x^2$

```
> f:=x->x^(1/2);
```

$$f := x \to \sqrt{x}$$

```
> g:=x->x^2;
```

$$g := x \rightarrow x^2$$

```
> plot({f(x),g(x)},x=-2..2);
> solve(f(x)=g(x),x);
```

$$0, 1, \left(-\frac{1}{2} + \frac{1}{2}I\sqrt{3}\right)^2, \left(-\frac{1}{2} - \frac{1}{2}I\sqrt{3}\right)^2$$

```
> int(f(x)-g(x),x=0..1);
```

$$\frac{1}{3}$$

Exercise 20.5 Solve the following differential equations, first investigating them with the **fieldplot** function.

(a) $\frac{dy}{dx} = x, y(0) = 2$

```
> with(plots):
> fieldplot( [x,x],x=-2..2,y=2..4);
> diff(y(x),x)=x;
```

$$\frac{\partial}{\partial x} y(x) = x$$

```
> dsolve({",y(0)=2},y(x));
```

$$y(x) = \frac{1}{2}x^2 + 2$$

(b) $\frac{dy}{dx} = x + y, y(0) = 0$

```
> fieldplot( [x,x+y],x=-2..2,y=-2..2);
> diff(y(x),x)=x+y(x);
```

$$\frac{\partial}{\partial x} y(x) = x + y(x)$$

```
> dsolve({",y(0)=0},y(x));
```

$$y(x) = -x - 1 + e^x$$

(c) $\frac{dy}{dx} = x^2 + y, y(0) = 1$

```
> fieldplot( [x,x^2+y],x=-2..2,y=0..2);
> diff(y(x),x)=x^2+y(x);
```

$$\frac{\partial}{\partial x} y(x) = x^2 + y(x)$$

```
> dsolve({",y(0)=1},y(x));
```
$$y(x) = -x^2 - 2x - 2 + 3e^x$$

Exercise 20.6 Using the method of integration by parts, integrate the following functions.

(a) $y(x) = xe^x$
```
> y:=x->x*exp(x);
```
$$y := x \rightarrow x e^x$$

```
> intparts(Int(y(x),x),x);
```
$$x e^x - \int e^x \, dx$$

```
> intparts(Int(y(x),x),exp(x));
```
$$\frac{1}{2} e^x x^2 - \int \frac{1}{2} e^x x^2 \, dx$$

(b) $y(x) = \frac{1}{x} e^x$
```
> y:=x->1/x*exp(x);
```
$$y := x \rightarrow \frac{e^x}{x}$$

```
> intparts(Int(y(x),x),1/x);
```
$$\frac{e^x}{x} - \int -\frac{e^x}{x^2} \, dx$$

```
> intparts(Int(y(x),x),exp(x));
```
$$e^x \ln(x) - \int e^x \ln(x) \, dx$$

(c) $y(x) = \ln(x)$
```
> y:=x->ln(x);
```
$$y := \ln$$

```
> intparts(Int(y(x),x),1);
```
$$x \ln(x) - x - \int 0 \, dx$$

```
> intparts(Int(y(x),x),ln(x));
```
$$x \ln(x) - \int 1 \, dx$$

Exercise 20.7 Convert the following functions into partial fractions and then integrate them.

(a) $y(x) = \frac{x^2-4}{x^2+3x+2}$

```
> y:=x->(x^2-4)/(x^2+3*x+2);
```

$$y := x \rightarrow \frac{x^2 - 4}{x^2 + 3x + 2}$$

```
> convert(y(x),parfrac, x);
```

$$1 - 3\frac{1}{x+1}$$

```
> int(",x);
```

$$x - 3\ln(x+1)$$

(b) $y(x) = \frac{2x+8}{18x^2+52x-6}$

```
> y:=x->(2*x+8)/(18*x^2+52*x-6);
```

$$y := x \rightarrow \frac{2x+8}{18x^2 + 52x - 6}$$

```
> convert(y(x),parfrac, x);
```

$$-\frac{1}{28}\frac{1}{x+3} + \frac{37}{28}\frac{1}{9x-1}$$

```
> int(",x);
```

$$-\frac{1}{28}\ln(x+3) + \frac{37}{252}\ln(9x-1)$$

Exercise 20.8 Approximate the area under the curve $y(x) = -\frac{1}{3}x^2 + 4x + 9, -13 \leq x \leq 1$

(a) using the Trapeziod rule with $n = 20$

```
> y:=x->-1/3*x^2-4*x+9;
```

$$y := x \rightarrow -\frac{1}{3}x^2 - 4x + 9$$

```
> plot(y(x),x=-13..1);
> trapezoid(y(x),x=-13..1,20);
```

$$\frac{49}{15} + \frac{7}{10}\left(\sum_{i=1}^{19}\left(-\frac{1}{3}\left(-13+\frac{7}{10}i\right)^2 + 61 - \frac{14}{5}i\right)\right)$$

```
> evalf(");
```

$$217.3966667$$

(b) using Simpson's rule with $n = 10$

```
> simpson(y(x),x=-13..1,20);
```

$$\frac{98}{45} + \frac{14}{15} \left(\sum_{i=1}^{10} \left(-\frac{1}{3} \left(-\frac{137}{10} + \frac{7}{5}i \right)^2 + \frac{319}{5} - \frac{28}{5}i \right) \right)$$
$$+ \frac{7}{15} \left(\sum_{i=1}^{9} \left(-\frac{1}{3} \left(-13 + \frac{7}{5}i \right)^2 + 61 - \frac{28}{5}i \right) \right)$$

```
> evalf(");
```
$$217.7777778$$

c)
```
> int(y(x),x=-13..1);
```
$$\frac{1960}{9}$$

```
> evalf(");
```
$$217.7777778$$

438

Appendix B Solutions to Part II Exercises

B.7 Trigonometry

Exercise 21.1 Animate a family of ellipses $x^2 + \frac{y^2}{n} = 1$, for $n = 1, 2, 3, \ldots$ using *Maple's* functions plot and implicitplot.

```
> with(plots):
> animate([cos(t),n*sin(t),t=0..2*Pi],n=1..10,scaling=CONSTRAINED);
> display({seq(implicitplot(x^2+y^2/(n)=1,x=-1..1,y=-5..5,
> scaling=CONSTRAINED),n=1..10)},insequence=true);
```

Exercise 21.2 Given that $\sin 30° = \frac{1}{2}$, use the approximation

$$\sin(x + h) \approx \sin x + h\frac{d\sin x}{dx}$$

with $x = \frac{\pi}{6}$ radians and $h = 0.0175$ radians to estimate $\sin 31°$.

```
> sin(x)+h*diff(sin(x),x);
```
$$\sin(x) + h\cos(x)$$

```
> subs({x=Pi/6,h=0.0175},");
```
$$\sin\left(\frac{1}{6}\pi\right) + .0175\cos\left(\frac{1}{6}\pi\right)$$

```
> evalf(");
```
$$.5151554446$$

Exercise 21.3 Differentiate the following trigonometric functions.

(a) $y = \sin(\frac{x}{2})$
```
>   y:=x->sin(x/2);
```
$$y := x \to \sin\left(\frac{1}{2}x\right)$$

```
> diff(y(x),x);
```
$$\frac{1}{2}\cos\left(\frac{1}{2}x\right)$$

(b) $y = \cos(3x)^2$
```
>   y:=x->cos(3*x)^2;
```
$$y := x \to \cos(3x)^2$$

```
> diff(y(x),x);
```
$$-6\cos(3x)\sin(3x)$$

(c) $y = \cot(x)$
```
>   y:=x->cot(x);
```
$$y := \cot$$

```
> diff(y(x),x);
```
$$-1 - \cot(x)^2$$

(d) $y = \sin(x^{-1})$
```
> y:=x->sin(x^(-1));
```
$$y := x \rightarrow \sin\left(\frac{1}{x}\right)$$

```
> diff(y(x),x);
```
$$-\frac{\cos\left(\frac{1}{x}\right)}{x^2}$$

Exercise 21.4   Define, for constant a, b,
$$f(x) = \cos(a + x)\cos(b - x) - \sin(a + x)\sin(b - x)$$
and show that $f'(x) = 0$. Hence f must be a constant function and in particular, $f(0) = f(b)$. Hence deduce the formula for $\cos(a + b)$ and similarly derive that for $\sin(a + b)$.
```
> f:=x->cos(a+x)*cos(b-x)-sin(a+x)*sin(b-x);
```
$$f := x \rightarrow \cos(a + x)\cos(b - x) - \sin(a + x)\sin(b - x)$$

```
> diff(f(x),x);
```
$$0$$

```
> f(0);
```
$$\cos(a)\cos(b) - \sin(a)\sin(b)$$

```
> f(b);
```
$$\cos(a + b)$$

Exercise 21.5   Integrate the following trigonometric functions.

(a) $y = \sin(3x)$
```
> y:=x->sin(3*x);
```
$$y := x \rightarrow \sin(3x)$$

```
> int(y(x),x);
```

$$-\frac{1}{3}\cos(3x)$$

(b) $y = \cos(x)^{3/2}\sin(x)$

```
> y:=x->cos(x)^(3/2)*sin(x);
```

$$y := x \rightarrow \cos(x)^{3/2}\sin(x)$$

```
> int(y(x),x);
```

$$-\frac{2}{5}\cos(x)^{5/2}$$

(c) $y = x^2\cos(x^3 + 1)$

```
> y:=x->x^2*cos(x^3+1);
```

$$y := x \rightarrow x^2\cos(x^3 + 1)$$

```
> int(y(x),x);
```

$$\frac{1}{3}\sin(x^3 + 1)$$

(d) $y = \sin(x) + \sin(2x)$

```
> y:=x->sin(x)+sin(2*x);
```

$$y := x \rightarrow \sin(x) + \sin(2x)$$

```
> int(y(x),x);
```

$$-\cos(x) - \frac{1}{2}\cos(2x)$$

B.8   Exponents and Logarithms

Exercise 22.1   Advertising of a certain product has been discontinued; the company plans to resume advertising when sales have declined to 75% of their initial rate. (This phenomenon actually occurred when the Sony Corporation first introduced home videotape recorders in the United States in 1976.)   If after 1 week without advertising, sales have declined to 95% of their initial rate, when should the company expect to resume advertising? (p395, Edwards and Penney [6])

```
> diff(S(t),t)=-k*S(t);
```

$$\frac{\partial}{\partial t} S(t) = -k\, S(t)$$

```
> dsolve({",S(0)=1},S(t));
```

$$S(t) = e^{(-kt)}$$

```
> assign(");
> solve({S(t)=0.95,t=7});
```

$$\{\, t = 7., k = .007327613486 \,\}$$

```
> assign(op(2,"));
> solve(S(t)=0.75,{t});
```

$$\{\, t = 39.25999551 \,\}$$

Exercise 22.2   Differentiate the following logarithmic functions.

(a) $y = \log(3x)$

```
> y:=x->log(3*x);
```

$$y := x \rightarrow \log(3x)$$

```
> diff(y(x),x);
```

$$\frac{1}{x}$$

(b) $y = \log(\sqrt{2 - x})$

```
> y:=x->log((2-x)^(1/2));
```

$$y := x \rightarrow \log\left(\sqrt{2 - x}\right)$$

```
> diff(y(x),x);
```

$$-\frac{1}{2}\frac{1}{2 - x}$$

(c) $y = \sin(\log(x))$

```
> y:=x->sin(log(x));
```

$$y := x \rightarrow \sin(\log(x))$$

```
> diff(y(x),x);
```

$$\frac{\cos(\ln(x))}{x}$$

Exercise 22.3　Differentiate the following exponential functions.

(a) $y = e^{2x-x^2}$

```
> y:=x->exp(2*x-x^2);
```

$$y := x \rightarrow e^{(2x-x^2)}$$

```
> diff(y(x),x);
```

$$(2 - 2x)e^{(2x-x^2)}$$

(b) $y = \frac{e^x}{x^2}$

```
> y:=x->exp(x)/x^2;
```

$$y := x \rightarrow \frac{e^x}{x^2}$$

```
> diff(y(x),x);
```

$$\frac{e^x}{x^2} - 2\frac{e^x}{x^3}$$

(c) $y = e^{\sin(x)}$

```
> y:=x->exp(sin(x));
```

$$y := x \rightarrow e^{\sin(x)}$$

```
> diff(y(x),x);
```

$$\cos(x)e^{\sin(x)}$$

Exercise 22.4　Use the *Maple* function **series** to obtain quadratic approximations to e^x and $\frac{1}{1+x}$. Hence deduce that the slope of the tangent to $\frac{e^x}{1+x}$ is approximately x when x is small.

```
> series(exp(x),x,9);
```

$$1 + x + \frac{1}{2}x^2 + \frac{1}{6}x^3 + \frac{1}{24}x^4 + \frac{1}{120}x^5 + \frac{1}{720}x^6 + \frac{1}{5040}x^7 + \frac{1}{40320}x^8 +$$
$$O(x^9)$$

```
> series(1/(1+x),x,9);
```
$$1 - x + x^2 - x^3 + x^4 - x^5 + x^6 - x^7 + x^8 + O(x^9)$$

```
> (1+x+1/2*x^2+1/6*x^3+1/24*x^4)*(1-x+x^2-x^3+x^4-x^5);
```
$$\left(1 + x + \frac{1}{2}x^2 + \frac{1}{6}x^3 + \frac{1}{24}x^4\right)\left(1 - x + x^2 - x^3 + x^4 - x^5\right)$$

```
> expand(");
```
$$\frac{1}{2}x^2 - \frac{1}{3}x^3 + \frac{3}{8}x^4 - \frac{3}{8}x^5 - \frac{5}{8}x^6 - \frac{3}{8}x^7 - \frac{1}{8}x^8 - \frac{1}{24}x^9 + 1$$

```
> diff(",x);
```
$$x - x^2 + \frac{3}{2}x^3 - \frac{15}{8}x^4 - \frac{15}{4}x^5 - \frac{21}{8}x^6 - x^7 - \frac{3}{8}x^8$$

By inspection, the lowest order term here is x, as required.

Exercise 22.5 Use `series` to show that for small x the slope of the tangent to $\left(\frac{1+x}{1-x}\right)^{1/3}$ is $\frac{2}{3}$.

```
> series(((1+x)/(1-x))^(1/3),x,5);
```
$$1 + \frac{2}{3}x + \frac{2}{9}x^2 + \frac{22}{81}x^3 + \frac{38}{243}x^4 + O(x^5)$$

```
> diff(1+2/3*x+2/9*x^2+22/81*x^3+38/243*x^4,x);
```
$$\frac{2}{3} + \frac{4}{9}x + \frac{22}{27}x^2 + \frac{152}{243}x^3$$

For small x, all of the terms except the constant $\frac{2}{3}$ are negligible.

Exercise 22.6 Integrate the following functions.

(a) $y = \frac{2}{x+3}$
```
> y:=x->2/(x+3);
```
$$y := x \to 2\frac{1}{x+3}$$

```
> int(y(x),x);
```
$$2\ln(x+3)$$

(b) $y = \frac{x}{x^2-1}$
```
> y:=x->x/(x^2-1);
```
$$y := x \to \frac{x}{x^2-1}$$

```
> int(y(x),x);
```

$$\frac{1}{2}\ln(x-1) + \frac{1}{2}\ln(1+x)$$

(c) $y = e^{-x+2}$

```
> y:=x->exp(-x+2);
```

$$y := x \rightarrow e^{(2-x)}$$

```
> int(y(x),x);
```

$$-e^{(2-x)}$$

(d) $y = x^3 e^{x^4+1}$

```
> y:=x->x^3*exp(x^4+1);
```

$$y := x \rightarrow x^3 e^{(x^4+1)}$$

```
> int(y(x),x);
```

$$\frac{1}{4}e^{(x^4+1)}$$

B.9 Polar Coordinates

Exercise 23.1 Convert the following Cartesian coordinates into polar coordinates.

(a) $(2, -5)$

```
> eq1:=r=sqrt(x^2+y^2);
```

$$eq1 := r = \sqrt{x^2 + y^2}$$

```
> eq2:=tan(theta)=y/x;
```

$$eq2 := \tan(\theta) = \frac{y}{x}$$

```
> p:=[2,-5];
```

$$p := [2, -5]$$

```
> x:=p[1];
```

$$x := 2$$

```
> y:=p[2];
```

$$y := -5$$

```
> eq1;
```

$$r = \sqrt{29}$$

```
> eq2;
```

$$\tan(\theta) = \frac{-5}{2}$$

```
> convert(solve(",theta),degrees);
```

$$-180\,\frac{\arctan\left(\frac{5}{2}\right)\,degrees}{\pi}$$

```
> evalf(");
```

$$-68.19859053 \; degrees$$

(b) $(-1, -3)$

```
> p:=[-1,-3];
```

$$p := [-1, -3]$$

```
> x:=p[1];
```

$$x := -1$$

```
> y:=p[2];
```
$$y := -3$$

```
> eq1;
```
$$r = \sqrt{10}$$

```
> eq2;
```
$$\tan(\theta) = 3$$

```
> convert(solve(",theta),degrees);
```
$$180\,\frac{\arctan(3)\,degrees}{\pi}$$

```
> evalf(");
```
$$71.56505115\,degrees$$

(c) $(0,6)$
```
> p:=[0,6];
```
$$p := [0,6]$$

```
> x:=p[1];
```
$$x := 0$$

```
> y:=p[2];
```
$$y := 6$$

```
> eq1;
```
$$r = \sqrt{36}$$

```
> assign(");
> sin(theta)=r/y;
```
$$\sin(\theta) = \frac{1}{6}\sqrt{36}$$

```
> convert(solve(",theta),degrees);
```
$$90\,degrees$$

Exercise 23.2   Convert the following polar coordinates into Cartesian coordinates.

(a) $(3, \frac{\pi}{4})$
```
> eq1:=x=r*cos(theta);
```
$$eq1 := x = r \cos(\theta)$$

```
> eq2:=y=r*sin(theta);
```
$$eq2 := y = r \sin(\theta)$$

```
> p:=[3,Pi/4];
```
$$p := \left[3, \frac{1}{4}\pi\right]$$

```
> r:=p[1];
```
$$r := 3$$

```
> theta:=p[2];
```
$$\theta := \frac{1}{4}\pi$$

```
> solve({eq1,eq2},{x,y});
```
$$\left\{y = \frac{3}{2}\sqrt{2}, x = \frac{3}{2}\sqrt{2}\right\}$$

(b) $(\sqrt{2}, -\frac{\pi}{6})$
```
> p:=[sqrt(2),-Pi/6];
```
$$p := \left[\sqrt{2}, -\frac{1}{6}\pi\right]$$

```
> r:=p[1];
```
$$r := \sqrt{2}$$

```
> theta:=p[2];
```
$$\theta := -\frac{1}{6}\pi$$

```
> solve({eq1,eq2},{x,y});
```
$$\left\{y = -\frac{1}{2}\sqrt{2}, x = \frac{1}{2}\sqrt{2}\sqrt{3}\right\}$$

(c) $(1, 50°)$
```
> p:=[1,50*Pi/180];
```
$$p := \left[1, \frac{5}{18}\pi\right]$$

```
> r:=p[1];
```
$$r := 1$$

```
> theta:=p[2];
```
$$\theta := \frac{5}{18}\pi$$

```
> solve({eq1,eq2},{x,y});
```
$$\left\{ x = \cos\left(\frac{5}{18}\pi\right), y = \sin\left(\frac{5}{18}\pi\right) \right\}$$

```
> evalf(");
```
$$\{\, x = .6427876095, y = .7660444432 \,\}$$

Exercise 23.3   Convert the following functions into polar form.

    (a) $x + y + 1 = 0$

```
> theta:='theta':r:='r':
> eq:=x+y+1=0;
```
$$eq := x + y + 1 = 0$$

```
> subs({x=r*cos(theta),y=r*sin(theta)},eq);
```
$$r\cos(\theta) + r\sin(\theta) + 1 = 0$$

```
> with(student):
> isolate("",r);
```
$$r = -\frac{1}{\cos(\theta) + \sin(\theta)}$$

    (b) $x^2 + y^2 - 4 = 0$

```
> eq:=x^2+y^2-4=0;
```
$$eq := x^2 + y^2 - 4 = 0$$

```
> subs({x=r*cos(theta),y=r*sin(theta)},eq);
```
$$r^2\cos(\theta)^2 + r^2\sin(\theta)^2 - 4 = 0$$

```
> isolate(",r^2);
```
$$r^2 = 4\,\frac{1}{\cos(\theta)^2 + \sin(\theta)^2}$$

(c) $x^2 + xy + 8 = 0$

```
> eq:=x^2+xy+8=0;
```
$$eq := x^2 + xy + 8 = 0$$

```
> subs({x=r*cos(theta),y=r*sin(theta)},eq);
```
$$r^2 \cos(\theta)^2 + xy + 8 = 0$$

```
> isolate(",r^2);
```
$$r^2 = \frac{-xy - 8}{\cos(\theta)^2}$$

Exercise 23.4   In each of the following cases, find the area enclosed by the loop of the graph.

(a) $r = 3\sin(2\theta)$ Hint: $0 \le \theta \le \frac{\pi}{2}$

```
> with(plots):
> polarplot(3*sin(2*theta),theta=0..2*Pi);
> polarplot(3*sin(2*theta),theta=0..Pi/2);
> int(1/2*(3*sin(2*theta))^(2),theta=0..Pi/2);
```
$$\frac{9}{8}\pi$$

(b) $r = 5\cos(3\theta)$

```
> polarplot(5*cos(3*theta),theta=0..Pi);
> solve(5*cos(3*theta)=5,{theta});
```
$$\{\theta = 0\}$$

```
> solve(5*cos(3*theta)=0,{theta});
```
$$\left\{\theta = \frac{1}{6}\pi\right\}$$

```
> polarplot(5*cos(3*theta),theta=-Pi/6..Pi/6);
> int(1/2*(5*cos(3*theta))^2,theta=-Pi/6..Pi/6);
```
$$\frac{25}{12}\pi$$

(c) $r = \cos(3\theta/2)$

```
> polarplot(cos(3*theta/2),theta=-2*Pi..2*Pi);
> polarplot(cos(3*theta/2),theta=-Pi/3..Pi/3);
> int(1/2*(cos(3*theta/2))^2,theta=-Pi/3..Pi/3);
```
$$\frac{1}{6}\pi$$

Exercise 23.5   Find the area of the circle $r = \sin(x) + \cos(x)$ by integration in polar coordinates. Check the answer by writing the equation of the circle in rectangular coordinates, finding its radius, and then using the familiar area formula. (p510, Edwards and Penney [6])

First plot the function $r = \sin(x) + \cos(x)$ in polar coordinates and note that r is measured from the origin and not from the centre of the circle. When this happens, you will find that your range of integration will need to be $0 \le \theta \le \pi$.

```
> r:=sin(theta)+cos(theta);
```
$$r := \cos(\theta) + \sin(\theta)$$

```
> polarplot(r,theta=0..Pi/2);
> Int(1/2*r^2,theta=0..Pi);
```
$$\int_0^\pi \frac{1}{2}\left(\cos(\theta) + \sin(\theta)\right)^2 d\theta$$

```
> expand(");
```
$$\frac{1}{2}\int_0^\pi \cos(\theta)^2\, d\theta + \int_0^\pi \cos(\theta)\sin(\theta)\, d\theta + \frac{1}{2}\int_0^\pi \sin(\theta)^2\, d\theta$$

```
> value(");
```
$$\frac{1}{2}\pi$$

```
> r:='r':
> subs({cos(theta)=x/r,sin(theta)=y/r},r=sin(theta)+cos(theta));
```
$$r = \frac{x}{r} + \frac{y}{r}$$

```
> isolate(",r);
```
$$r = -\sqrt{x + y}$$

```
> assign(");
> x^2+y^2=r^2;
```
$$x^2 + y^2 = x + y$$

```
> a1:=completesquare(x^2-x);
```
$$a1 := \left(x - \frac{1}{2}\right)^2 - \frac{1}{4}$$

```
> a2:=completesquare(y^2-y);
```
$$a2 := \left(y - \frac{1}{2}\right)^2 - \frac{1}{4}$$

> a1+a2;

$$\left(x - \frac{1}{2}\right)^2 - \frac{1}{2} + \left(y - \frac{1}{2}\right)^2$$

> r:='r':
> subs(r^2=1/2,area=Pi*r^2);

$$area = \frac{1}{2}\pi$$

Exercise 23.6

(a) Show that the area enclosed by the cardioid $r = a(1 + \cos\theta)$ is
$\frac{1}{2}\int_0^{2\pi} r^2 d\theta = \frac{3\pi a^2}{2}$.

> r:=theta->a*(1+cos(theta));

$$r := \theta \rightarrow a\left(1 + \cos(\theta)\right)$$

> 1/2*int(r(theta)^2,theta=0..2*Pi);

$$\frac{3}{2}\pi a^2$$

(b) Show that the area enclosed by one loop of the curve $r = a\cos 3\theta$ is
$\frac{\pi a^2}{2}$.

> r:=theta->a*cos(3*theta);

$$r := \theta \rightarrow a\cos(3\theta)$$

> 1/2*int(r(theta)^2,theta=0..2*Pi);

$$\frac{1}{2}\pi a^2$$

(c) Show that the area enclosed by the loops of the curve $r^2 = a^2\sin 2\theta$
is $\frac{1}{2}\int_0^{\frac{\pi}{2}} r^2 d\theta + \frac{1}{2}\int_0^{\frac{3\pi}{2}} r^2 d\theta = a^2$.

> 1/2*int(a^2*sin(2*theta),theta=0..Pi/2)+
> 1/2*int(a^2*sin(2*theta),theta=0..3*Pi/2);

$$a^2$$

Appendix C

Maple **Procedures**

C.1   angles

```
> angles:=proc(T)
> local u1,u2,u3,A,B,C,unk;
> # Written by E.A. Gonzalez
>          if type(T[given][1],table) then
>               u1 := T[given][1];
>               u2 := T[given][2];
>               u3 := T[given][3];
>               T[vertices] := T[given];
>               T[u1] := sqrt(u2[x]^2+u3[x]^2+u2[y]^2+u3[y]^2-
>               2*u2[x]*u3[x]-2*u2[y]*u3[y]);
>               T[u2] := sqrt(u1[x]^2+u3[x]^2+u1[y]^2+u3[y]^2-
>               2*u1[x]*u3[x]-2*u1[y]*u3[y]);
>               T[u3] := sqrt(u2[x]^2+u1[x]^2+u2[y]^2+u1[y]^2-
>               2*u2[x]*u1[x]-2*u2[y]*u1[y]);
>               T[u4] := solve(T[u1]^2=T[u2]^2+T[u3]^2-
>               2*T[u2]*T[u3]*cos(A),A);
>               T[u5] := solve(T[u2]^2=T[u1]^2+T[u3]^2-
>               2*T[u1]*T[u3]*cos(B),B);
>               T[u6] := solve(T[u3]^2=T[u1]^2+T[u2]^2-
>               2*T[u1]*T[u2]*cos(C),C);
>               T[angles] := [convert(T[u4],degrees),
>               convert(T[u5],degrees),convert(T[u6],degrees)]
>          elif type(T[given][2],algebraic) and type(T[given][1],
>          algebraic)
>          then
>               T[u1] := T[given][1];
>               T[u2] := T[given][2];
>               T[u3] := T[given][3];
>               T[u4] :=solve(T[u1]^2=T[u2]^2+T[u3]^2-
>               2*T[u2]*T[u3]*cos(A),A);
>               T[u5] :=solve(T[u2]^2=T[u1]^2+T[u3]^2-
>               2*T[u1]*T[u3]*cos(B),B);
>               T[u6] :=solve(T[u3]^2=T[u1]^2+T[u2]^2-
>               2*T[u1]*T[u2]*cos(C),C);
>               T[angles] :=[convert(T[u4],degrees),convert(T[u5],
>               degrees),convert(T[u6],degrees)]
>          else
>               T[u1] := T[given][1];
>               T[u3] := T[given][3];
>               unk := sqrt(T[given][1]^2+T[given][3]^2-
>               2*T[given][1]*T[given][3]*cos(op(2,T[given][2])));
>               T[u4] :=solve(T[u1]^2=unk^2+T[u3]^2-2*unk*
>               T[u3]*cos(A),A);
```

```
>                    T[u5]  :=solve(unk^2=T[u1]^2+T[u3]^2-2*T[u1]*T[u3]*
>                    cos(B),B);

>                    T[u6]  :=solve(T[u3]^2= T[u1]^2+unk^2-2*T[u1]*unk*
>                    cos(C),C);

>                    T[angles] := [convert(T[u4],degrees),convert(T[u5],
>                    degrees),convert(T[u6],degrees)]
>           fi;
>           RETURN(T[angles])
> end;
```

C.2   showAntiderivatives

```
> # This function is based on a function of the same name in
> D.C.M. Burbulla and C.T.J Dodson, "Self-Tutor for Computer
> Calculus Using Maple", Prentice-Hall Canada Inc., Scarborough
> 1993.  It has been modified to include plot options such as
> title.

> # Modified by E.A. Gonzalez

> showAntiderivatives:=proc(expression,plotrange);

> if nargs=2 then plot({expression}union{seq(int(expression,x)+i,
> i=-5..5)},plotrange)

>    else plot({expression}union{seq(int(expression,x)+i,i=-5..5),
>    plotrange,args[3 .. nargs])

> fi

> end:
```

Bibliography

[1] D.C.M. Burbulla and C.T.J. Dodson: *Self-Tutor for Computer Calculus Using Maple*. Prentice-Hall Canada Inc., Scarborough 1993.

[2] B.W. Char, K.O. Geddes, G.H. Gonnet, B.L. Leong, M.B. Monagan and S.M. Watt: *Maple V Library Reference Manual*. Springer, New York 1991.

[3] B.W. Char, K.O. Geddes, G.H. Gonnet, B.L. Leong, M.B. Monagan and S.M. Watt: *Maple V Language Reference Manual*. Springer, New York 1991.

[4] B.W. Char, K.O. Geddes, G.H. Gonnet, B.L. Leong, M.B. Monagan and S.M. Watt: *Maple V First Leaves: A Tutorial Introduction to Maple V*. Springer, New York 1991.

[5] J.S. Devitt: *Calculus With Maple V*. Brooks/Cole, Pacific Grove, California 1993.

[6] C.H. Edwards, Jr. and D.E. Penney: *Calculus and Analytic Geometry*. 2nd edn. Prentice-Hall, Inc., Englewood Cliffs, New Jersey 1986.

[7] S. Goldenberg and H. Greenwald: *Calculus Applications in Engineering and Science*. D.C. Heath and Company, Toronto 1990.

[8] J. Hahn: LaTeX *For Everyone*. Personal TeX Inc., Mill Valley, California 1991.

[9] A. Heck: *Introduction to Maple*. Springer, New York 1993.

[10] M.H. Holmes, J.G. Ecker, W.E. Boyce, and W.L. Siegmann: *Exploring Calculus with Maple*. Addison-Wesley Publishing Company, Don Mills, Ontario 1993.

[11] L. Lamport: LaTeX *A Document Preparation System*. Addison-Wesley, California 1986.

[12] *The Maple Technical Newsletter*. Birkhäuser, Boston

Index

D. Redfern

The Maple Handbook

Maple V, Release 4

In preparation:

3rd ed. 1995.
Hardcover
ISBN 0-387-94538-5

This up-to-date handbook covers
the latest release of Maple and will
be an indispensible guide for all Maple users.

It is a complete reference and learning tool, covering
Maple's built-in mathematical, graphic, and system-based
commands in a precise and consistent manner.

For each function, the author includes information on
common parameter sequences, what type of output to
expect, additional hints, and cross references to the program's documentaion manuals.

The GETTING STARTED WITH MAPLE tutorial will help
those first-time users get off to a quick start.

■ ■ ■ ■ ■ ■ ■ ■ ■ ■

Springer

Springer-Verlag, Postfach 31 13 40, D-10643 Berlin, Fax 0 30 / 82 07 - 3 01 / 4 48   e-mail: orders@springer.de   tm.BA95.09.19

A. Heck

Introduction to Maple

1993. XIII, 497 pp.
84 figs. Hardcover
ISBN 3-540-97662-0

Introduction to Maple provides the reader with a hands-on, in-depth presentation of the most important aspects of the Maple V system. It teaches not only **what** can be done by sthe system, but also **how and why** it can be done.

By presenting a variety of problems and showing how Maple can be used to solve them, the reader will learn both the capabilities of the system and how to utilize it efficiently in his or her work. The many examples and exercises range from elementary to more sophisticated and will encourage the exploration and use of the Maple system.

The author has written the book with an emphasis on developing a good understanding of Maple data structures. This understanding will prove invaluable in effectively manipulating and simplifying expressions and will create a good foundation for learning Maple as a programming language.

Springer

Springer-Verlag, Postfach 31 13 40, D-10643 Berlin, Fax 0 30 / 82 07 - 3 01 / 4 48   e-mail: orders@springer.de

R.M. Corless

Essential
Maple

An Introduction for
Scientific Programmers

1995. XIV, 218 pp.
32 figs. Softcover
ISBN 3-540-94209-2

Essential Maple provides a quick introduction to programming in Maple, with an
overview of the most commonly-used commands and constructs. It summarizes basic
material, highlights slippery points, and gives tips on programming.
It also covers more subtle topics unique to Maple:

- option remember,
- the assume facility,
- the use of packages in Maple,
- evaluation rules,
- data structures,
- computation sequences,
- simplification,
- solution of equations,
- sequence accelleration,
- the Maple model of floating-point evaluation,
- calling other programs from Maple,
- operators,
- structured types,
- local, global, and environment variables,
- tracing and debugging.

Springer

Springer-Verlag, Postfach 31 13 40, D-10643 Berlin, Fax 0 30 / 82 07 - 3 01 / 4 48   e-mail: orders@springer.de

Springer-Verlag
and the Environment

We at Springer-Verlag firmly believe that an international science publisher has a special obligation to the environment, and our corporate policies consistently reflect this conviction.

We also expect our business partners – paper mills, printers, packaging manufacturers, etc. – to commit themselves to using environmentally friendly materials and production processes.

The paper in this book is made from low- or no-chlorine pulp and is acid free, in conformance with international standards for paper permanency.